T0230029

ASTROPHYSICAL APPLICATIONS OF GRAVITATIONAL LENSING

INTERNATIONAL ASTRONOMICAL UNION

UNION ASTRONOMIQUE INTERNATIONALE

ASTROPHYSICAL APPLICATIONS OF GRAVITATIONAL LENSING

PROCEEDINGS OF THE 173RD SYMPOSIUM OF THE
INTERNATIONAL ASTRONOMICAL UNION,
HELD IN MELBOURNE, AUSTRALIA,
9–14 JULY, 1995

EDITED BY

C. S. KOCHANEK

Harvard-Smithsonian Center for Astrophysics,
Cambridge, MA, U.S.A.

and

J. N. HEWITT

MIT, Cambridge, MA, U.S.A.

KLUWER ACADEMIC PUBLISHERS

DORDRECHT / BOSTON / LONDON

A C.I.P. Catalogue record for this book is available from the Library of Congress

ISBN-13:978-0-7923-3955-7   e-ISBN-13:978-94-009-0221-3
DOI:10.1007/978-94-009-0221-3

Published on behalf of
the International Astronomical Union
by
Kluwer Academic Publishers, P.O. Box 17, 3300 AA Dordrecht, The Netherlands.

Kluwer Academic Publishers incorporates
the publishing programmes of
D. Reidel, Martinus Nijhoff, Dr W. Junk and MTP Press.

Sold and distributed in the U.S.A. and Canada
by Kluwer Academic Publishers,
101 Philip Drive, Norwell, MA 02061, U.S.A.

In all other countries, sold and distributed
by Kluwer Academic Publishers Group,
P.O. Box 322, 3300 AH Dordrecht, The Netherlands.

Printed on acid-free paper

All Rights Reserved
©1996 International Astronomical Union
Softcover reprint of the hardcover 1st edition 1996
No part of the material protected by this copyright notice may be reproduced or utilized in any form or by any means, electronic or mechanical including photocopying, recording or by any information storage and retrieval system, without written permission from the publisher.

TABLE OF CONTENTS

Preface

Chapter 1: Classical Cosmology

Chapter 2: The Hubble Constant & Time Delays

Chapter 3: Large Scale Structure

Chapter 4: Quasar Absorption Lines

Chapter 5: Galaxy Clusters

Chapter 6: Galaxies

Chapter 7: Microlensing In the Galaxy

Chapter 8: Quasar Structure & Microlensing

Chapter 9: Observational Developments

Chapter 10: Emerging Applications

Chapter 11: Lens Surveys

Summary

Appendices

Index

PREFACE

EDWIN TURNER AND RACHEL WEBSTER
Co-Chairs, Scientific Organizing Committee

IAU Symposium 173, Astrophysical Applications of Gravitational Lenses, was held in Melbourne, Australia from July 9-14, 1995. The Symposium was sponsored by IAU Commissions 47 and 40.

With the discovery by Walsh and collaborators of the first instance of a gravitational lens, the multiply imaged quasar 0957+561, the area of gravitational lensing moved from speculative theory to a major astrophysical tool. Since that time, there have been regular, approximately biennial international meetings both in Europe and in North America, which have specifically focussed on gravitational lensing.

On this occasion, with the blessing of the IAU, the meeting was held at the University of Melbourne in Australia. It was the first international astronomical meeting to be held at the University of Melbourne, and hopefully has given the astronomical community some enthusiasm for trekking half-way round the globe to Australia to discuss their latest work.

Although there are still interesting fundamental questions in the theory of gravitational lensing, we felt that gravitational lensing has now 'come-of-age'. It provides a powerful new method for the study of the mass distributions of a wide range of cosmological objects, for the determination of cosmological parameters and also the possibility of using a 'natural telescope' to study background objects, as envisaged by Zwicky over half a century ago. The meeting was therefore organized to highlight areas of astrophysics where gravitational lensing is currently making an impact, or where we anticipate new results over the next few years.

Undoubtably the greatest impression was created by the new images from the the Hubble Space Telescope. These images show morphological details in the faint arcs of background galaxies which are blurred by seeing at ground level. These images have driven new theoretical investigations to determine optimal mass inversion techniques for clusters of galaxies. However there was also considerable discussion of the remarkable successes of the searches for gravitational lensing by compact objects in our own galaxy. It is likely that the model of our galaxy will be substantially revised to take account of these new results. The final talk of the conference was given by Bill Press, who introduced the possibility of gravitationally lensing some of the indigenous Australian fauna. Bill entertained the conference with speculations of future applications of gravitational lensing.

The success of the meeting owes much to the generous financial assistance of some of the major astronomical institutions in Australia: the Anglo-Australian Telescope, the Australia Telescope National Facility, Mount Stromlo and Siding Springs Observatory and the Research Centre for Theoretical Astrophysics. Assistance was also received from the IAU, SUN Microsystems Australia and the Department of Industry, Science and Technology in the Australian Government. We are grateful to the School of Physics at the University of Melbourne, and in particular to Janice Long, Margaret McGregor, Peter Cairns, Victoria Ibbetson, and Catherine Trott for ensuring that the meeting ran smoothly.

Finally, the task of editing the proceedings has been ably and efficiently undertaken by Chris Kochanek and Jackie Hewitt.

Scientific Organizing Committee

- Charles Dyer
- Masataka Fukugita
- Paul Hewett
- Jacqueline Hewitt
- Bohdan Paczynski
- Genevieve Soucail
- Jean Surdej
- Sjur Refsdal
- Kandu Subramanian
- Ed Turner
- Rachel Webster

Local Organizing Committee

- Michael Drinkwater
- Paul Francis
- David Jauncey
- Lucyna Kedziora-Chudczer
- Margaret Mazzolini
- Peter Quinn
- Rachel Webster

LENSING LIMITS ON THE COSMOLOGICAL CONSTANT

HANS-WALTER RIX
Max-Planck-Institut für Astrophysik
Postfach 1523, D-85740 Garching, Germany

Abstract. The statistics of gravitationally lensed quasars requires that $\lambda < 0.7$ at 90% confidence. This limit is weakly affected by the evolution of galaxies through merging. If E/S0 galaxies have much higher dust opacities at redshifts of 0.5-1.0 than they do locally, these limits can be evaded.

1. Why care about λ ?

After spending much of its life in the corner-cabinet for "unorthodox cosmologies", the cosmological constant, λ (or Ω_λ) has become socially acceptable in the last years (see Caroll, Press and Turner (1991), for a review). Cosmologists have reminded themselves that the equation governing the expansion of the universe

$$\mathrm{H}^2 \equiv \left(\frac{\dot{R}}{R}\right)^2 = \frac{8\pi G}{3}\rho_{matter} - \frac{k}{R^2} + \frac{\Lambda}{3}$$

indeed may contain the constant $\Lambda \neq 0$[1]. The aesthetic arguments for *expecting* $\lambda = 0$ are weak and therefore the value of λ should be determined empirically. However, this realization in itself was not sufficient to prompt widespread interest in measuring the actual value of λ. This interest has been sparked largely by the persistent conflict (see, e.g. Freedman et al.1995), between the age determination of globular clusters ($\gtrsim 15$Gyrs) and the estimates of the Hubble constant (≈ 65km s^{-1} Mpc^{-1}), which lead to a universe younger than some globular clusters for cosmologies with $\lambda = 0$. Yet, such an age conflict would not exist in a flat universe with $\Omega_{matter} = 0.3$ and $\lambda = 0.7$.

---

[1]This constant is usually expressed as its current epoch contribution to Ω: $\lambda \equiv \Omega_\lambda \equiv \Lambda/(2H_0^2)$

C. S. Kochanek and J. N. Hewitt (eds), Astrophysical Applications of Gravitational Lensing, 1–6.

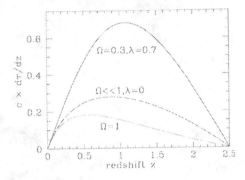

Figure 1. The relative optical depth to lensing in different cosmologies. All other things equal, many more lenses are expected in λ -dominated cosmologies

2. λ determinations from lensing statistics

It was suggested about six years ago (Fukugita et al.1990, Turner 1990) that the statistics of multiply lensed objects, such as QSOs, might provide a useful test for λ -dominated cosmologies. The differential optical depth for multiple imaging, $d\tau/dz$, due to a lens population of local space density n_l, can be written as the probability of intercepting a target of area unity, dP/dz, times the cross-section for image splitting, A_{crit}. These quantities are given by $dP/dz = n_l(z=0)\,(1+z)^3\,dl/dz$ and $A_{crit} = f_l(\sigma)D_{OL}D_{LS}/D_{OS}$. The factor $f_l(\sigma)$ depends on the mass profile of the lens (mostly the velocity dispersion σ) and it is for now assumed that the lens population does not evolve in co-moving coordinates. The optical depth can be written as

$$\frac{d\tau}{dz} = n_l(0)\,f_l(\sigma) \times (1+z)^3\,\frac{dl}{dz}\,\frac{D_{OL}D_{LS}}{D_{OS}}.$$

While the first two factors only depend on the properties of the lenses, the last three factors only depend on the geometry of the universe. These geometrical factors are plotted in Figure 1 for a source population at $\langle z \rangle = 2.5$ for three cosmologies: (a) $\Omega_{matter} \ll 1$, $\lambda = 0$, (b) $\Omega_{matter} = 1$, $\lambda = 0$ and (c) $\Omega_{matter} = 0.3$, $\lambda = 0.7$. The optical depth to lensing for a given lens population is 4-5 times higher in a flat universe with $\lambda = 0.7$ than with $\lambda = 0$. Therefore, with a statistically well defined sample of sources, e.g. distant QSOs, and with an independently characterized population of lenses, one can hope to constrain λ from the statistics (mostly the frequency and separation distribution[2]) of multiple images.

---

[2]But see Kochanek (1992) for a different statistical λ test, employing the distribution of lens-redshifts. In general, all observational constraints can and should be included in a likelihood analysis

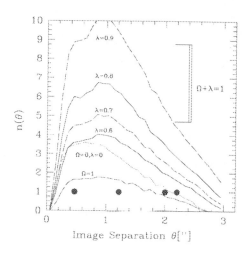

Figure 2. Lens Models for the HST Snapshot-Survey. Shown is the number of expected lenses per separation interval, n(θ), *vs.* the separation θ. The solid dots indicate the separations of the observed lenses.

3. Application to the HST Snapshot Survey

The practical application of this cosmological test became possible with the advent of a suitable and statistically well defined sample of distant sources[3], the HST Gravitational Lens Snapshot survey (Maoz et al.1993, and references therein). About 500 luminous QSOs with redshifts $>$ 1 were surveyed for multiple images, down to separations of $\Delta\theta \gtrsim 0.1$" for magnitude differences of $\Delta m \lesssim 2.5$. This survey found only one new lens at 0.47" separation (Bahcall et al.1992) in addition to four previously known lenses.

This data set was modeled independently by Maoz and Rix (1993) and Kochanek (1993). In these models the relevant lens population was taken to be the locally observed elliptical galaxies; these were assumed to have constant co-moving density and fixed mass profiles. The mass normalization of the lenses (with dark halos) can be determined both by a combination of independent stellar-dynamical constraints and by the separations of the observed multiple images. With a conventional cosmology (either $\Omega = 1$ or $\Omega \ll 1, \lambda = 0$), such empirical, non-evolving lens models, can explain the observed abundance and separation distribution $n(\theta)$, once the amplification bias and the survey detection limits are accounted for (see Maoz & Rix 1993 and Kochanek 1993 for details). In contrast, for a λ-dominated cosmology these models predict many more lenses than observed. Formally, Maoz and Rix (1993) found $\lambda < 0.7$ at 90% confidence (Figure 2). Kochanek (1993) found $\lambda \lesssim 0.8$ with less constrained lens models; the formal limits of

---

[3]The absence of statistically complete information about the source redshifts in radio surveys complicate the lensing analysis. Here we focus on optically selected samples, despite their drawbacks (Section 4.2)

his likelihood analysis have improved to $\lambda < 0.55\ (2\sigma)$ with the inclusion of further data (Kochanek 1995). For determining limits on λ it is conservative to account only for the well understood lens population of elliptical galaxies; if other objects (groups, dark potential wells, etc..) also contribute, the limits on λ will become tighter.

4. Loop-Holes and Caveats

These limits on the cosmological constant were (and had to be) derived in the context of a model, predicting the abundance and separation of multiple images. For this result to be useful, it must not be very sensitive to the model assumptions. In the following we discuss the two most critical assumptions that enter the models.

4.1. EVOLUTION OF THE LENS POPULATION

The models by Maoz and Rix (1993) and Kochanek (1993) assume that the locally observed lens population is uniformly (in co-moving co-ordinates) sprinkled throughout the universe. In this respect the models ignore the vast body of theoretical prediction and observational evidence that galaxies evolve with cosmological epoch. It may seem hopeless to implement our fragmentary knowledge of galaxy evolution into such lensing models. Can they make meaningful predictions nonetheless? For several reasons these models may indeed place robust upper limits on λ : First, we only have to consider the evolution of the assumed lens population, elliptical galaxies with $l \gtrsim 0.3L_*$ and only to redshifts ≈ 1; this is a small and relatively well understood subset of all galaxies. The observed small scatter in the color-magnitude relation for ellipticals (e.g. Bower et al.1992), and the evolution of the fundamental plane with redshift (e.g. Franx 1993) indicate that the bulk of stars in these galaxies has formed at redshifts beyond unity. However, these galaxies may still evolve by merging. Assuming simple merging histories, Mao and Kochanek (1994) and Rix et al.(1994) addressed the effect of such evolution on the lensing statistics. Both N-body simulations and the existence of the "Fundamental Plane" indicate that the merger of two progenitor galaxies with characteristic dispersion, σ_i, leads to a product with $\sigma_f = 2^{1/4}\sigma_i$. Since for an isothermal sphere the cross section increases as σ^4, the total cross-section is approximately conserved under merging: $2A_{crit}^{init} \approx A_{crit}^{final}$. As galaxies are "split-up" into smaller pieces, going back in time, the observable distribution of image splittings becomes skewed towards smaller separations. Quantitatively, Rix et al.(1994) found that none of the merging scenarios explored resulted in a "softer" upper limit on λ than the no-evolution model.

4.2. THE ABUNDANCE OF DUSTY LENSES

Multiply imaged QSOs are found in very special directions, namely where their light passes through the inner parts (a few kpc) of a galaxy. During this passage the QSO light may become partially extincted by dust in the galaxy. This would lead to the under-representation of multiple images in a magnitude limited sample, which in turn may permit a dominant λ to explain the observed lensing frequency. This is indeed the most serious concern in the modeling procedure, as a good number of radio-selected lens-system show indications of severe reddening (Lawrence et al.1994). However, at this point in time it is not yet conclusively settled in which fraction of these "red radio lenses" the reddening occurs at the redshift of the lens (as opposed to the source).

There is observational evidence that the optical depth due to dust in present-day elliptical is small ($\tau \ll 1$) at the impact parameters of lensed images (a few kpc). The dust content of elliptical has been studied extensively through its IR emission (see Goudfrooij (1995) and references therein); these investigations suggest a smoothly distributed dust component, but its optical depth at R_{eff} is only a few hundredth. A more direct argument comes from the photometry of superimposed cluster ellipticals (Lauer 1988): assume that the foreground galaxy in the superimposed pair contained dust, $\tau_f(R_f)$, with a radial profile comparable to the stars', $I_f(R_f)$. The centers of the two galaxies are separated by D. Now consider the intensity, I_1, at a distances R_b ($< D$) and I_2 at $-R_b$ from the center of the background galaxy. The intensities are given by

$$
\begin{aligned}
I_1 &= && I_f(R_f) + I_b(R_b)\exp\left(-\tau(R_f)\right) \\
I_2 &= && I_f(R_f + 2D) + I_b(R_b)\exp\left(-\tau(R_f + 2D)\right) &\approx I_b(R_b),
\end{aligned}
$$

where the exponential factors are unity in the dust free case. Lauer found that in the majority of cases the light distributions could be decomposed into two sets of nearly concentric ellipses, as expected for the superposition of symmetric, dust-free galaxies. Compared to transparent galaxies, the intensity of the background galaxy at $\pm R_b$ in the dusty case would differ by $\Delta I_b(R_b) = I_b(R_b)[1 - \exp\left(-\tau(R_f)\right)]$ between the two opposite, but equidistant, point from the center. When fitting isophotes this would lead to a center shift, ΔR_b, of $\Delta R_b/R_b \approx \tau(R_f) \left(\partial ln I_b/\partial ln R_b\right)^{-1}$ for $\tau(R_f) \ll 1$, where $\partial ln I_b/\partial ln R_b \sim 1 - 3$ is the logarithmic luminosity gradient of the background galaxy. Quantitatively, the upper limits on ΔR_b in Lauer's data indicate $\tau(R_f) < 0.04$, with R_f of order kpc. These data indicate that at least in local ellipticals, dust extinction is unimportant at the level relevant to lensing statistics.

5. Future Prospects

On the face of it, the statistics of gravitational lensing provide a limit on λ that makes the cosmological constant almost uninteresting for solving the "age problem" and other outstanding cosmological issues. What can be done in the near future to tighten these limits and to make them less model dependent? On the one hand, we can expect the completion of the MIT-Greenbank survey and the ongoing CLASS survey. These surveys will greatly increase the amount of radio data for lensing statistics. Similarly, the Sloan-Survey will increase tremendously the database of optically selected QSO. Maybe even more important, it should be possible to close some of the loop-holes in the lens modeling: (1) Detailed modeling of a few individual lenses (see Kochanek in this volume) will improve our understanding of how the lensing properties of (E/S0) galaxies, are related to their photometric and stellar kinematic properties; in turn this will constrain better the statistical lensing properties of the photometrically defined lens population. (2) Direct observations with HST will also lead to rapid improvement in our understanding of the galaxy evolution for $0 < z < 1$. (3) HST observations will further allow us to study the dust content of elliptical galaxies at intermediate redshifts. These observations may lay to rest the of the most worrisome systematic uncertainty in the modeling, preferential dust extinction of multiply imaged objects. Taken together, these arguments give reason to be optimistic that within a few years lensing statistics will settle whether λ is cosmologically important, or not.

References

Bahcall, J., et al., 1992, ApJL, 392, L1
Bower, R., Lucey, J. & Ellis, R., 1992, MNRAS, 254, 601
Carroll, S., Press, W. & Turner, E., 1992, ARAA, 30, 499
Franx, M., 1993, ApJL, 407, L5
Freedman, W., et al., 1995, Nature, 371, 757
Fukugita, M., Futamase, T. & Kasai, M., 1990, MNRAS, 246, 25p
Goudfrooij, P., 1995, PASP, 107, 502
Kochanek, C.S., 1992, ApJ, 384, 1
Kochanek, C.S., 1993, ApJ, 419, 12
Kochanek, C.S., 1995, ApJ, submitted.
Lauer, T., 1988, ApJ, 325, 49
Lawrence, C., et al., 1994, ApJL, 420, L9
Mao, S. & Kochanek, C.S., 1994, MNRAS, 268, 569
Maoz, D., et al., 1993, ApJ, 409, 28
Maoz, D. & Rix, H.-W., 1993, ApJ, 416, 425
Rix, H.-W., Maoz, D., Turner, E. & Fukugita, M., 1994, ApJ, 435, 49
Turner, E., 1990, ApJL, 365, L43

A TECHNICAL MEMORANDUM ON CORE RADII IN LENS STATISTICS

CHRISTOPHER S. KOCHANEK
Harvard-Smithsonian Center for Astrophysics
60 Garden Street, Cambridge, MA 02138, USA

Abstract. Quantitative estimates of lensing probabilities must be self-consistent. In particular, for asymptotically isothermal models: (1) using the $(3/2)^{1/2}$ correction for the velocity dispersion overestimates the expected number of lenses by 150% and their average separations by 50%, thereby introducing large cosmological errors; (2) when a core radius is added to the SIS model, the velocity dispersion must be increased; and (3) cross sections and magnification bias cannot be separated when computing the lensing probability. When we self-consistently calculate the effects of finite core radii in flat cosmological models, we find that the cosmological limits are independent of the core radius.

1. Introduction

Asymptotically isothermal potentials are consistent with most data on gravitational lenses. They explain the observed numbers of lenses (Maoz & Rix 1993, Kochanek 1993), fit most observed image configurations (e.g. Kochanek 1991a), and are consistent with stellar dynamics (Kochanek 1994, Franx 1993). Photometry of early type galaxies (Tremaine et al. 1994), and the absence of central images in most lenses (Wallington & Narayan 1993, Kassiola & Kovner 1993) suggest that the lens potentials have a small or vanishing core radius. There is, however, a persistent myth (generated in part by the author) that small core radii can dramatically alter the expected number of lenses without other observational consequences. We can trace the current versions of this myth to inconsistencies in either the dynamical normalization of the models or the calculation of the magnification bias. In this technical memorandum we briefly explore these consistency re-

7

C. S. Kochanek and J. N. Hewitt (eds), Astrophysical Applications of Gravitational Lensing, 7–12.
© 1996 IAU

quirements and the resulting effects of a finite core radius on cosmological limits.

We confine ourselves to a circular isothermal density distribution with $\rho = \sigma_{DM}^2 / 2\pi G(r^2 + s^2)$ where σ_{DM} is the velocity dispersion of the dark matter, and s is the core radius (Hinshaw & Krauss 1987). The lens deflects rays by

$$\frac{\partial \phi}{\partial r} = b \frac{(r^2 + s^2)^{1/2} - s}{r} \qquad (1)$$

where $b = 4\pi(\sigma_{DM}/c)^2 D_{LS}/D_{OS}$, D_{LS} and D_{OS} are proper motion or angular diameter distances between the lens and the source and the observer and the source respectively, and ϕ is the two-dimensional lensing potential. (The alternate softened isothermal model with $\phi = b(r^2 + s^2)^{1/2}$ introduced by Blandford & Kochanek (1987) is almost indistinguishable from the Hinshaw & Krauss (1987) model if its core radius is twice as large.)

The lens is supercritical and able to generate multiple images if $\beta = s/b < 1/2$, with a tangential critical line at $r_+ = b(1 - 2\beta)^{1/2}$ and a radial critical line at $r_- = b[\beta - \beta^2/2 - \beta^{3/2}(4 + \beta)^{1/2}]^{1/2}$. The caustics are at $u_+ = 0$ and u_- where the cross section is $\tau = \pi u_-^2 = \pi b^2[1 + 5\beta - \beta^2/2 - \beta^{1/2}(4+\beta)^{3/2}/2]$ (Hinshaw & Krauss 1987). The cross section declines very rapidly with β and near the threshold of $\beta = 1/2 - \epsilon$ it declines as $\tau \propto \epsilon^3$ if b is held fixed. If we assume a *constant comoving core radius* the cross section can be integrated analytically to compute the optical depth (Krauss & White 1992).

2. Dynamical Normalizations

The first question we must address is the normalization of the singular model ($s \to 0$). Historically, Turner, Ostriker, & Gott (1984) argued that if the central velocity dispersion of the stars is σ_c then the dark matter should have velocity dispersion $\sigma_{DM} = (3/2)^{1/2}\sigma_c$. However, Franx (1993), Kochanek (1993, 1994), and Breimer and Sanders (1993) show convincingly that real galaxies do not satisfy the assumptions used by Turner, Ostriker, & Gott (1984), and that for real galaxies $\sigma_{DM} \simeq \sigma_c$. Kochanek (1994) fit a sample of 37 early type galaxies from van der Marel (1991) and found that the best fit estimate was $\sigma_{DM*} = 225 \pm 10$ km s^{-1} for an L_* galaxy.

All existing studies of the effects of a core radius on lens statistics have added a core radius while leaving the velocity dispersion σ_{DM} or b unchanged. It is clear, however, that s and σ_{DM} must be correlated. Adding a core radius reduces the mass near the center of the galaxy, and the velocity dispersion must increase compared to its value in a singular model to maintain either the stellar velocity dispersions or the average image separations fixed. As a model calculation, we compute the average line-of-sight velocity dispersion inside one effective radius R_e assuming a Hernquist (1990) distribution ($\nu \propto r^{-1}(r + a)^{-3}$) for the stars with $a \simeq 0.45 R_e$. With the assumption that the velocity dispersion tensor of the stars is isotropic, the dark matter dispersion increases as $\sigma_{DM} \propto 1 + 2(s/R_e)$ with the addition of a core radius.

For a more realistic model we use van der Marel (1991) sample and fit isotropic dynamical models to each galaxy, assuming the core radius is a constant fraction of the estimated effective radius for each galaxy, and that the velocity dispersion scales as $L/L_* = (\sigma_{DM}/\sigma_{DM*})^4$. The χ^2 surface of the fit to the observed velocity dispersion profiles is shown in Figure 1, and the dashed line is the scaling law estimated from the Hernquist (1990) model. Models with large core radii cannot fit the data because of the contradiction between a homogeneous core and a steeply rising luminosity profile. The formal 95% confidence upper limit on the core radius is $s_* \lesssim 0.08 R_e$ or $s_* \lesssim 300 h^{-1}$ pc for $R_{e*} = 4 h^{-1}$ kpc. For a core radius of $s_* = 100 h^{-1}$ pc the fractional increase in the velocity dispersion is 7.5% or 17 km s^{-1}, less than the uncertainty in the value of σ_{DM*}. Nonetheless, its effects on models with a finite core radius are striking; it produces a 33% increase in the expected number of lenses if we keep the ratio of the core radius to the critical radius fixed (s/b constant).

Self-consistency in lens models also requires a velocity dispersion that increases as the core radius becomes larger since the average image separation must stay fixed as the core radius increases. The image separation is approximately twice the tangential critical radius of the lens ($\Delta\theta \simeq 2r_+ = 2b(1 - 2\beta)$), so that if the core radius increases (larger β), the only way to maintain constant average image separations is to also increase the average velocity dispersion (larger b). If we model this by keeping the tangential critical radius $r_+ = b(1 - 2\beta)$ fixed, then the lens cross section τ decreases as $\tau \propto \epsilon^2$ instead of ϵ^3.

3. Magnification Bias

Self-consistent calculations of the lensing probability such as Kochanek & Blandford (1987), Kochanek (1991b, 1993), Wallington & Narayan (1993), Kassiola & Kovner (1993), and (in most regimes) Maoz & Rix (1993) automatically include the effects of the core radius on the magnification bias, but most treatments of softened isothermal models examined only the effects of a core radius on lensing cross sections (e.g. Dyer 1984, Blandford & Kochanek 1987, Hinshaw & Krauss 1987, Krauss & White 1992, Fukugita & Turner 1991, Fukugita et al. 1992, Bloomfield-Torres & Waga 1995). Core radii have a powerful effect on the cross section for multiple imaging. However, using the change in the cross section grossly overestimates the effects

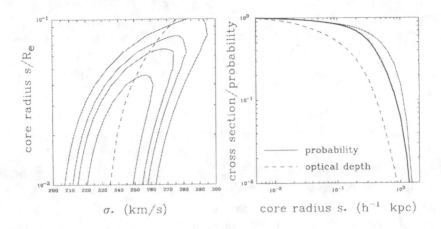

Figure 1. (Left) Contours of the χ^2 for the dynamical fits. The light solid lines show the 68%, 90%, 95%, and 99% confidence limit changes on $\Delta\chi^2$ for the fit to the sample. The dashed line shows the expected scaling of σ_{DM*} with s/R_e if we keep the average velocity dispersion interior to R_e fixed in the Hernquist/softened isothermal sphere dynamical model.

Figure 2. (Right) Variation in cross section (dashed line) and lensing probability (solid line) with core radius s for a lens with $\sigma_{DM} = 250$ km s^{-1}. The values are normalized to unity at the minimum core radius. The heavy solid line shows the lensing probability excluding image systems with detectable central images using the same normalization as for the total probability. The results are given for the average over the quasar data sample including selection effects.

of a finite core radius on the lensing probability in bright quasar samples. The core radius first eliminates images with low total magnifications, but the bright quasar samples are dominated by highly magnified images of fainter quasars and magnification bias significantly reduces the effects of adding a core radius on the probability.

We can understand this analytically in the Hinshaw & Krauss (1987) model. Inconsistent models of the effects of a core radius estimate the lensing probability by using the optical depth multiplied by the magnification bias for the singular model. When two images are merging on the radial caustic, the third image is located at $r_{out} = 2\beta u_-/r_-^2$, and the average magnification produced by the lens is $\langle M \rangle = r_{out}^2/u_-^2 = 4\beta^2/r_-^4$. When the core radius is small, the average magnification is 4, but near the threshold the average magnification diverges as $\langle M \rangle \propto \epsilon^{-2}$. The magnification probability distribution is approximately $P(> M) = (\langle M \rangle/2M)^2$ when $M > \langle M \rangle/2$ for fold caustic statistics (e.g. Blandford & Narayan 1986). If we assume a single power law quasar number counts distribution with $dN/dm \propto 10^{\gamma(m-m_0)}$ then the magnification bias varies with the average magnification as $B(m) \propto \langle M \rangle_0^{2.5\gamma}$ for $\gamma < 0.8$. As the core shrinks, the average magnification increases, which drives up the magnification bias. The

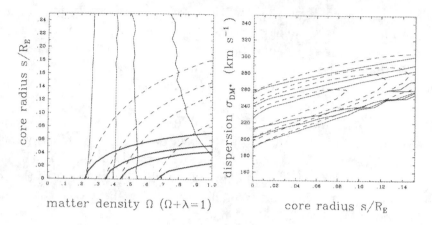

Figure 3. Cosmological effects of finite core radii. The left panel shows contours as a function of the ratio of the core radius to the effective radius and the cosmological model. The light solid lines are the constraints from lensing alone, the dashed lines adds the velocity dispersion prior as a function of core radius estimated in §2, and the heavy solid lines add the goodness of fit to the dynamical data. The right panel shows dependence of σ_{DM*} on the core radius in an $\Omega_0 = 1$ cosmology. The light contours use only the lens data, and the dashed lines include the prior probability distribution for σ_{DM*} estimated from dynamical models. The contours are drawn at the 68%, 90%, 95%, and 99% confidence levels for one parameter.

lensing probability, including the change in the magnification bias, varies as $\tau B(m) \propto \epsilon^{3-5\gamma}$ not $\tau \propto \epsilon^3$. For large average magnifications the effective value of γ is the faint slope of the quasar number counts, $\gamma \simeq 0.27 \pm 0.07$, and $\tau B(m) \propto \epsilon^{1.65 \pm 0.35}$. For bright quasars the increase in the bias is greater because of the steeper number counts slope, and the effects of the core radius are still smaller.

To emphasize this point, Figure 2 shows the relative variation of the cross section and the true lensing probability including magnification bias for a lens with $\sigma_{DM} = 250$ km s^{-1} as a function of the core radius averaged over the full quasar data sample. For a core radius of $s = 100h^{-1}$ pc using the cross section instead of the true probability underestimates the lensing probability by about 40%. This comparison overestimates the effect of a finite core radius because we did not include the dependence of σ_{DM} on s.

4. The Cosmological Effects of Softened Isothermal Spheres

We assume that the core radii of galaxies are proportional to their effective radii $s = s_*(L/L_*)^{1.2}$, and the models are characterized by a fixed ratio of s/R_e. To simplify the calculations, we set the "Faber-Jackson" exponent to be $\gamma = 4$, so the core radius varies with the velocity dispersion as $s = s_*(\sigma_{DM}/\sigma_{DM*})^{4.8} = s_*(L/L_*)^{1.2}$ (see Kochanek 1995 for details).

The right panel of Figure 3 shows the best fit value of σ_{DM*} as a function of s_* in an $\Omega_0 = 1$ cosmology. As expected from §2, the velocity dispersion increases as the core radius increases with $\sigma_{DM*} \propto 1 + s/R_e$. This is shallower than the slope seen in the dynamical models. The left panel of Figure 3 shows the dependence of the cosmological limits for a flat universe ($\Omega_0 + \lambda_0 = 1$) on the core radius using only the lens data, the lens data combined with the prior probability distribution for σ_{DM*} derived from the dynamical model of §2, and finally the lens data, the dynamical velocity dispersion prior, and the likelihood of the dynamical model. If we use only the lensing data, the cosmological limits are nearly independent of the core radius – this is a radically different picture of the effect of a core radius than that found in inconsistent calculations.

Acknowledgements: This research was supported by the Alfred P. Sloan Foundation.

References

Blandford, R.D., & Narayan, R. 1986, ApJ, 310, 568
Blandford, R.D., & Kochanek, C.S., 1987, ApJ, 321, 658
Bloomfield-Torres, L.F.B., & Waga, I., 1995, this volume
Breimer, T.G., & Sanders, R.H., 1993, A&A, 274, 96
Dyer, C.C., 1984, ApJ, 287, 26
Franx, M., 1993, in Galactic Bulges, ed. H. Dejonghe & H.J. Habing, (Dordrecht: Kluwer) 243
Fukugita, M., & Turner, E.L., 1991, MNRAS, 253, 99
Fukugita, M., Futamase, T., Kasai, M., & Turner, E.L., 1992, MNRAS, 393, 3
Hernquist, L., 1990, ApJ, 356, 359
Hinshaw, G., & Krauss, L.M., 1987, ApJ, 320, 468
Kassiola, A., & Kovner, I., 1993, ApJ, 417, 450
Kochanek, C.S., 1991a, ApJ, 373, 354
Kochanek, C.S., 1991b, ApJ, 379, 517
Kochanek, C.S., 1993, ApJ, 419, 12
Kochanek, C.S., 1994, ApJ, 436, 56
Kochanek, C.S., 1995, submitted to ApJ
Kochanek, C.S., & Blandford, R.D., 1987, ApJ, 321, 676
Krauss, L.M., & White, M., 1992, ApJ, 394, 385
Maoz, D., & Rix, H.-W., 1993, ApJ, 416, 425
Tremaine, S., Richstone, D.O., Yong-Ik Byun, Dressler, A., Faber, S.M., Grillmair, C., Kormendy, J., & Lauer, T.R., 1994, AJ, 107, 634
Turner, E.L., Ostriker, J.P., & Gott, J.R., 1984, ApJ, 284, 1
van der Marel, R.P., 1991, MNRAS, 253, 710
Wallington, S., & Narayan, R., 1993, ApJ, 403, 517

GRAVITATIONAL LENSES AMONG HIGHLY LUMINOUS QUASARS: LARGE OPTICAL SURVEYS

J.-F. CLAESKENS

European Southern Observatory (Chile), and
Aspirant au FNRS (Belgium),
Institut d'Astrophysique, Université de Liège,
5 Avenue de Cointe, B-4000 Liège, Belgium

A.O. JAUNSEN

Nordic Optical Telescope,
Ap. 474, S/C de La Palma,
E-38700 Canarias, Spain

AND

J. SURDEJ

STScI, 3700 San Martin Drive, Baltimore, MD 21218, USA,
Member of the Astrophysics Division,
Space Science Department of the European Space Agency, and
Directeur de Recherche au FNRS, Belgium

Abstract. The search for multiply imaged quasars among highly luminous quasars (HLQs) is a very good strategy to determine the fundamental parameters of the Universe. We report on the present observational status of a combined sample of HLQs, including new observations obtained with the Nordic Optical Telescope (NOT) and at ESO. This combined sample of HLQs now contains 1178 distinct HLQs. A complete list of the total sample will be soon made available, through a World-Wide-Web page. Preliminary maximum likelihood results are also presented, using a simple statistical model to constrain the values of galactic parameters, of the number counts of QSOs, and of the cosmological constant.

13

1. Introduction

The systematic search for multiply imaged quasars among HLQs ($M_V \leq$ -27 for $H_o = 50$ km s^{-1} Mpc^{-1}, $q_o = 0.5$) has been in progress for 9 years, since its initial start in 1986 (Surdej et al. 1987). This technique has proven to be very efficient in order to find new gravitational lenses (GLs) (Turner et al. 1984, Surdej et al. 1987, 1988, 1993a). Today, the large number of observed HLQs also allows us to perform meaningful maximum likelihood analyses to constrain the parameter values of the lens and of the cosmological models.

In Section 2, we present the properties of the present merged sample of HLQs, together with those of the sub-samples. We then give in Section 3 a short overview of the main relations used to estimate the probability for lensing. Finally, we present, in Section 4 preliminary constraints on some galactic and cosmological parameters of the model, using the maximum likelihood analysis and give a short conclusion in Section 5.

2. Survey sample

Previous observations of HLQs have been conducted at ESO (Surdej et al. 1993a), with HST (Maoz et al. 1993), the CFHT (Crampton et al. 1992, Yee et al. 1993), and with the NOT (Jaunsen et al. 1995). The statistics of the merged sample has been improved by the observation of new HLQs, carried out with the NOT and at ESO, and by a careful re-analysis of the HST frames (see Surdej et al. in these proceedings).

The new observations have been incorporated in the respective sub-samples presented in Table 1. The present merged sample contains 1528 observations of 1178 different HLQs. The last column of Table 1 shows the number (and the fraction) of HLQs kept in each subsample, after selecting the best observation among multiple ones of the same HLQ. It can be seen from this table that the average brightness of the observed samples is not declining from a lack of bright HLQs.

There are 6 lenses in the merged sample: UM673 (Surdej et al. 1987), PG1115+080 (Weymann et al. 1980), Q1009-025 (Surdej et al. 1993b), HE1104-1805 (Wisotski et al. 1993), Q1208+1011 (Magain et al. 1992) and H1413+117 (Magain et al. 1988). Also a few new promising GL candidates are being found, like J03.13 (Claeskens et al. 1995). The observed properties of these lenses can be found in Surdej et al. (1995), and also by connecting to the GL World-Wide-Web page at STScI (the URL address is given below).

The good correlation between intrinsically bright QSOs and high lensing probability is illustrated in Figure 1, which represents the probability for

TABLE 1. Individual properties of the sub-samples

Sub-sample	$< z >$	$< V >$	$< M_V >$	<FWHM> [arc-sec.]	N_q	$N_{q,best}$ (%)
HST (Maoz et al.)	2.1	17.6	−27.7	-	495	408(82%)
ESO (Surdej et al.)	2.2	17.6	−27.8	1.00	396	279(70%)
CFHT (Crampton et al.)	2.4	18.1	−27.4	0.66	101	81(80%)
CFHT (Yee et al.)	2.2	17.7	−27.8	0.76	104	32(31%)
NOT (Jaunsen et al.)	2.0	17.5	−27.6	0.90	432	378(87%)
TOTAL	2.2	17.7	−27.6	0.90	1528	1178

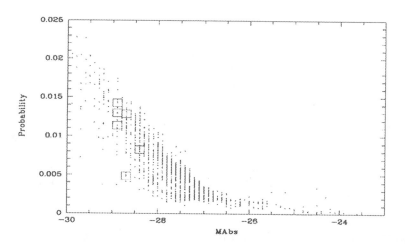

Figure 1. Plot of the absolute V magnitude versus the lensing probability for the 1178 HLQs. The 6 accepted GL candidates in the sample are marked by the squares

each HLQ to be lensed, versus their absolute magnitude. The 6 lensed QSOs in the sample are marked with a square.

A complete list of the total sample will be soon made available and updated through a World-Wide-Web page. Preliminary URL addresses are:

http://www.uio.no/~ajaunsen/index.html
http://www.stsci.edu/ftp/stsci/library/grav_lens/grav_lens.html.

3. Statistical model

As in previous statistical works on optical samples of QSOs (Surdej et al. 1993a, Kochanek 1993, Jaunsen et al. 1995), we have used a simple statistical model, where the lens galaxy is described by means of a Singular Isothermal Sphere (SIS). We recall hereafter the main formalism, but refer the reader to previous publications for further details.

The optical depth or probability for lensing, in an Einstein-de-Sitter Universe is, for a SIS lens model, given by:

$$\tau = \frac{1}{30} F D_s^3(z_s),\tag{1}$$

where F is the lens effectiveness parameter ($F \propto n_0 \sigma_*^4$ with n_0 being the local spatial density of galaxies (E/S0-type), σ_* is the one-dimensional velocity dispersion of the galaxy, and D_s is the proper distance in a flat universe).

An important effect that must be taken into consideration when computing the frequency of lensing in the universe is the well known *magnification bias*. The bias stems from the fact that QSOs dimmer than the target QSO are amplified to brighter luminosities due to the effects of lensing. The expected frequency of lensing is therefore higher than that directly estimated from the general optical depth. The magnification bias can be estimated for each QSO from its magnitude m_i by the following expression:

$$Bias(m_i, M_{Min}(r_s), \infty) = \int_{M_{Min}(r_s)}^{\infty} \frac{8}{M^3} \frac{n_q(m_i + 2.5 log M(r))}{n_q(m_i)} dM.\tag{2}$$

The total integrated bias is computed within the range of the survey search angular radius, r_s. For each radius, a corresponding minimum magnification $M_{Min}(r_s)$ (or maximum magnitude difference) is found using the survey angular selection function (ASF). The ASF is determined for each instrumental configuration (i.e. each survey sample) and is determined through simulations (cf. Surdej et al. 1993a).

Finally, the probability for observing a lensed QSO, taking into account the general optical depth, magnification bias and detectability, is expressed by

$$p_i^{SF}(m_i, z_i) = \frac{S_{cat} F}{30} D_s^3(z_i) \int_0^{r_s/2} p_c(b_{cr}) Bias(m_i, M(2b_{cr}), M_2) db_{cr}.\tag{3}$$

$S_{cat} = 0.7$ is the correction factor for incompleteness of the sample, due to the fact that QSO surveys loose about 30% of GL-systems (Kochanek 1991). $p_c(b_{cr})$ represents the probability for a SIS lens galaxy to make a lens with angular separation $2b_{cr}$, and M_2 is a finite (large) constant to prevent any divergence in equation (2), when the slope of the QSO number counts is changed.

4. Maximum likelihood analysis

4.1. INTRODUCTION

The maximum likelihood (ML) technique is a powerful tool to set constraints on parameters of the statistical model, using all the observed properties of the sample. So far, we have taken into account the *frequency* of lenses in the optical sample and the *observed angular separation* between their images (described by the configuration probability p_{Conf}; see Kochanek 1993, Jaunsen et al. 1995). We have also included a configuration probability, $p_{Conf,Rad}$, for the $N_{L,Radio}$ radio lenses whose ASF is known. $p_{Conf,Rad}$ is the probability of the observed angular separation between the lensed radio images, normalized by all the *detectable* angular separations. So, using a Poissonian statistics for the occurrence of lenses, we can write the logarithmic likelihood function to be maximized as:

$$\ell = \ln L = -\sum_{i=1}^{N_U} p_i + \sum_{j=1}^{N_L} \ln p_j + \sum_{j=1}^{N_L} \ln p_{Conf,Obs,j} + \sum_{j=1}^{N_{L,Radio}} \ln p_{Conf,Rad,j}, \quad (4)$$

where N_L and N_U are the number of lensed and unlensed objects in the sample.

4.2. INPUT DATA

The input data in the model are listed hereafter. These are also the default or reference values when the parameters are kept constant in the ML analysis.

- $n_{Ell} = 0.005h^3Mpc^{-3}$ (Fukugita & Turner 1991)
- $n_{Spiral} = 0.015h^3Mpc^{-3}$ (Fukugita & Turner 1991)
- $\sigma^*_{Ell} = 220$ km/s (Fukugita & Turner 1991)
- $\sigma^*_{Spiral} = 143$ km/s (Fukugita & Turner 1991)
- $\alpha = -1.1$ exponent of the Schechter luminosity function
- Tully-Fisher exponent $\gamma_{Ell} = 4$
- Tully-Fisher exponent $\gamma_{Spiral} = 2.6$

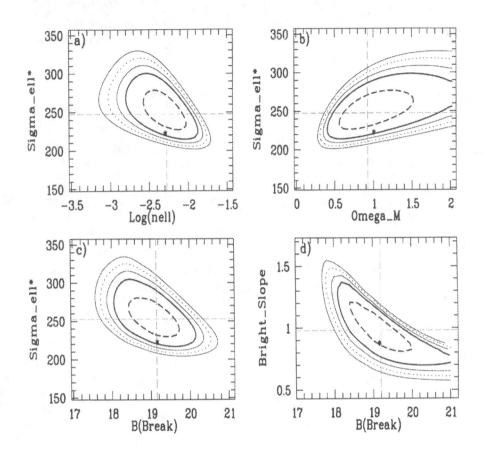

Figure 2. Maximum likelihood results: a) $(n_{Ell}, \sigma^*_{Ell})$ plane; b) $(\Omega_M, \sigma^*_{Ell})$ plane (flat universe); c) (B(Break), σ^*_{Ell}) plane; d) (B(Break), Bright end slope) plane.

- Magnitude break, bright end slope and faint end slopes of the differential number counts of QSOs (B band): 19.15, 0.86 & 0.28 resp. (Hartwick & Schade 1990)
- Cosmological Model: Einstein - de- Sitter ($\Omega_M = 1$, $\lambda_o = 0$)
- Angular radius for the search of secondary images around the QSOs: $r_s = 3''$
- Observed sample: 1528 observations of 1178 distinct HLQs
- 6 optical lenses (with $\Delta\theta \leq 3''$) (see Section 2)
- 10 radio lenses (with $\Delta\theta \leq 3''$): MG0414+053, B1422+231, B1938+666, B0218+356, MG1131+045, MG1654+134, MG1549+304 (Kochanek 1993, Refsdal & Surdej 1994), and F10214+4724 (Lehár & Broadhurst 1995), Q1600+434 (Jackson et al. 1995), Q1608+656 (Myers 1995).

4.3. RESULTS

Figure 2 displays the likelihood contours in 4 different parameter spaces. The most inner solid (dashed) heavy contours are such that their projections onto the one dimensional axes give the 90% (68%) confidence level intervals for the corresponding parameters. The dashed cross indicates the ML estimate of the parameters, and the star shows their reference values (given in Section 4.2).

We infer the following 90% confidence intervals for the parameters:

- Figure 2a): $n_{Ell}^* \in [0.0017, 0.012]$ $h_{100}^3 \mathrm{Mpc}^{-3}$, and $\sigma_{Ell}^* \in [215, 300]$ km/s. The resulting estimate of the efficiency parameter F for the ellipticals is 0.037, and is well inside the [0.012-0.070] range at the 90% confidence.
- Figure 2b): The upper limit on the cosmological constant in a flat universe is 0.55 at the 90% confidence.
- Figure 2c) & d) The parameters B(Break) and the bright end slope exponent of the double power law for the surface density of QSOs belong respectively to $[18, 20]$ and $[0.7, 1.4]$ (at the 90% confidence level).

5. Conclusions

We have reviewed in this paper the present status of large optical samples of HLQs. The total merged sample contains 1178 distinct objects among which there are 6 lenses and a few more very promising GL candidates with small angular separations (e.g. the doubly imaged QSO J03.13), which are being found using precise PSF subtraction techniques. The ML analysis shows that the lensing statistics are fully compatible with independent estimates of galactic parameters (without any correction factor to account for dark matter in the halo), and with the standard number counts of QSOs (double power law). The upper value on the cosmological constant is 0.55 at the 90% confidence level. A better knowledge of our sample will still improve these constraints. For that, we need i) to identify *all* the lenses present in the sample; ii) to know the redshift of all the lensing galaxies (with the Keck and VLT telescopes).

References

Claeskens, J.F., Surdej, J., & Remy, M., 1995, A&A, submitted
Crampton, D., McClure, R., & Fletcher, J., 1992, ApJ, 392, 23
Fukugita, M., & Turner, E., 1991, MNRAS, 253, 99
Hartwick, F., & Schade, D., 1990, ARA&A, 28, 437
Jackson, N., De Bruyn, A.G., Myers, S., Bremer, M.N., et al., 1995, MNRAS, 274, L25
Jaunsen, A.O., Jablonski, M., Pettersen, B.R., & Stabell, R., 1995, A&A, in press

Kochanek, C.S., 1991, ApJ, 379, 517

Kochanek, C.S., 1993, ApJ, 419, 12

Lehár, J., & Broadhurst, T., 1995, in these proceedings

Magain, P., Surdej, J., Swings, J.-P., Borgeest, U., Kayser, R., Kühr, H., Refsdal, S., & Remy, M., 1988, Nature, 334, 325

Magain, P., Surdej, J., Vanderriest, C., et al., 1992, A&A, 253, L13 (Erratum: 1993, A&A, 272, 383)

Maoz, D., Bahcall, J. N., Schneider, D. P., Bahcall, N. A., Djorgovski, S., et al., 1993, ApJ, 409, 28

Myers, S.T., 1995, in these proceedings

Refsdal, S. & Surdej, J., 1994, Report on Progress in Phys, 128, 295

Surdej, J., Magain, P., Swings, J.-P., Borgeest, U., Courvoisier, T.J.-L., Kayser, R., Kellermann, K.I., Kühr, H., & Refsdal, S., 1987, Nature, 329, 695

Surdej, J., Magain, P., Swings, J.P., et al., 1988, in Large Scale Structures: Observations and Instrumentations, eds. C. Balkowski & S. Gordon

Surdej, J., Claeskens, J.-F., Crampton, D., Filippenko, A.V., Hutsemékers, D., et al., 1993a, AJ, 105, 2064

Surdej, J., Remy, M., Smette, A., et al., 1993b, in Gravitational lenses in the Universe, eds. J. Surdej, D. Fraipont-Caro, E. Gosset, S. Refsdal & M. Remy, (Liège: Université de Liège) 153

Surdej, J., et al., 1995, in these proceedings

Turner, E., Ostriker, J., & Gott, J., 1984, ApJ, 284, 1

Weymann, R., Latham, D., Angel, J., Green, R., Liebert, J., Turnshek, D., & Tyson, J., 1980, Nature, 285, 641

Wisotski, L., Koehler, T., Kayser, R., & Reimers, D., 1993, A&A, 278, L15

Yee, H., Filippenko, A., & Tang, D., 1993, AJ, 105, 7

PREDICTED LENS REDSHIFTS AND MAGNITUDES

P. HELBIG
Hamburger Sternwarte
Gojenbergsweg 112, D-21029 Hamburg, Germany

Abstract. For gravitational lens systems with unknown lens redshifts, the redshifts and brightnesses of the lenses are predicted for a variety of cosmological models. Besides providing hints as to which systems should be observed with a realistic chance of measuring the lens redshifts, which are needed for detailed lensing statistics and for modelling the lenses, these calculations give a visual impression of the influence of the cosmological model in gravitational lensing.

1. Introduction and Basic Theory

The singular isothermal sphere produces an image separation twice the radius of the Einstein ring *independent* of the relative angular positions of source and lens (Turner et al. 1984) and allows one to define a cross section for multiply imaging a background source by a single (galaxy) lens. The relative probability p of the lens being at a given redshift is proportional to this cross section, the relative numbers of lenses of the appropriate mass and to the volume element dV/dz at the given redshift. The cross section for a single lens is not constant as a function of redshift since the cross section depends on redshift-dependent angular size distances and since the mass ($\rightarrow$ velocity dispersion) needed to produce the observed image separation depends on the redshift. The relative numbers of such galaxies one can get from the Schechter function (Schechter 1976) after converting the velocity dispersion to an absolute magnitude *via* the Faber-Jackson relation (Faber & Jackson 1976) for ellipticals or the Tully-Fisher relation (Tully & Fisher 1977) for spirals (Kochanek 1992). The volume element can be calculated in the standard way (Feige 1992). Both the angular size distances and dV/dz depend on the cosmological model. Since the Faber-Jackson/Tully-Fisher relation provides the absolute magnitude, the apparent magnitude can be calculated with standard methods.

21

2. Calculations, Results and Discussion

Values of $(-0.5, 0.3)$, $(0.0, 0.3)$, $(0.0, 1.0)$, $(1.0, 0.0)$, $(0.7, 0.3)$ and $(1.0, 1.0)$ were used for (λ_0, Ω_0). These values were chosen to satisfy the majority of the following constraints: (a) compatibility with all relatively certain and well-understood observations (b) maximization of the differences due to the cosmological model within the above area, (c) inclusion of several 'standard models' for purposes of comparison, (d) limitation of the size of the poster. The cosmological models examined here are thus not meant to be exhaustive but merely illustrative and somewhat representative.

For each lens system, for each cosmological model $p(z_d)$ as well as the lens brightness were calculated (separately for elliptical and spiral galaxies). Except in the case of $1104 - 1805$, since $m(z_d)$ is so steep, selection effects will probably cause those lenses which happen to have a low redshift to be found, regardless of the cosmological model. Neglecting the extreme de Sitter model $(1.0, 0.0)$ one can make a relatively robust prediction for the redshift and brightness of the lens galaxy in $1104 - 1805$, since in this case $p(z_d)$ peaks at approximately the same redshift in all cosmological models, the width of the probability distribution is small, and $m(z_d)$ is comparatively not very steep. (All are consequences of the relatively large image separation in this system, larger than that in any of the comparable systems with *known* redshifts. This system also has the smallest source redshift of the systems with unknown lens redshifts, which also contributes somewhat to the effect.) The lens should lie in the range $0.3 < z_d < 0.7$ and be brighter than about 21.5 in R, which means that it could be detected. Since here there is not a strong selection effect in favor of spirals over ellipticals, one would expect the lens to be an elliptical. This is probably the case, and a spiral lens would be so bright that it probably would have been found already. The fact that no lens has yet been found in this system can have one (or more) of three reasons: our cosmological model is near the de Sitter model, there is an unseen cluster responsible for the large image separation and thus the approximations used here break down, or the brightness of the images makes measuring the lens redshift difficult.

The complete poster text and figures can be obtained from
`ftp://ftp.uni-hamburg.de/pub/misc/astronomy/aus_poster.uu`

References

Turner, E. L., Ostriker, J. P. & Gott, J. R., 1984, ApJ, 284, 1
Schechter, P., 1976, ApJ, 203, 297
Tully, R. & Fisher, J., 1977, A&A, 54, 661
Faber, S. M. & Jackson, R. E., 1976, ApJ, 204, 668
Feige, B., 1992, Astr. Nachr., 313, 139
Kochanek, C. S., 1992, ApJ, 384, 1

VACUUM DECAYING COSMOLOGICAL MODELS AND GRAVITATIONAL LENSING

I. WAGA[1,2] AND L. F. BLOOMFIELD TORRES[2]
[1] NASA/Fermilab Astrophysics Center
Fermi National Accelerator Laboratory, Batavia, IL 60510
[2] Universidade Federal do Rio de Janeiro,
Instituto de Física
Rio de Janeiro - RJ - Brasil -21943

We investigate the statistical properties of gravitational lenses for flat models in which a cosmological term decreases with time as $\Lambda \propto a^{-m}$, where a is the scale factor and m is a parameter ($0 \leq m < 3$) (Bloomfield Torres and Waga 1995).

We use two different approximations in our computations. In the first one, that we call "A1", we follow Fukugita and Turner (1991) and use a simplified approach to take into account magnification bias (as in Fukugita & Peebles 1995). In the second approach, named "A2", we model E/S0 galaxies as singular isothermal spheres, use a velocity dispersion of $\sigma_\star = 220$ km s^{-1} (no correction factor), and the magnification bias is computed with the Wallington and Narayan (1993) quasar luminosity function (QLF).

In Table 1 we display, for three flat models and for the open FRW model with $\Omega_{m0} = 0.2$ (model D), the predicted number of lensed quasars for the Hubble Space Telescope (HST) Snapshot Survey (Maoz et al. 1993). By assuming Poisson statistics we also display, for all cases, the probability of detecting the observed number (four) of lensed quasars.

We applied Kochanek's (1993) maximum likelihood method to the flat models. Contours of constant likelihood (95.4%, 68% and 50% confidence levels) are plotted in Figure 1. If $m \simeq 0$, regions with a high value of Λ have low likelihood. However if, for instance, $m \gtrsim 1$ and $\Omega_{m0} \gtrsim 0.2$ or $m \gtrsim 1.5$ this constraint does not exist and models with low Ω_{m0} have high likelihoods for reproducing the observed lens statistics in the HST Snapshot Survey.

C. S. Kochanek and J. N. Hewitt (eds), Astrophysical Applications of Gravitational Lensing, 23–24.
© 1996 IAU

TABLE 1. **Predicted number of lensed quasars for the HSTSS and model probabilities**

	model A $k = 0$ $\Omega_{m0} = 1$	model B $k = 0, m = 2$ $\Omega_{m0} = 0.2$	model C $k = 0, m = 0$ $\Omega_{m0} = 0.2$	model D $k = -1$ $\Omega_{m0} = 0.2$
"A1"	2.6	4.7	10.7	4.1
probability	18%	17%	0.3%	19%
"A2"	3.2	5.7	12.7	5.1
probability	20%	12%	0.03%	15%

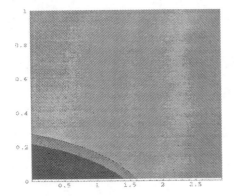 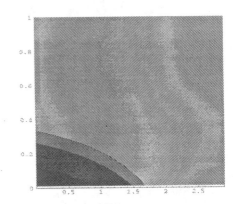

Figure 1. Contours of constant likelihood (95.4%, 68% and 50% confidence levels) for flat models are plotted in the Ω_{m0} (y-axis) and m (x-axis) parameter space. Regions with larger likelihood are represented by lighter shades. The left panel was obtained following Fukugita and Turner (1991) and Fukugita and Peebles (1995). For the right panel we used the Wallington and Narayan (1993) QLF.

Acknowledgements: We would like to thank Chris Kochanek for several discussions on lens statistics, core radius, magnification bias and other selection effects . This work was supported in part by the Brazilian agency CNPq and by the DOE and NASA at Fermilab through grant NAG5-2788.

References

Bloomfield Torres, & L. F., Waga, I., 1995, submitted to MNRAS
Fukugita, M., & Peebles, P. J. E., 1995, submitted to ApJL
Fukugita, M., & Turner, E.L., 1991, MNRAS, 253, 99
Kochanek, C. S., 1993, ApJ, 419, 12
Maoz, D., et al., 1993, ApJ, 409, 28
Wallington, S., & Narayan, R., 1993, ApJ, 403, 517

NOTE ON A SUPER-HORIZON-SCALE
INHOMOGENEOUS COSMOLOGICAL MODEL

K. TOMITA
Yukawa Institute for Theoretical Physics,
Kyoto University, Kyoto 606-01, Japan

1. Introduction

Many observations of large-scale and cosmological structures in the universe have been collected, but so far there is no consistent theoretical explanation. In the region within 100 Mpc from us, the observed two-point correlations of galaxies and clusters of galaxies can be described well by low-density homogeneous cosmological models (Bahcall & Cen 1993; Suto 1993). On the other hand, the observed anisotropies of the cosmic microwave background radiation have been explained well by comparatively high-density cosmological models such as the Einstein-de Sitter model (Bunn & Sugiyama 1994). In the intermediate scale, the angular sizes of the cores of quasars have been measured and their redshift dependence has been shown to be more consistent with the Einstein-de Sitter model than with the low-density models (Kellermann 1993). The number count-magnitude relation for remote galaxies supports low-density models with a nonzero cosmological constant (for example, Fukugita et al. 1990), but these models may be inconsistent with the observed distribution of Lyα clouds (Fukugita & Lahav 1991).

To avoid this contradictory situation it may be useful to drop homogeneous models and examine the observational consequences of an inhomogeneous cosmological model that consists of a local, low-density, negative curvature region, an inhomogeneous region, and a nearly flat, homogeneous region. This model is analogous to the void models, which have been studied by many people, and it is well-known that there is a narrow region with comparatively high density in these models. In view of the large-scale structure of the universe, we shall be confronted with severe observational contradictions if the high density shell lies at redshifts $z < 1.5$. However, it may be possible to have the shell between the redshifts of $z_1 = 1.5$-2.0 and

25

C. S. Kochanek and J. N. Hewitt (eds), Astrophysical Applications of Gravitational Lensing, 25–26.
© 1996 *IAU*

$z_2 \sim 5$, which corresponds to the epoch when the number of quasars and Lyα clouds changed dramatically (Hartwick & Schade 1990; Bechtold 1994). In a recent paper (Tomita 1995a) I derived such a spherically-symmetric inhomogeneous cosmological model and some of its observational properties.

2. An Inhomogeneous Cosmological Model

The Universe is assumed to consist of three regions: an inner low-density homogeneous region (a), a self-similar region (b), an outer, nearly-flat, homogeneous region (c). Its evolution is described in terms of the Tolman-Bondi solution without a cosmological constant. The present densities of the regions are $\Omega_0(a) = 0.1\text{-}0.2$, $\Omega_0(c) \simeq 1.0$, and the shell region has $\Omega_0(b) > \Omega_0(c)$. The observer must be near the origin of the inner region, to be consistent with the small dipole-anisotropy of CMB (Tomita 1995b).

In this model we derived the redshift dependence of angular, d_A, and luminosity distances, d_L, the number density N_u of Lyα clouds per unit redshift, and the number counts $N(m)$ of faint galaxies. The angular diameter of quasar cores is proportional to $1/d_A$, and we found that it can reproduce the observed behavior (Kellermann 1993), because it is approximately the Einstein-de Sitter model for $z > z_1$. The number density N_u can reproduce the observed rapid changes in the numbers of quasars and Lyα clouds. Moreover, number counts in the present model can produce the observed excess over the counts in homogeneous models with $\Lambda = 0$, because of the larger number density of galaxies. Thus, this inhomogeneous model may play an important role in observational cosmology.

References

Bahcall, N.A., & Cen, R., 1993, ApJL, 407, L49
Bechtold, J., 1994, ApJ, 91, 1
Bunn, E. F., & Sugiyama, N., 1995, ApJ, 446, 49
Fukugita, M., & Lahav, O., 1991, MNRAS 253, 17p
Fukugita, M., Takahara, F., Yamashita, K., & Yoshii, Y., 1990, ApJL, 361, L1
Hartwick, F.D.A., & Schade, D., 1990, ARA&A, 28, 437
Kellermann, K.I., 1993, Nature, 361, 134
Suto, Y., 1993, Prog Theor Phys, 90, 1173
Tomita, K., 1995a, ApJ, 451, 1
Tomita, K., 1995b, preprint YITP/U-95-15

THE HUBBLE CONSTANT: PRESENT STATUS

MASATAKA FUKUGITA

*Yukawa Institute, Kyoto University, Kyoto 606, Japan, and
Institute for Advanced Study, Princeton, NJ08540, U.S.A.*

Abstract. The current status on the value of the Hubble constant is reviewed with the emphasis given to the origin of the discrepancy among authors. I argue that the situation is not too controversial and straightforward reading of observations indicates a high value ($H_0 \simeq 75$ km s^{-1} Mpc^{-1}).

1. Introduction

The value of the Hubble constant has been an issue of controversy over many years, the current version being summarized as whether $H_0 = 50$ km s^{-1} Mpc^{-1} or 80 km s^{-1} Mpc^{-1} (see Jacoby et al. 1992; van den Bergh 1992, Fukugita, Hogan & Peebles 1993 for general reviews). The great advances that have been made over the last five years, however, lead us to the conclusion, in my opinion, that we have no serious controversy on H_0.

In this talk I shall discuss the key points that gave rise to the longstanding controversy and summarize the recent advances concerning these points. In particular, I will address the reasons why I believe that the controversy is basically resolved, although I do not mean that all problems in the distance scale were resolved.

There are two paths to estimating the Hubble constant: one is to measure the distance to the Virgo cluster and derive H_0 either by estimating the Hubble recession velocity of the Virgo cluster, after correcting for large peculiar motions, or by using the relative distances of the Virgo and Coma clusters plus the recession velocity of the Coma cluster. The other path is to circumvent the Virgo cluster and to tie distant clusters directly to nearby galaxies with securely determined distances. In the first approach the controversy is ascribed to the distance to the Virgo cluster, and whether

C. S. Kochanek and J. N. Hewitt (eds), Astrophysical Applications of Gravitational Lensing, 27–36.
© 1996 IAU

it is 16 or 22 Mpc. In the second approach, the controversy is whether the Tully-Fisher (TF) relation, which has been believed to be the most reliable distance indicator that reaches the distances beyond the Virgo cluster, gives a correct distance or suffers from a strong selection bias: a straightforward reading of the TF relation, when calibrated with nearby galaxies, leads to a high value of H_0. There are a few other distance indicators that reach beyond the Virgo cluster. They occasionally lead to a low value of H_0. The credibility of these results is also a subject for this talk.

2. Where People Agree

2.1. THE DISTANCES TO NEARBY GALAXIES

There are no debates over the distances to galaxies determined with the Cepheid period-luminosity relation. The most fundamental of these is the distance to the Large Magellanic Cloud (LMC). The Cepheid distance (Feast 1991), obtained both in the optical and the near infrared, is confirmed by the expanding photosphere method using SN1987A (Schmidt et al. 1992) and the time delay of the ring echo associated with this supernova (Panagia et al. 1991), both of which are physical methods that do not need local calibrations. We conclude that the distance to LMC is 50±3 kpc, although RR Lyraes give a value a little smaller than this.

About 20 galaxies have distances measured with Cepheid observations, an area where great advances have been made with the *Hubble Space Telescope*. Among these galaxies M31, M33, NGC300, N2403 and M81 (and NGC3109) have been used as local calibrators for a variety of secondary distance indicators. No doubt has been cast on these distances up to the error of the measurement, which is 5-10% relative to the LMC distance. The most recent advances are the determination of the distances to two galaxies in the Virgo cluster (Pierce et al. 1994, Freedman et al. 1994) and M96 in the Leo group (Tanvir et al. 1995)

2.2. THE RELATIVE DISTANCES AMONG CLUSTERS

In contrast to the case of the distance to the Virgo cluster, there is little dispute over the relative distances of the Virgo and Coma clusters. De Vaucouleurs (1993) compiled the estimates in the literature and concluded that the ratio is 5.60±0.30 (5%), and few authors disagree with this value. After correcting for the proper motion of our Galaxy, one obtains $v_H = 7200 \pm 80\text{km s}^{-1}$ (1.1% error). Taking these values, we obtain $H_0 = 80$ if $d(\text{Virgo}) = 16$ Mpc, and 58 if $d(\text{Virgo}) = 22$ Mpc.

We note that little controversy is seen for the distance ratios between other clusters. A classical example is given by the Hubble diagram of first rank elliptical galaxies (Kristian, Sandage & Westphal 1978), which extends linearly at least to $z = 0.4$. The scatter is small, and if $H_0 = 80$ is obtained from nearby clusters, it should represent a global value up to at least 1000 Mpc. We stress that the present controversy about H_0 comes from the calibration of the absolute distance to galaxies near the Virgo distance.

3. Local Calibrations and Checks Between Distance Indicators

Distance ladders require calibrations to determine the zero point, and local calibrations are made primarily using galaxies with Cepheid distances. The TF relation can be calibrated using M31, M33, NGC300, NGC2403 and M81 (and NGC3109). The scatter is 0.3 mag for these galaxies even for the B band TF relation. There are two novel secondary distance indicators developed over the last five years: one technique using surface brightness fluctuations (SBF) (Tonry & Schneider 1988) and one using planetary neb- ula luminosity functions (PLNF) (Jacoby et al. 1990). These two indicators are applied to old objects, elliptical galaxies or spiral galaxy bulges, where the composition of the stellar population is reasonably uniform. Calibra- tions are made using M31, and M81 gives an additional check.

The expanding photosphere method for type II supernovae (EPM) is a physical method that does not need local calibrations (Schmidt et al. 1992). A check against the Cepheid distances to the LMC, M101 and M100, however, is valuable, since it verifies the elaborate procedures involved in the EPM.

Most of secondary distance indicators do not have a solid physical basis, but are based on empirical grounds; this has been taken to be a weak point in the distance ladder argument. For this reason it is of crucial importance to check the distances obtained by various indicators with each other, which would justify the validity of indicators and allows us to document the error.

It was shown that the rms scatter between the Tully-Fisher distance and that estimated with SBF is 10-15% for clusters, where ellipticals and spirals are supposed to be well mixed. A remarkable agreement is seen for a number of galaxies between the distance obtained with SBF and that with PNLF: the rms scatter is 5-7% for 16 galaxies up to 17 Mpc (Ciardullo, Jacoby & Tonry 1993). Such excellent agreement is unlikely to be fortuitous. Another interesting test is between EPM and Tully-Fisher distances (Pierce 1994; Schmidt et al. 1994). Good agreement is found between the two for 11 galaxies after allowing for a 10% offset (the EPM distance is longer). The rms scatter is on the order of 15%. This test not only gives us an additional verification of EPM, but also justifies many steps needed to obtain the TF distance. It is unlikely that the TF relation gives a wrong distance beyond 20% error for majority of spiral galaxies.

TABLE 1. Distances to Leo I galaxies

	Type	SBF	PNLF	TF	Cepheid
NGC3377	E	9.72	10.61		
NGC3379 (M105)	E	9.42	10.05		
NGC3384	S0	9.46	10.42		
NGC3368 (M96)	Sab			11.96	11.59
NGC3351 (M95)	Sb			10.94	

The new observation of Cepheids in M96 in the Leo I group (Tanvir et al. 1995) gives an interesting testing ground for a number of indicators. In Table 1 we present distances (in Mpc) to five galaxies of the Leo I group. All distances agree to 20%. However, we see that the distance to M96 (both Cepheids and TF) is systematically 10-20% larger than the distance to E and S0 galaxies estimated with SBF and PNLF. Two possible explanations are: (i) that calibrations of SBF (and also PNLF) for early type galaxies with the bulges of spirals (M31 and M81) have 20% (10%) errors, and (ii) that M96 is actually located behind the elliptical-S0 system, although Leo I is usually supposed to be a small system from its dimension on the sky. Allowing for possibility (i), we conclude that the error of these four indicators is at most 20% for Leo I located at 10 Mpc from the Milky Way.

4. The Distance to the Virgo Cluster

Application of the PNLF to five ellipticals in the Virgo cluster (M87, M86, M84, M49 and NGC4649) gives 15.2±0.2 Mpc, and SBF yields 15.6±0.6 Mpc for the same five galaxies (Ciardullo et al. 1993). These values agree with the 15.8±1.3 Mpc obtained from the TF relation applied to 26 spiral galaxies (Pierce & Tully 1988; see also Mould et al. 1983). Cepheid measurements for two spiral galaxies also give consistent results: 14.9±1.2 Mpc for NGC4571 (Pierce et al. 1994) and 17.1±1.8 Mpc for M100 (Freedman et al. 1994). These distances (summarized in Table 2) support the high value of H_0.

On the other hand, Sandage & Tammann (Sandage 1995; Sandage & Tammann 1994; 1990) have been claiming differently, as summarized in Table 3. The six methods listed in Table 3 consistently give 20–23.5 Mpc. I believe only two methods, the distances from the TF relation and from supernovae, deserve serious attention, since the other indicators are not well qualified. The distance from globular cluster luminosity functions (GCLF) depends on how it is derived. Secker & Harris (1993) discussed how the

TABLE 2. Distances to the Virgo cluster (the short scale)

method	distance (Mpc)	ref.
TF(bright spirals)	15.8 ± 1.3	Pierce & Tully 1988
PNLF	15.2 ± 0.2	Ciardullo et al. 1993
SBF	15.6 ± 0.6	Ciardullo et al. 1993
Cepheid (M100)	17.8 ± 1.2	Freedman et al. 1994
Cepheid (NGC4571)	15.6 ± 1.3	Pierce et al. 1994

GCLF distances to NGC4365, M49, and M60 are consistent with those inferred from the PNLF and SBF methods, while Sandage & Tammann, using the same data, but a different manipulation of the data, find the longer distance. The problems are that the position of the peak of the GCLF does not quite agree between Milky Way and M31, and that the positions of the peak for Virgo galaxies are not well determined due to a flat feature and poor statistics. This allows a distance that depends on how the data are manipulated. The method also lacks a cross-check with other distance indicators. The data for novae (Pritchet & van den Bergh 1987) are too poor to infer any accurate result: the result is basically derived from one nova in NGC4472, which is given the smallest error bar. Also, the data are too poor to constrain the form of the maximum luminosity-decline rate relation, so the result depends largely on how to parametrize the relation. I would ignore the result of $D_n - \sigma$ relation, since it lacks a good local calibration. When a distance indicator relation with a large scatter is calibrated with a single galaxy, the resulting distance is largely uncertain. If the PNLF/SBF distance to E and S0 galaxies of Leo I is used for a calibration, the Virgo distance becomes 16Mpc instead of 23Mpc obtained with a calibration using the M31 or M81 bulge. Cross checks that allow us to know the error are not made for $D_n - \sigma$.

The Origin of the controversy in the TF distance to the Virgo cluster. This is a serious issue, since it could discredit the use of the TF relation to estimate extragalactic distance scales. An important point, however, is that there is no serious disagreement on the distance to each galaxy; the difference between the two schools arises from the different choice of the sample. Kraan-Korteweg, Cameron & Tammann (1987) (as quoted in Table 3) and Fouqué et al. (1990) have used complete spiral samples in the Virgo cluster region of the sky. On the other hand, Pierce & Tully (1988) have chosen only bright spiral galaxies and discarded some galaxies which give larger distances. Mould et al. (1983)'s sample shares the same characteristics as Pierce & Tully's. This selection procedure has naturally aroused

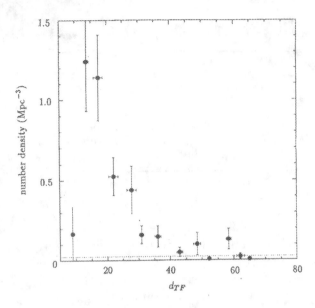

Figure 1. The distribution of spiral galaxy number density as a function of the TF distance. The dotted line is the base line for the field.

the suspicion that their samples suffer from a very strong selection bias. On the other hand, it is suspected that the Virgo cluster has a substantial depth, and that fainter spirals in the "complete sample" are those actually located in the background of the core.

A further study explicitly demonstrated that the Virgo cluster is elongated from 10 to 30 Mpc almost along the line of sight, and that the peak of the galaxy number density is located at about 15 Mpc (Fukugita et al. 1993; Yasuda et al. 1995). This means that an unusually large scatter of the TF relation for the complete sample of the Virgo spirals is caused by the actual depth, and not by the intrinsic scatter of the TF relation. The best evidence for the depth effect comes from the fact that HI deficient galaxies are located only in the range 14-20 Mpc, and this coincides with the region where the density is very high. We show in Fig. 1 the distribution of spiral-galaxy number density as a function of the distance. The position of the density peak also agrees with the positions of elliptical galaxies given by PNLF and SBF. If the density peak is identified as the Virgo core, the distance is about 15 Mpc. On the other hand, a simple average of all spiral galaxies gives 20 Mpc in agreement with the long distance listed in Table 3. (This also explains 20Mpc from the size of spiral galaxies in Table 3.)

We remark here that the present Cepheid observation does not give a compelling distance to the Virgo core, since we do not know the relative position of these spiral galaxies to the core. It is unfortunate that these

TABLE 3.   The Distance to the
Virgo cluster by Sandage & Tammann (the long scale)

method	distance (Mpc)
TF	20.9 ± 1.5
SNe Ia and II	20.9 ± 2.6
GCLF	21.3 ± 2.6
Novae	20.6 ± 4.5
$D_n - \sigma$	23.4 ± 2.1
size of Galaxy	20.0 ± 1.9

galaxies cannot be used to calibrate the TF distance, since they are too face-on to obtain a reliable distance with the TF relation.

5.   The Use of Supernovae as Distance Indicators

5.1.   TYPE IA SUPERNOVAE

The conventional use of type Ia SNe is to take the maximum brightness as a universal standard candle. Now, it is well recognized that there are some type Ia SNe with absolute brightnesses that are clearly dimmer. The scatter in the Hubble diagram varies from 0.2 mag (Branch & Miller 1993; Vaughan et al. 1995) to 0.6 mag (Leibundgut & Pinto 1992), depending on the selection of the sample. The zero point of the Hubble diagram is then determined with SN1937C in IC4182 (4.8 Mpc) or SN1972E in NGC5253 (4.1 Mpc). In this way Sandage & Tammann (1993) obtained $H_0 = 47 \pm 5$ or $H_0 < 55$ (Sandage & Tammann 1994) and Branch & Khokhlov (1995) obtained $H_0 = 55 \pm 5$.

Riess, Press & Kirshner (1995) have recently reconsidered the issue of the standard candle. They have shown that the maximum luminosity is correlated with the luminosity decline rate, as indicated earlier by Phillips (1993). Using the light curve shape (LCS) as a control parameter, they found that the scatter of the Hubble diagram is reduced to as small as 0.21 mag without any selection of the sample. This correction affects significantly both Hubble diagram and calibrator SN1972E, and brings $H_0 = 53 \pm 11$ with the conventional method up to $H_0 = 67 \pm 9$.

I admit, however, that the value of H_0 from type Ia SNe is still controversial. Branch and collaborators (Nugent et al. 1995) claim that it is difficult to reconcile the high H_0 with current models of SNeIa for the amount of radioactive ^{56}Ni.

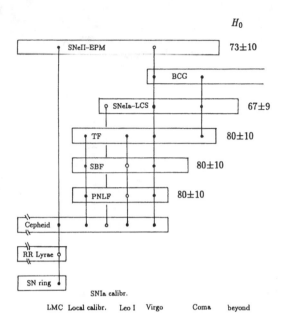

Figure 2. The distance ladder that leads to $H_0 = 70 - 80$km s^{-1} Mpc^{-1}. The vertical lines connected with solid dots mean that the relevant ladders are tightly constrained with each other, and those with open dots are those constrained allowing for $\approx 10\%$ offset. Typical H_0 resulting from each ladder is also indicated.

5.2. TYPE II SUPERNOVAE

Type II SNe do not give a standard candle. A variant of the Baade-Wesselink method, named EPM, has been developed to estimate the distance to type II SNe (Schmidt et al. 1992). SNeII do not emit photons like a black body, because the flux is substantially diluted by a scattering dominated atmosphere (Wagoner 1981). The distance obtained with this method agrees well with that from the TF relation, allowing for a 10% offset. The Hubble constant derived by Schmidt et al. (1994) is 73±6, which is about 10% smaller than the value obtained with TF, PNLF and SBF.

Sandage & Tammann have given long distances to the Virgo cluster with both type Ia and type II SNe. The distance with type II SNe is derived from two methods, radiation flux calculated assuming a black body (which overestimates the flux) and the radio sphere parallax measurement for SN1979C in M100 with VLBI. The latter assumes that the radio sphere and optical sphere are identical, which yields 22Mpc to M100, compared to 17Mpc from Cepheids.

6. The Global value of the Hubble constant

The linearity of the distance-recession velocity relation for type II SNe extends to 180 Mpc (Schmidt et al. 1994), and the Hubble diagram of type Ia SNe shows excellent linearity up to 300 Mpc (Riess et al. 1995). A classical example of the Hubble diagram for first rank ellipticals in clusters indicates linearity between the apparent V (or R) magnitude and the recession velocity to $z \simeq 0.4$, where evolution may affect the results (Kristian et al. 1978). The scatter in the diagram is small enough to exclude a change of the intercept (=Hubble constant) by a factor of 1.5. A modern version is provided by Lauer & Postman (1992) up to $cz \simeq 15000$ km s^{-1}, and their diagram does not allow a change of H_0 by more than 15%. The results make the suspicion that a high H_0 is a local effect unlikely: once H_0 is determined with nearby galaxies, the same value describes the expansion of the Universe up to at least $z \sim 0.4$.

Another interesting physical method, which has recently attracted our attention, is the use of the Zeldovich-Sunyaev (ZS) effect (Birkinshaw et al. 1991). All earlier attempts gave small values of H_0. A general caution to be made, however, is that clusters are selected on the basis of surface brightness and such a selection method induces a bias towards clusters elongated along the line of sight. Since the argument using the ZS effect assumes spherical symmetry for clusters, this readily causes a bias towards a low H_0. This bias should be stronger for distant clusters, such as those used for the ZS test. Recent observations (Herbig et al. 1995; Meyers et al. 1995) of nearby clusters allow this point to be examined. Meyers et al. (1995) derived $H_0 = 74^{+29}_{-24}$ for the Coma cluster, consistent with the value obtained from other methods albeit with a large error. The lesson is that it is too premature to take H_0 measurements using the ZS effect in distant clusters seriously.

7. Conclusions

The most important advance in recent years is that the errors of a number of distance indicators are now well documented using cross-checks among distances from a variety of methods, and we are able to discriminate between reliable and unreliable indicators. We summarize that TF, SBF and PNLF give $H_0 = 80 \pm 10$, and SNeII (EPM) and SNeIa (LCS) give $H_0 = 70 \pm 10$. We conclude that the current best value is

$$H_0 = 75 \pm 10 \text{ km s}^{-1}\text{Mpc}^{-1}.$$

We note that many different distance indicators are so tightly correlated (allowing for occasional 10% error), as shown in Fig. 2, that it seems difficult to break the chain to obtain $H_0 \approx 50$. We are now tempted to take the difference between $H_0 = 80$ from the conventional distance ladders and $H_0 = 70$ from supernovae more seriously. I believe that there is not much controversy as to the value of the Hubble constant as far as its observational status is concerned. The reason why many people quote "Hubble constant is highly uncertain" comes from its notorious history during its infancy, and more importantly that "theorists want the controversy".

References

Birkinshaw, M., Hughes, J. P. & Arnaud, K. A., 1991, ApJ, 379, 466

Branch, D. & Miller, D. L., 1993, ApJL, 405, L5

Branch, D. & Khokhlov, A., 1995, Phys Repts, in press

Ciardullo, R., Jacoby, J. H. & Tonry, J. L., 1993, ApJ, 419, 479

de Vaucouleurs, G., 1993, ApJ, 415, 10

Feast, M. W., 1991, in Observational Tests of Cosmological Inflation (Dordrecht: Kluwer), 147

Fouqué, P., Bottinelli, L., Gougenheim, L. & Paturel, G., 1990, ApJ, 349, 1

Freedman, W. L., et al., 1994, Nature, 371, 757

Fukugita, M., Hogan, C. J. & Peebles, P. J. E., 1993, Nature, 366, 309

Fukugita, M., Okamura, S. & Yasuda, N., 1993, ApJL, 412, L13

Herbig, T., Lawrence, C. R., Readhead, A. C. S. & Gulkis, S. 1995, ApJ, submitted

Jacoby, G. H., Ciardullo, R & Ford, H. C. 1990, ApJ, 356,332

Jacoby, G.H., et al., 1992, PASP, 104, 599

Kraan-Korteweg, R. C., Cameron, L. M. & Tammann, G. A., 1988, ApJ, 331, 620

Kristian, J., Sandage, A & Westphal, J., 1978, ApJ, 221, 383

Lauer, T. & Postman, M., 1992, ApJ, 221, 383

Leibundgut, B. & Pinto, P. A., 1992, ApJ, 401, 49

Meyers, S. T., Baker, J. E., Readhead, A. C. S., et al., 1995, Caltech preprint

Mould, J., Aaronson, M. & Huchra, J., 1980, ApJ, 238, 458

Nugent, P. et al., 1995, PhysRevLett 75, 394

Panagia, N. et al., 1991, ApJL, 380, L23

Phillips, M.M., 1993, ApJL, 413, L105

Pierce, M.J., 1994, ApJ, 430, 53

Pierce, M.J. & Tully, B.R., 1988, ApJ, 330, 579

Pierce, M. J. et al. 1994, Nature, 371, 385

Pritchet, C. & van den Bergh, S., 1987, ApJ, 318, 507

Riess, A.G., Press, W.H. & Kirshner, R.P., 1995, ApJL, 438, L17

Sandage, A., et al., 1995, in Particles, Strings & Cosmology (Johns Hopkins University, March 1995), in press

Sandage, A. & Tammann, G.A. 1990, ApJ, 365, 1

Sandage, A. & Tammann, G.A., 1993, ApJ, 415, 1

Sandage, A. & Tammann, G.A., 1994, in Proc. of the 3rd Chalonge School, in press

Schmidt, B.P., Kirshner, R.P. & Eastman, R.G., 1992, ApJ, 395, 366

Schmidt, B.P., et al., 1994, ApJ, 432, 42

Secker, J. & Harris, W.E., 1993, AJ, 105, 1358

Tanvir, N.R., et al., 1995, Nature, submitted

Tonry, J. L. & Schneider, D. P. 1988, AJ, 96, 807

van den Bergh, S., 1992, PASP, 104, 861

Vaughan, T.E., Branch, D., Miller, D. L. & Perlmutter, S., 1995, ApJ, in press

Wagoner, R.V., 1981, ApJL, 250, L65

Yasuda, N., Fukugita, M. & Okamura, S., 1995, preprint

RADIO MEASUREMENT OF THE TIME DELAY IN 0218+357

E.A. CORBETT, I.W.A. BROWNE AND P.N. WILKINSON
University of Manchester, Nuffield Radio Astronomy Laboratories
Jodrell Bank, Macclesfield, Cheshire, SK11 9DL, U.K.

AND

A.R. PATNAIK
Max-Planck-Institut für Radioastronomia
Auf dem Hügel 69, D-53121 Bonn, Germany

Abstract. A time delay of 12 ± 3 days has been measured for the lens system 0218+357 using VLA 15GHz polarization observations.

1. Introduction

The potential for using a measurement of the time-delay in a gravitational lens system to give a value for the Hubble constant has long been appreciated (Refsdal 1964). However, only a few systems suitable for such a determination have been found. The 0218+357 system (Patnaik et al. 1993), found in JVAS (The Jodrell/VLA Astrometric Survey – see Patnaik et al. 1992), is one of the most promising. It consists of 2 images of a strong flat spectrum radio core plus a low brightness ring of emission (Figure 1). The 2 core images are compact, have high radio polarizations ($\sim 10\%$) and show evidence for radio variability. Both the ring and VLBI images of the compact core images (Patnaik et al. 1995) give useful constraints on the mass distribution of the lensing galaxy. The lensing galaxy redshift has been measured to be 0.6847 from an [OII] 3727Å emission line and optical absorption lines (Browne et al, 1993). It has been confirmed by the detection of 21cm neutral hydrogen absorption (Carilli et al. 1993) and recently by CO and other molecular absorption lines (Wicklind & Combes 1995). Lawrence et al. (in these proceedings) report the detection of a broad emission line and associated absorption from which they propose a redshift

37

C. S. Kochanek and J. N. Hewitt (eds), Astrophysical Applications of Gravitational Lensing, 37–42.
© 1996 IAU

of 0.96 for the lensed object. In this contribution we report the successful determination of a time delay in 0218+357.

2. VLA Observations

Observations were made with the VLA[1] in its A-configuration at frequencies of 15GHz and 8.4GHz on 25 irregularly spaced occasions during 3-months in 1992/3. The resolution of the VLA at these 2 frequencies is 0.12 arcsec and 0.2 arcsec respectively, adequate to separate the compact images which are 0.335 arcsec apart. We set out to monitor the total flux densities, the polarized flux densities and polarization position angles of the compact images. As well as 0218+357, three calibration sources and a "control source" were observed during the 45 minutes allocated to each epoch.

3. Data Analysis

Owing to the short duration of the observations at each epoch, some non-standard calibration methods were necessary. For flux density calibration we assumed that the VLA primary calibrator 3C84 had constant flux density over the period of observation. When all epochs had been calibrated, the validity this assumption was checked by seeing if the flux densities derived in this way for the CSS sources 3C48, 3C119 and 3C147 were constant with time. This proved to be the case. The antenna polarization residual terms were also derived from the 3C84 observations on the assumption that 3C84 is unpolarized – something supported by the long-term monitoring program of Aller & Aller (personal communication). The zero of polarization position angles was set arbitrarily with respect to that of the position angle of 3C119.

The target, the calibration sources and the control source (3C119) were all mapped using an AIPS procedure involving 2 cycles of phase self-calibration. Stokes I, Q and U flux densities for all but the target source were derived using the AIPS task JMFIT. In the case of 0218+357, fitting in the image plane proved not to be a satisfactory method to derive the core image flux densities because the core images are blended with the extended ring emission. Instead we exploited the fact that the ring emission is almost completely resolved at baselines $\geq 400\text{k}\lambda$. We excluded all baselines shorter than 400 kλ from the self-calibrated data-set and used the task UVFIT to model-fit 2 point components to the visibility data. The component separation and position angle were fixed to the values found from high resolution MERLIN and VLBI maps. In this way I, Q and U flux densities

---

[1]The VLA is operated by Associated Universities, Inc., under a cooperative agreement with the National Science Foundation.

were obtained for the two images, uncontaminated by the low brightness emission.

4. Results

The peak-to-peak errors in total flux density are ~7% at 15GHz and ~3% at 8.4GHz due to uncertainties in calibration. These uncertainties masked any clear signal that could be used to determine a time delay, but it was evident that the ratio of the flux densities of the 0218+357 images varied, indicating real variability on a time scale comparable to, or shorter than, the time delay. Small residual errors in the polarization position angle calibration compromise the usefulness of this parameter for time delay measurements. On the other hand the percentage polarizations of the two images show clear variations. The 15GHz polarization light curves are shown in Figure 2. The 8.4GHz results show similar trends with the percentage polarizations of both the A and B images increasing with time. The formal errors on the measured percentage polarizations are $\leq 0.2\%$. Moreover, systematic flux density calibration errors cancel. Hence percentage polarization is a much more robust parameter to use in a time delay determination. In Figure 3 we show the 15GHz percentage polarization light curve for the control source 3C119. In contrast to the steady increase in the percentage polarization seen in 0218+357, 3C119 has constant polarization. We note that the percentage polarization of the strongest image (A) of 0218+357 shows significant changes on time scales as short as 2 or 3 days – something that should enable the time delay to be measured to an accuracy of ~1 day if similar variations can be detected in B.

The polarization properties of the images of a point object should be identical. There are, however, complications. The first is that, if the object being imaged has internal polarization substructure, different magnification gradients across the images could give rise to each image having different integrated polarizations. In the case of 0218+357 we know that the 15GHz substructure has a size of only ~ 1mas (Patnaik et al. 1995) – a scale over which the models indicate that the magnifications are essentially constant. Therefore this is not a problem in the present case.

An additional complication is that we view the images through the ISM lensing galaxy and the two ray paths suffer different degrees of Faraday rotation, and even Faraday depolarization. Both effects occur in the 0218+357 system; the rotation measures of the two images differ by 814 radian m^{-2} and, at frequencies of 5GHz and lower, the percentage polarization of the A image is much less than that of the B image, something that we attribute to depolarization caused by the Faraday depth changing across the face of the A-image. In the case of the Faraday rotation it is possible to determine the rotation measure difference and correct for it on the assumption that a λ^2 law is obeyed. The way to avoid the effects of depolarization is to observe at high frequencies. We think 15GHz is high enough to be free from this problem and hence we base most of our conclusions on observations at this frequency. We also note that the A-image is the most depolarized. It is this image that has the higher average percentage polarization during the period of our observations. Thus any small residual depolarization in A will result in the separation in the light curves for the two images (Figure 2) being underestimated. This would also lead to our measured time delay being an underestimate of the true time delay.

5. Time Delay Analysis

We slide the polarization light curve for one image relative to the other and minimize the difference to give the delay. We have done this visually and, more formally, by linearly interpolating between the measurements in one light curve and adjusting the time delay to minimize χ^2 (normalized for the decreasing number of epochs compared for increasing delays) between the two time series. The plot of the normalized χ^2 against delay is shown in Figure 4. There is a distinct minimum at a delay of 12 days. The formal result is that the delay is 12 days with a 1σ error of ± 3 days.

6. Conclusions

- (1) We have determined the time delay in the gravitationally lensed system 0218+357 from the variation of the radio percentage polarization of the images with time. The delay is 12 ± 3 days.

- (2) Given the redshift of the lensed object measured by Lawrence et al. (this conference), the value for the Hubble constant obtained is $\sim$60 km s^{-1} Mpc^{-1} (Nair, this conference).

- (3) With more frequent and regular sampling of the polarization light curve we can greatly improve the accuracy of our measurement of time delay in the future.

References

Browne, I.W.A., et al., 1993, MNRAS, 263, L32
Carilli, C.L., et al., 1993, ApJL, 412, L59
Patnaik, A.R., et al., 1992, MNRAS, 254, 655
Patnaik, A.R., et al., 1993, MNRAS, 261, 435
Patnaik, A.R., Porcas, R.W. & Browne, I.W.A., 1995, MNRAS, 274, L5
Refsdal, S., 1964, MNRAS, 128, 307
Wiklind, T. & Combes, F., 1995, A&A, 299, 382

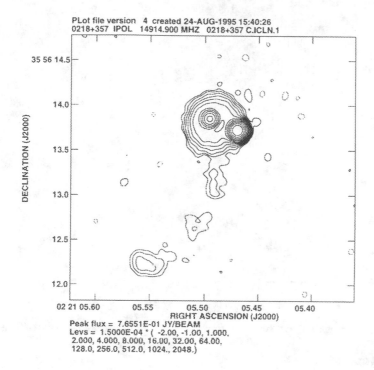

PLot file version 4 created 24-AUG-1995 15:40:26
0218+357 IPOL   14914.900 MHZ   0218+357 C.ICLN.1

Peak flux = 7.6551E-01 JY/BEAM
Levs = 1.5000E-04 * (-2.00, -1.00, 1.000,
2.000, 4.000, 8.000, 16.00, 32.00, 64.00,
128.0, 256.0, 512.0, 1024., 2048.)

Figure 1. VLA 2cm map of 0218+357 showing the two core images, the ring and a weak jet.

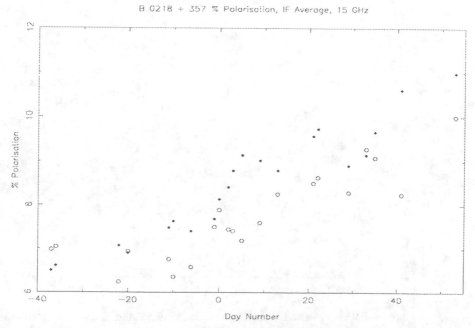

Figure 2. Percentage polarization of the A-image (filled circles) and B-image (open circles) as a function of time.

E.A. CORBETT ET AL.

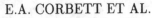

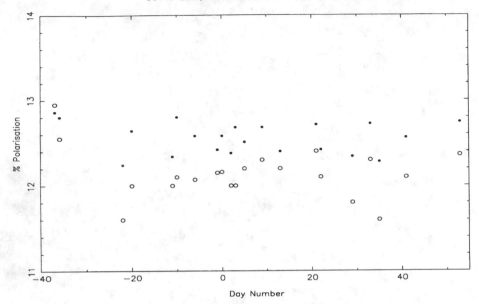

Figure 3. The percentage polarization as a function of time for the control source 3C119.
The two VLA IFs are distinguished by different symbols.

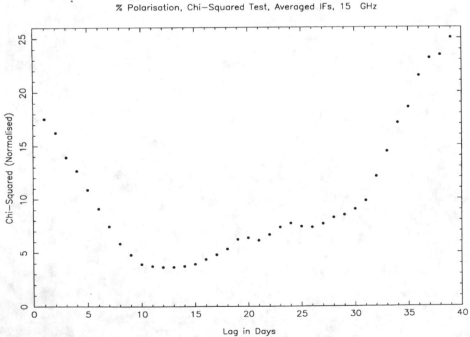

Figure 4. The Normalized χ^2 as a function of time lag between the A and B polarization
light curves.

THE VLA LIGHT CURVES OF 0957+561, 1979-1994

D. B. HAARSMA, J. N. HEWITT AND B. F. BURKE
Massachusetts Institute of Technology

AND

J. LEHÁR
Harvard-Smithsonian Center for Astrophysics

Abstract. The gravitational lens 0957+561 has been monitored by our group for 16 years, using the VLA at 5 GHz. Since our last report in 1992, both images have returned to their quiescent flux levels, and the A image has brightened again. We apply various analysis methods to the light curves, and obtain a preliminary estimate of $\tau = 455 \pm 40$ days for the time delay.

Since its discovery (Walsh *et al.* 1979), we have monitored 0957+561 using the VLA at 5 GHz. There are currently 111 good observations up to December 1994, taken approximately every month. We map the VLA data and measure the fluxes of the two quasar images using the techniques of Lehár *et al.* (1992). The new light curves are presented in Figure 1. Since 1990, the fluxes of both the A and B images have decreased back to their quiescent levels, and the A image has risen again in flux. The strong fluctuation in the B image in the spring of 1990 appears to be anomalous, but these data calibrated without problems and produced good quality maps, so they are not bad data points. When the curves are lined up at delays greater than 500 days, it can be seen that the A image does not decrease fast enough in 1991 to match the B image flux, so long delays seem to be excluded by the new feature.

The method of Press *et al.* (1992a, 1992b) uses a maximum likelihood estimator based on a structure function analysis of the light curves. Press *et al.* found a χ^2 minimum to the 1979-1990 data of $\tau = 548^{+10}_{-16}$ days (95% C.L.) and flux ratio $R = 0.694 \pm 0.002$; there is also a secondary minimum around $\tau = 455$ days and $R = 0.698$. Using the same analysis on the 1979-1994 data, we find that the χ^2 minimum has shifted to $\tau = 455$ days and $R = 0.698$ (see Figure 2). Note that there is still a local minimum

C. S. Kochanek and J. N. Hewitt (eds), Astrophysical Applications of Gravitational Lensing, 43–44.
© 1996 IAU

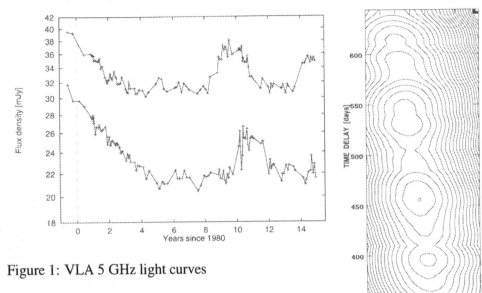

Figure 1: VLA 5 GHz light curves

Figure 2: Chi-squared surface for new light curves:
contours start at 280 and increase by 10 in chi-squared

around $\tau = 530$ days, as well as around $\tau = 400$ days. A Monte Carlo analysis that assumes a gaussian process gives an uncertainty for the new result of ± 10 days. The Monte Carlo χ^2 surfaces do not have the structure of the χ^2 surface of Figure 2, so we suspect that this analysis underestimates the uncertainty in the time delay. From inspection of Figure 2, we estimate an uncertainty of ± 40 days. Rapid variations, especially during the B-image rise in 1990, are responsible for the ambiguity in the delay. Interstellar scintillation may be important, and an analysis that takes this into account is needed.

Pelt et $al.$ (1994) determine the time delay by minimizing the dispersion of the combined light curve. For the 1979-1990 data, the dispersion using only AB pairs has a minimum at $\tau = 533 \pm 40$ days. When the observations on 90Apr10 and 90May07 are removed, the minimum shifts to $\tau = 409 \pm 23$ days. Applying the same analysis to the 1979-1994 data, we find the dispersion minimum is again at 533 days, and it also shifts to 409 days when the 90Apr10 and 90May07 observations are removed.

References

Lehár J., Hewitt J.N, Roberts D.H., & Burke B.F., 1992, ApJ, 384, 453
Pelt J., Hoff W., Kayser R., Refsdal S., & Schramm T., 1994, A&A, 286, 775
Press W.H., Rybicki G.B., Hewitt J.N., 1992a, ApJ, 384, 404
Press W.H., Rybicki G.B., Hewitt J.N., 1992b, ApJ, 384, 416
Walsh D., Carswell R.F., & Weymann R.J. 1979, Nature, 279, 381

WHY IS THE "TIME DELAY CONTROVERSY IN Q0957+561 NOT YET DECIDED"?

V.L. OKNYANSKIJ
Sternberg Astronomical Ins.
13 Universitetskij prospekt, 119899, Moscow, Russia

Abstract. The time delay in Q0957+561 remains indeterminate due to differences between statistical methods, irregular data spacings, and (possibly) microlensing.

Until now, the first gravitational lens, Q0957+561 A B, has been the most attractive object for time delay determinations (see Beskin & Oknyanskij 1995). Yet in spite of intensive efforts to measure the correct value of the time delay using long-term optical and radio monitoring, we have no time delay value that would be accepted by all specialists working in the field. The published time delays for Q 0957+561 can be divided into three sets:

1. The time delay value is about 400-430 days (Schild and Cholfin 1986, Schild 1990, Vanderriest et al. 1989, Pelt et al. 1994, 1995).

2. The time delay value is about 520-555 days (Beskin and Oknyanskij 1992,1995, Press et al. 1992).

3. A definite time delay value cannot be found due to gaps in the data sets and possible microlensing effects (Falco et al. 1991).

Pelt et al. (1994) published a paper entitled "The Time Delay Controversy On QSO 0957+561 Is Not Yet Decided". They discussed two possible values for time delay about 410 and 540 days, but really preferred the first of them. Then, in the next paper (Pelt et al. 1995), using new extended data (Schild and Thomson 1994, below ST94), they rejected values near 540 days as a possible time delay and concluded that the correct value is about 423 ± 6 days. The purpose of present paper is to explain why we cannot be sure of the Pelt et al. (1995) conclusion and that the "time delay controversy is (still) not yet decided."

The following is a brief summary of our arguments:

C. S. Kochanek and J. N. Hewitt (eds), Astrophysical Applications of Gravitational Lensing, 45–48.
© 1996 IAU

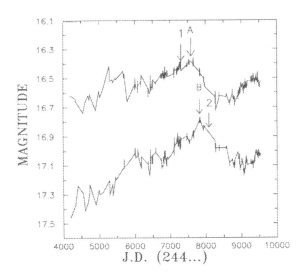

Figure 1. Optical light curves (only Schild's data) of Q0957+561 A,B. The light curve for image B is shifted down by 0.5 magnitude. The 1 marks the expected location of a maximum in A light curve given the location of the maximum in the B light curve and a time delay of 540 days. The 2 marks the same point for the B light curve.

- It was shown in Beskin and Oknyanskij (1992, 1995) that time delay values near 415 days may be artifacts. We should be very careful before admitting that the real time delay has about the same value expected from aliasing.
- The PDS method (Pelt et al. 1994) cannot eliminate the problem of irregular data spacing. It is simple to show that the signal/noise ratio in the PDS function depends on the time shift and must be higher for a time shift near 423 days than for one near 540 days.
- Removing trends from data sets can introduce additional noise, and give an incorrect value of the time delay.
- The spacing of weights and the density of observations in the new data set (Schild & Thomson 1994) is very different in the first and second parts of data. The first part of the data (before J.D. 2447120) covers more time, but it is practicality negligible in the time delay analysis.
- We note (see Figure 1) that the absolute maxima and the beginning of falloffs in A and B light curves cannot be explained by either the 540-day nor the 423-day time delay. They can be explained by a time delay of about 200-300 days. We will call this interval (J.D. 2447950-2449170) "irregular".

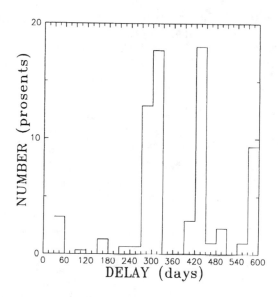

Figure 2. The distribution of time delays determined by the MCCF method (Oknyanskij 1994) for 500 Monte-Carlo simulations using the real B light curve and an artificial A light curve (see text) in the "irregular" interval.

— The autocorrelation function of the B image in the "irregular" interval has a very strong maximum of about 0.9 for a time shift near 400-440 days. We found from Monte-Carlo simulations (using the real B light curve and simulated data consisting of white noise with some random trend in place of the real A light curve) that there is a high probability of finding a time delay of near 420 days (see Figure 2).

In conclusion, the time delay controversy for Q 0957+561 remains undecided for two kinds of reasons:

1) *subjective* reasons (using different statistical methods including several methods that are not quite correct for irregularly spaced data), and

2) *objective* reasons (low signal/noise ratios for the time delay determinations due to periodic gaps in the data sets, possible microlensing and the small amplitude of the source variability).

Acknowledgements: The author is very thankful to R.Schild for sending photometric data before publication. This investigation has been in part financially supported by the Science Education Centre "COSMION" within the "Cosmoparticle Physics" project.

References

Beskin, G.M., & Oknyanskij, V.L., 1992, in Gravitational Lenses, eds. R. Kayser, T. Schramm & L. Nieser, (Berlin: Springer-Verlag) 67

Beskin, G.M., & Oknyanskij, V.L. 1995, A&A in press

Falco, E.E., Wambsganss, J., & Schneider P., 1991, MNRAS, 251, 698

Oknyanskij, V.L., 1994, Astrophys & Space Sci, 222, 157

Pelt, J. et al., 1994, A&A, 256, 775

Pelt, J. et al., 1995, A&A in press

Schild R.E., Cholfin, B., 1986, ApJ, 300, 209

Schild, R.E., 1990, AJ, 100, 1771

Schild, R.E., & Thomson D.J., 1994, AJ in press

Vanderriest, C., Schneider, J., Herpe, G., et al., 1989, A&A, 215, 1

GRAVITATIONAL LENSING OF QUASAR 0957+561 AND THE DETERMINATION OF H_0

GEORGE RHEE
Department of Physics, University of Nevada, Las Vegas

GARY BERNSTEIN
Department of Astronomy, University of Michigan

AND

TONY TYSON AND PHIL FISCHER
AT&T Bell Labs

1. Introduction

The double quasar 0957+561 was the first discovered instance of multiple imaging via gravitational lensing. The galaxy cluster is an important deflector as well as the first ranked galaxy. This has so far precluded construction of a unique model of the lens, reducing the accuracy of the derived H_0 value. We have obtained deep images of the system at CFHT. The cluster is sufficiently massive to cause distortions on distant background galaxy images. We have used a mass map derived from lensing distortions to improve the accuracy of the cluster center location and place new limits on H_0.

2. The Time Delay

Vanderriest et al. (1989) obtain a value of 415 ± 20 days based on optical data. Schild (1991) obtains a value of 404 ± 10 days also based on optical monitoring. From radio monitoring studies Lehar et al. (1991) conclude that the time delay is 513 ± 40 days.

3. The Mass Distribution

The galaxy principally responsible for the lensing (G1) has a redshift of 0.36. Bernstein et al. (1993) find that the surface brightness profile is well

C. S. Kochanek and J. N. Hewitt (eds), Astrophysical Applications of Gravitational Lensing, 49–50.
© 1996 IAU

fit by a power law with index $n = -1.94 \pm 0.06$. The position angle of the galaxy may twist slightly about a mean value of 55° . From spectroscopic observations Rhee (1991) derives a velocity dispersion 300 ± 50 km s^{-1} for G1.

The parameters for the galaxy model are: the velocity dispersion, the ellipticity, the galaxy position angle and the power law index. The cluster is modeled as a pseudo-isothermal sphere. The cluster maps show that the cluster potential is not perfectly smooth. This can be taken into account by adding a shear component to the two components listed above. The shear component has two parameters the magnitude of the shear and its position angle. The effect of the shear may be taken as some indication of the changes in the models that would result from the addition of lumps to the mass distribution.

The observations were made with FOCAM and the SAIC CCD at CFHT in January 1994. We obtained a 3 hour integration in B band and a 4 hour integration in I-band. By inverting the estimated shear field inferred from the observed shapes and orientations of background galaxies we have obtained a map of the surface mass density of the matter lensing Q0957+561.

4. Conclusion

From the map of the mass distribution we can locate the center of the cluster relative to the lensing galaxy. We measure the cluster center position to be at a distance (r_c) of 9 arcseconds from the lensing galaxy in the direction $\theta_c = 193°$. By limiting the range in values that these two parameters can have we eliminate all models except those having H_0 less than 70 km s^{-1} Mpc^{-1}.

Acknowledgements

George Rhee acknowledges travel support from the UNLV Physics department Bigelow fund and accommodation support from an IAU grant.

References

Bernstein, G.M., Tyson, J.A., & Kochanek, C.S. 1993, AJ, 105, 816
Lehar, J., Hewitt, J.N., & Roberts, D.H., 1989, in Gravitational Lenses, ed. J.M. Moran, J.N. Hewitt, & K.Y. Lo (Berlin: Springer) 84
Rhee, G. 1991, Nature, 350, 211
Schild, R., AJ, 1990, 100, 1771
Vanderriest, C., Schneider, J., Herpe, G., Chevreton, M., Moles, M., & Wlerick, G. 1989, A&A, 215, 1

THE Q0957+561 TIME DELAY

RUDOLPH E. SCHILD
Harvard-Smithsonian Center for Astrophysics
60 Garden Street, Cambridge MA 02138

AND

DAVID J. THOMSON
AT&T Bell Labs
Murray Hill, NJ 07974-2079

Abstract. The Q0957+561 time delay remains controversial, and we present four new time delay calculations based upon the intensively sampled data of the last three years.

To avoid problems due to sampling in time delay estimation, the TwQSO was scheduled for observations each night of the 9-month observing window that begins on Oct. 1, and on average 130 nights of data have been obtained each season for the past three years.

In Figure 1 we show the data plotted for a 404-day time delay, but with no other alterations. The crosses are for image A, and the open squares are for image B. Note that the B brightness record extends to JD 249500 only, because of the time delay correction. The B magnitudes have been shifted with a small magnitude offset to show the agreement with the correctly plotted A data. A 0.1 magnitude brightness increase was seen near JD2948950, and a complex brightness change with decreasing, increasing, and then decreasing flux was seen near JD2949400. These features repeated 1.1 years later in the B image, and give a secure time delay signature.

In Figure 2 we show four cross-correlation calculations for just the 3-year data subset shown in Figure 1. All four calculations peak near 404 days, and show no correlation amplitude for the alternative 540 day value. Of particular interest is the approximately 100 day widths of the symmetrical parts of the peaks, since one would have expected much sharper autocorrelation peaks with daily sampled data.

51

C. S. Kochanek and J. N. Hewitt (eds), Astrophysical Applications of Gravitational Lensing, 51–52.
© 1996 IAU

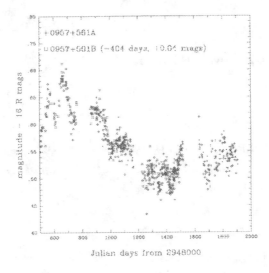

Figure 1.   The Q0957+561 data plotted for a 405-day time delay.

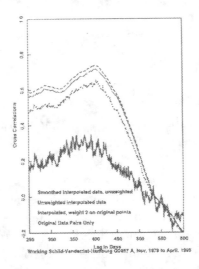

Figure 2.   Four autocorrelation calculations for the 3-year data subset

The upper curve in Figure 2 is a autocorrelation for the 3-year sub-sample with 7-day median smoothing filter first applied to the data, giving a cross-correlation peak of 0.73 for a 404-day lag. If the data are not smoothed, the curve of connected dots shows a slightly lower level of cross-correlation and a peak for 405 days. Attaching double weight to the real data points in the interpolated data set produces a 405-day peak lag and lower level of correlation (third curve from top). The Edelson-Krolick method produces a shorter, 360-day peak and a lower amplitude (solid curve).

PHOTOMETRIC MONITORING

OF THE LENSED QSO 0957+561

D. SINACHOPOULOS, M. BURGER AND E. VAN DESSEL
Royal Observatory of Belgium
Ringlaan 3, B - 1180 Brussels, Belgium

M. GEFFERT AND M. THIBOR
Sternwarte der Univ. Bonn
Auf dem Hügel 71, D - 53121 Bonn, Germany

AND

J. COLIN AND CH. DUCOURANT
Observatoire de Bordeaux
BP 89 - 33270 Floirac, France

We present our first results of a photometric monitoring project of the twin quasar 0957+561. This project aims mainly at the improvement of the determination of the time delay $\Delta T(A,B)$ for this gravitational lens, since the "time delay controversy on QSO 0957+561 (is) not yet decided" (Pelt et al. 1994). In addition, the quite large field of the CCD used allows also a long-term astrometric and photometric study of stars and galaxies in the field within a radius of about 10 arcminutes around the lens.

The present results are based on observations made with the 1m Cassegrain telescope of the Hoher List Observatory, its focal reducer (about f/4) and its 2048 × 2048 pixel CCD camera (15 × 15 μ^2), with a useful field of 20 × 20 arcmin^2 around this GL. A Bessel R filter has been used for the observations, but some V exposures have been taken as well in order to determine the color of the objects in the field. The exposure times were 3, 4 or 5 minutes. Longer exposures were excluded in order to avoid the guiding errors of the telescope.

Useful CCD data were obtained in two nights in April and three more nights in May 1995. Seeing was between 2.5 and 3.6 arcsecs. We have used DAOPHOT as well as ESO - MIDAS data fitting procedures for the data reduction. We fitted simultaneously two two - dimensional Moffat functions to the QSO images A and B, determining the PSF defining parameters

C. S. Kochanek and J. N. Hewitt (eds), Astrophysical Applications of Gravitational Lensing, 53–54.
© 1996 IAU

during fitting. One gets in this way the magnitude difference between the two QSO images. We combined this with aperture photometry of the total flux of the two QSO images and derived the magnitudes for each one of the images. We used star S3 (or ϵ) as a local standard star, adopting R = 14.77 mag from Schild and Weekes (1984, see there for a detailed discussion of the accuracy of this value). The magnitude difference between the two images and the astrometric results obtained from the two methods are the same at a 99% confidence level. To calculate the magnitude of the image B we adopted $R(G_1)$ = 18.7 mag (Vanderriest et al. (1989)) for the lensing galaxy G1. According to them the accuracy of this value is about 0.1 mag.

Results of our observations are presented in Table 1. Accuracies listed correspond to the mean error of the nightly mean.

The pixel size is 0.81 ± .01 arcsecs; the accuracy of the photocenter determination of one exposure is about 0.04 arcsecs.

TABLE 1. TwQSO Nightly Mean R Magnitudes

Date	R_A	σ_{R_A}	R_B	σ_{R_B}	$N_{exp.}$
24-APR-95	16.43	.01	16.29	.01	19
28-APR-95	16.47	.02	16.33	.02	10
22-MAY-95	16.39	.01	16.29	.02	10
23-MAY-95	16.49	.03	16.43	.03	5
27-MAY-95	16.41	.02	16.29	.01	3

Acknowledgements. This research was carried out mainly within the framework of the project "Service Centres and Research Networks", initiated and financed by the Belgian Federal Scientific Services (DWTC/SSTC).

References

Pelt, J., Hoff, W., Kayser, R., Refsdal, S., & Schramm, T., 1994, A&A, 286, 775
Schild, R.E. & Weekes, T., 1984, ApJ, 277, 481
Schild, R.E. & Cholfin, B., 1986, ApJ, 300, 209
Vanderriest C., et al., 1989, A&A, 215, 1

STRUCTURE FORMATION:

MODELS, DYNAMICS AND STATUS

T. PADMANABHAN

*Inter-University Centre for Astronomy and Astrophysics,
Post Bag 4, Ganeshkhind, Pune - 411 007, INDIA.
Email: paddy@iucaa.ernet.in*

1. Recipe for the Universe

All the popular models for structure formation are based on three key ingredients: (a) a model for the background universe (b) some mechanism for generating small perturbations in the early universe and (c) specification of the nature of the dark matter.

The background universe is usually taken to be a Friedmann model with an expansion factor $a(t)$. Such a model is completely specified if the composition of the energy density and the Hubble constant are specified. We will take $H_0 = 100h$ kms^{-1} Mpc^{-1} and express the energy density of the various species in terms of the critical density $\rho_c = (3H_0^2/8\pi G) = 1.88h^2 \times 10^{-29}$ g cm^{-3}, by writing $\rho_i = \Omega_i \rho_c$ for the ith species. From various observations, we can impose the following constraints: (i) $0.011h^{-2} \lesssim \Omega_B \lesssim 0.016h^{-2}$ (ii) $\Omega_{\text{vac}} \lesssim 0.8$ (iii) $\Omega_{\text{lum}} \simeq 0.007h$ (iv) $\Omega_R = 4.85h^{-2} \times 10^{-5}$ (v) $\Omega_{\text{total}} \equiv \Omega \gtrsim 0.3$. Theoretical models strongly favor $\Omega = 1$ and it is usual to invoke either a cosmological constant and/or nonbaryonic dark matter to achieve this (e.g. Carney & Latham 1987, Pagel 1991). We shall denote by Ω_{DM} the total contribution due to all nonbaryonic energy densities.

Models for structure formation also need to assume that small perturbations in the energy density existed at very early epochs. These perturbations can then grow via gravitational instability leading to the structures we see today. In most of the models these perturbations are generated by processes which are supposed to have taken place in the *very* early universe (say, at $z \gtrsim 10^{18}$). Inflationary models – which are probably the most successful ones in this regard – can produce density perturbations with an initial power spectrum $P_{\text{in}}(k) \simeq Ak$. Since each logarith-

C. S. Kochanek and J. N. Hewitt (eds), Astrophysical Applications of Gravitational Lensing, 55–64.
© 1996 IAU

mic interval in k space will contribute to the energy density an amount $\triangle_\rho^2(k) \equiv d\sigma^2/d(lnk) = (k^3 P(k)/2\pi^2)$ we find that $\triangle_\rho^2 \propto k^4$ for $P \propto k$. The contribution to the gravitational potential from the same range will be $\triangle_\varphi^2 = \triangle_\rho^2 (9H_0^4/4k^4 a^2)$ which is independent of k if $\triangle_\rho^2 \propto k^4$. Such a "scale-invariant" spectrum is produced in some other seeded models as well. All these models need to be fine-tuned to keep the amplitude A of the fluctuations small up to, say, $z \gtrsim 10^3$.

Given a Friedmann model with small inhomogeneities described by a power spectrum $P(k, z_{in})$ at a high redshift $z = z_{in}$, we can predict *unambiguously* the power spectrum $P(k, z_D)$ at the epoch of decoupling $z_D \approx 10^3$. This is because the perturbations at all relevant scales are small at $z \gtrsim z_D$ and we can use linear perturbation theory during this epoch. The shape of the spectrum at $z = z_D$ will not be a pure power law since the gravitational amplifiction is wavelength-dependent. In general, the power at small scales is suppressed (relative to that at large scales) due to various physical processes and the exact shape of the spectrum at $z = z_D$ depends on the kind of dark matter present in the universe. In a universe dominated by "hot dark matter" particles of mass $m \simeq 30eV$, the power per logarithmic interval in $k-$space, $\triangle(k) = (k^3 P(k)/2\pi^2)^{1/2}$, is peaked at $k = k_{max} \equiv 0.11$ $Mpc^{-1}(m/30eV)$ and falls exponentially for $k > k_{max}$. Hence, in these models, the scale $k = k_{max}$ will go nonlinear first and smaller structures have to form by fragmentation. If the universe is dominated by "cold dark matter" particles with mass $m \gtrsim 35GeV$, then $\triangle(k)$ is a gently increasing function of k for small k. If we set $P(k) \propto k^n$ locally, the index n changes from 1 at $k^{-1} \gtrsim 200h^{-1}Mpc$ to 0at$k^{-1} \simeq 10h^{-1}Mpc$ and to about (-2) at $k^{-1} \simeq 1h^{-1}$ Mpc. In such models small scales will go nonlinear first and the structure will develop hierarchically (e.g. Padmanabhan 1993).

The situation is more complicated if two different kinds of dark matter are present or if the cosmological constant is nonzero. The presence of the cosmological constant adds to the power at large scales but suppresses the growth of perturbations at small scales. Similar effects take place if a small fraction of the dark matter is hot and the bulk of it is cold (eg. $\Omega_{HDM} \simeq 0.2, \Omega_{CDM} \simeq 0.8$). In both the cases there will be more power at large scales and less power at small scales, compared to standard CDM model. The spectrum $\triangle(k)$ is still a gently increasing function of k and small scales go nonlinear first.

The fact, that one can compute the power spectrum at $z \simeq z_D$ analytically, allows one to predict large scale anisotropies in CMBR unambiguously in any given model. Comparing this prediction with the anisotropy observed by COBE one can fix the amplitude A of the power spectrum. For a wide class (Padmanabhan 1992) of the models, $\triangle(k) \cong 10^{-3}(kL)^2$ with $L \simeq (24 \pm 4)h^{-1}$ Mpc for $k^{-1} \gtrsim 80h^{-1}$ Mpc. For CDM like models the function $\triangle(k)$ flattens out at larger k and is about unity around $k^{-1} \simeq 8h^{-1}$ Mpc. In pure HDM models, $\triangle(k)$ has a maximum value of $\triangle_m \simeq 0.42h^{-2}(m/30eV)^2$ at $k_m \simeq 0.11 \text{Mpc}^{-1}(m/30eV)$ and decreases exponentially at $k \gtrsim k_m$.

The evolution of the power spectrum after decoupling (for $z < z_D$) is more difficult to work out theoretically. In general, the power spectrum grows in amplitude (preserving the shape), as long as the perturbations are small (e.g. Padmanabhan 1993). In this case, we can write $\triangle(k, z) = [f(z)/f(z_D)]\triangle(k, z_D)$ for $z < z_D$. For example, in CDM models with $\Omega = 1, f(z) = (1 + z)^{-1}$; thus $\triangle(k)$ grows by a factor 10^3 at all scales between the epoch of decoupling ($z_D \simeq 10^3$) and the present epoch ($z = 0$), if we assume that linear theory is valid at all scales. The resulting $\triangle_0(k)$, obtained by linear extrapolation, is often used to specify the properties of the models. This spectrum correctly describes the power at large scales (say, $k^{-1} \gtrsim 30h^{-1}$ Mpc) where $\triangle_0 \lesssim 0.1$. The "density contrast" $\sigma(R)$ measures the rms fluctuations in mass within a randomly placed sphere of radius R; up to factors of order unity, $\sigma(R) \simeq \triangle(k \simeq R^{-1})$ in hierarchical models. For most of the, COBE normalized, CDM-like models $\sigma(R) \approx 1$ around $R \approx 8h^{-1}Mpc$. Clearly linear theory cannot be trusted at smaller scales.

There are two major difficulties in understanding the physics at these small scales. Firstly, the true power $\triangle_{\text{true}}(k)$ of dark matter will be larger than $\triangle_0(k)$ due to nonlinear effects which are difficult to model analytically. Since dark matter particles interacts only through gravity, it is, of course, possible to study the formation of dark matter structures by numerical simulations. But to gain insight into the dynamics, it will be helpful to have simple analytic models explaining the N-body results.

Secondly, it is important to understand gas dynamical processes before one can compare theory and observations at small scales. Since baryons can dissipate energy and sink to the minima of the dark matter potential wells, the statistical properties of visible galaxies and dark matter halos could be quite different. The situation is further complicated by the fact that in hierarchical models, considerable amounts of merging takes place at small scales. It is usual to quantify our ignorance at these scales by a 'bias' (acronym for 'Basic Ignorance of Astrophysical Scenarios') factor b and write $\xi_{\text{gal}}(r) = b^2\xi_{\text{mass}}(r)$. Such a parametrization is useful only if b is independent of scale and morphology of galaxies. This seems to be somewhat unlikely. Since small scale observations are based on galactic properties, while theoretical calculations usually deal with underlying mass distribution, any scale (or morphology) dependence of b could play havoc with predictive power of the theory.

Recently some amount of progress has been achieved as regards the first aspect, viz, understanding nonlinear clustering of dark matter (Hamilton et al. 1991, Nityananda & Padmanabhan 1994, Bagla & Padmanabhan 1993, Padmanabhan et al. 1995). This approach is based on the relationship between the mean correlation function $\bar{\xi}(x,a)$ and the mean relative pair velocity $v(x,a)$. These quantities are related by an *exact* equation.

$$\frac{\partial F}{\partial A} - h(A,X)\frac{\partial F}{\partial X} = 0$$

where $F = \ln\left[x^3(1+\bar{\xi})\right]$, $A = \ln a$, $X = \ln x$ and $h = -(v/\dot{a}x)$. The characteristics of this equation shows that, as the evolution proceeds, power from a large scale l is transferred to smaller scales up to $x = l(1+\bar{\xi})^{-1/3}$. By analyzing the behavior of h, it is possible to express $\bar{\xi}(x,a)$ in terms of the mean correlation function *in the linear theory*, $\bar{\xi}_L(l,a)$. It turns out that: $\bar{\xi}(a,x) = Q[\bar{\xi}_L(a,l)]^n$ with $l^3 = x^3(1+\bar{\xi})$ where $Q = n = 1$ for $\bar{\xi}_L \leq 1.2$; $Q = 0.7, n = 3$ for $1.2 \leq \bar{\xi}_L \leq 6.5$ and $Q = 11.7, n = 1.5$ for $\bar{\xi}_L \geq 6.5$. This relation shows that $\bar{\xi}$ is steeper than $\bar{\xi}_L$.

Unfortunately, no such simple pattern exists in the dynamics of baryons coupled to dark matter. The gas dynamical processes introduce several characteristic scales into the problem and the evolution becomes quite complicated. The only reliable way of probing these systems seems to be through massive hydro simulations which are still at infancy.

It is clear from the above discussions that our theoretical understanding is best at large scales ($k^{-1} \gtrsim 30h^{-1}$ Mpc) where linear theory is valid, $\Delta_0(k)$ is well determined and baryonic astrophysical processes are not important. At the intermediate scales ($3h^{-1}$ Mpc $\lesssim k^{-1} \lesssim 30h^{-1}$ Mpc), it is not too difficult to understand the dark matter dynamics by some approximation but the baryonic physics begins to be nontrivial. At still smaller scales, ($k^{-1} \lesssim 3h^{-1}$ Mpc) there is considerable uncertainty in our theoretical predictions. We shall now turn to the observational probes of the power spectrum at different scales.

2. Probing the power spectrum

One of the direct ways of constraining the models is to estimate the density contrast $\sigma_{obs}(R)$ from observations at different scales and compare it with the theoretically predicted values. Fortunately, we now have observational probes covering four decades of scales from 10^{-1} Mpc to 10^3 Mpc. We shall discuss the probes of different scales in decreasing order.

2.1. NEAR HORIZON SCALES: (300 - 3000) H^{-1}MPC

These scales are so large that the best way to probe them is by studying the MBR anisotropy at angular scales which correspond to these linear scales. Since a scale L subtends an angle $\theta(L) \cong 1°(L/100h^{-1} \text{ Mpc})$ at $z \simeq z_D$, the $(\Delta T/T)$ observations at $(3° - 30°)$ probe these scales. The COBE-DMR observations (Smoot et al. 1992) of $(\Delta T/T)_{\text{rms}}$ and $(\Delta T/T)_Q$ allow one to obtain the following conclusions: (i) $\sigma(10^3 h^{-1} \text{ Mpc}) \simeq 5 \times 10^{-4}$ (ii) The power spectrum at large scales is consistent with $P_{\text{in}}(k) \simeq Ak$ and, if we take $\Omega = 1$,, then $A^{1/4} \cong (24 \pm 4)h^{-1}$ Mpc (iii) In this range, $\sigma(R) \cong (24 \pm 4h^{-1} \text{ Mpc}/R)^2$.

2.2.   VERY LARGE SCALES : $(80 - 300)H^{-1}$ MPC

2.2.1.   *CMBR probes:*
These scales span $(0.8° - 3°)$ in the sky at $z \simeq z_D$. Several ground based and balloon-borne experiments to detect anisotropy in MBR probe this scale. For example, the UCSB South Pole experiment (Schuster et al. 1993) has reported a preliminary 'detection' of $(\Delta T/T) \simeq 10^{-5}$ at $1.5°$ scale, and a 95% confidence level bound of $(\Delta T/T) < 5 \times 10^{-5}$. This translates into the constraint of $\sigma(10^2 h^{-1} \text{ Mpc}) \lesssim 5 \times 10^{-2}$.

The angular anisotropy of CMBR is dominated by the gravitational potential wells of dark matter at large scales. However, at $\theta \simeq 1°$ baryonic process affect the pattern of anisotropy significantly. The precise determination of degree scale anisotropy can, therefore, help in distinguishing between different models (White, Scott & Silk 1994).

2.2.2.   *Galaxy surveys:*
Some galaxy surveys, notably CfA2 survey and pencil-beam surveys probe scales which are about $10^2 h^{-1}$ Mpc in depth (Broadhurst et al. 1990, Vogeley et al. 1992). Unfortunately, the statistics at these large scales is not good enough for one to obtain $\sigma(R)$ directly from these surveys.

2.3.   LARGE SCALES : $(40 - 80)$ H^{-1} MPC

2.3.1.   *CMBR probes:*
The scales correspond to $\theta_{MBR} \simeq (24' - 48')$ and are probed by the experiments looking for small angle anisotropies in MBR. The claimed detection (Cheng et al. 1994) by MIT-MASM of $(\Delta T/T) \cong (0.5 - 1.9) \times 10^{-5}$ at $\theta \simeq 28'$, if confirmed, will give a bound of $\sigma(50h^{-1} \text{ Mpc }) \lesssim 0.3$.

2.3.2.   *Galaxy Surveys:*
Several galaxy surveys, in particular the IRAS-QDOT and APM surveys, give valuable information about this range (Rowan-Robinson et al. 1990, Efstathiou et al. 1990, Saunders et al. 1991). The angular correlation of galaxies, measured by APM survey is $\omega(\theta) \simeq (1-5) \times 10^{-3}$ at $\theta \simeq 14°$. This corresponds to $\sigma(50h^{-1} \text{ Mpc}) \cong 0.2$. What is more important, these surveys

can provide valuable information about the shape of the power spectrum in this range if we assume that galaxies faithfully trace the underlying mass distribution.

2.3.3. *Large scale velocity field:*

Using distance indicators which are independent of Hubble constant, it is possible to determine the peculiar velocity field $v(R)$ of galaxies up to about $80h^{-1}$ Mpc or so. The motion of these galaxies can be used to map the underlying gravitational potential at these scales. Careful analysis of observational data shows (e.g. Dekel 1994) that $v(40h^{-1}$ Mpc$) \simeq (388 \pm 67)$ kms^{-1} and $v(60h^{-1}$ Mpc$) \simeq (327 \pm 82)$ kms^{-1}. From these values it is possible to deduce that $\sigma(50h^{-1}$ Mpc $) \simeq 0.2$. These observations also allow us to determine the value of the parameter $(\Omega^{0.6}/b_{IRAS})$ where b_{IRAS} is the bias factor with respect to IRAS galaxies. One finds that $(\Omega^{0.6}/b_{IRAS}) = 1.28^{+0.75}_{-0.59}$ which implies that if $\Omega = 1$, then $b_{IRAS} = 0.78^{+0.66}_{-0.29}$ and if $b_{IRAS} = 1$ then $\Omega = 1.51^{+1.74}_{-0.97}$.

2.3.4. *Clusters and voids:*

The cluster-cluster correlation function and the spectrum of voids in the universe can, in principle, tell us something about these scales. Unfortunately, the observational uncertainties are so large that one cannot yet make quantitative predictions.

2.4. INTERMEDIATE SCALES : $(8 - 40)H^{-1}$ MPC

2.4.1. *Galaxy Surveys:*

The galaxy–galaxy correlation function $\xi_{gg} \cong [r/5.4h^{-1}\text{Mpc}]^{-1.8}$ is fairly well determined at these scales. Direct observations suggest that $\sigma_{gal}(8h^{-1}$ Mpc$) \simeq 1$ but the σ_{DM} and σ_{gal} at these scales can be quite different because of possible biasing.

2.4.2. *Cluster Surveys:*

There have been several attempts to determine the correlation function of clusters of different classes. It is generally believed that $\xi_{cc} \simeq (r/L)^{-1.8}$ with $L \simeq 25h^{-1}$ Mpc. The index $n = 1.8$ is fairly well determined though the scale L is not; in fact, L seems to depend on the richness class of the cluster. The quantity $(\xi_{cc}/\xi_{gg})^{1/2}$ can be thought of as measure of the relative bias between cluster and galaxy scales. Observations suggest (Dalton et al. 1992, Nicol et al. 1992, Bahcall & West 1992, Postman et al. 1992) that this quantity depends on the cluster class and varies in the range $(2 - 8)$. The observational uncertainties are still quite large for this quantity to be of real use; but if the observations improve we will have valuable information from ξ_{cc}.

2.4.3. *Abundance of rich clusters:*

The scale $R = 8h^{-1}$ Mpc contain a mass of $1.2 \times 10^{13} \Omega h_{50}^{-1} M_{\odot}$. When this scale becomes nonlinear, it will reach an overdensity of about $\delta \simeq 178$, or – equivalently – it will contract to a radius of $R_f \simeq (8h^{-1}$ Mpc$)$ $/(178)^{1/3} \simeq 1.5h^{-1}$ Mpc. A mass of $10^{15} M_{\odot}$ in a radius of 1.5 Mpc is a good representation of Abell clusters we see in the universe. *This implies that the observed abundance of Abell clusters can be directly related to $\sigma(8h^{-1}$ Mpc$)$.* Several people have attempted to do this (White et al. 1993); the final results vary depending on the modeling of Abell clusters, and give $\sigma(8h^{-1}$ Mpc$) \simeq (0.5-0.7)$. Since $\sigma_{gal}(8h^{-1}$ Mpc$) \simeq 1$, this shows that $b \simeq (1.23-2)$ at $8h^{-1}$ Mpc.

It is possible to give this argument in a more general context (Subramanian & Padmanabhan 1994). Suppose that the contribution to critical density from collapsed structures with mass larger than M is $\Omega(M)$, at a given redshift z. Then one can show that

$$\Omega(M) = \mathrm{erfc}\left[\frac{\delta_c(1+z)}{\sqrt{2}\sigma_0(M)}\right]$$

where $\delta_c = 1.68$ and erfc(x) is the complementary error function. The Abell clusters (at $z = 0$) contribute in the range $\Omega \simeq (0.001 - 0.02)$. Even with such a wide uncertainty, we get $\sigma_{clus} \simeq (0.5 - 0.7)$.

2.5. SMALL SCALES : $(0.05 - 8)H^{-1}$ MPC

These scales correspond to structures with $M_{smooth} \simeq (3 \times 10^8 - 1.2 \times 10^{15}) \Omega h_{50}^{-1} M_{\odot}$ and we have considerable amount of observational data covering these scales. Unfortunately, it is not easy to make theoretical predictions at these scales because of nonlinear, gas dynamical, effects.

2.5.1. *Epoch of galaxy formation:*

Observations indicate that galaxy-like structures have existed even at $z \simeq 3$. This suggests that there must have been sufficient power at small scales to initiate galaxy formation at these high redshifts. Unfortunately, we do not have reliable estimate for the abundance of these objects at these redshifts and hence we cannot directly use it to constrain $\sigma(R)$.

2.5.2. *Abundance of quasars:*

The luminosity function of quasars is fairly well determined up to $z \approx 4$. If the astrophysical processes leading to quasar formation are known, then the luminosity function can be used to estimate the abundance of host objects at these redshifts. Though these processes are somewhat uncertain, most of the models for quasar formation suggest that we must have $\sigma(0.5h^{-1}$ Mpc$) \gtrsim 3$.

2.5.3. *Absorption systems:*

The universe at $1 \lesssim z \lesssim 5$ is also probed by the absorption of quasar light by intervening objects. These observations suggest that there exist significant amounts of clumped material in the universe at these redshifts with neutral hydrogen column densities of $N_{\text{HI}} \simeq (10^{15} - 10^{22})$cm^{-2}. We can convert these numbers into abundances of dark matter halos by making some assumptions about this structure. We find that (Subramanian & Padmanabhan 1994) in the redshift range of $z \simeq (1.7 - 3.5)$ damped Lyman alpha systems contribute a fractional density of $\Omega_{Ly} \simeq (0.06 - 0.23)$. This would require $\sigma(10^{12} M_\odot) \simeq (3 - 4.5)$.

2.5.4. *Gunn-Peterson bound:*

While we do see absorption due to *clumped* neutral hydrogen, quasar spectra do not show any absorption due to smoothly distributed neutral hydrogen. Since the universe became neutral at $z \lesssim z_D \simeq 10^3$, and since galaxy formation could not have made all the neutral hydrogen into clumps, we expect the IGM to have been ionized sometime during $5 \lesssim z \lesssim 10^3$. It is not clear what is the source for these ionizing photons. Several possible scenarios (quasars, massive primordial stars, decaying particles etc.) have been suggested in the literature though none of these appears to be completely satisfactory. In all these scenarios, it is necessary to form structures at $z \gtrsim 5$ so that an ionizing flux of about $J = 10^{-21}$ergs cm^{-2}s^{-1} Hz^{-1} sr^{-1} can be generated at these epochs. Once again, it is difficult to convert this constraint into a firm bound on σ though it seems that $\sigma(0.5h^{-1}$ Mpc$) \gtrsim 3$ will be necessary.

3. Gravitational lensing and large scale structure

In the above discussion we have not taken into consideration the constraints imposed by gravitational lensing effects on the structure formation models. This aspect will be discussed in detail in the other articles in this volume; here we shall contend ourselves with a brief mention of the possibilities.

Gravitational lensing probes the gravitational potential directly and can provide valuable information at very different scales. At the largest scales ($R \simeq 10^3$ Mpc) lensing can be used to probe the geometry of the universe. For example, it is possible to put firm bounds on the energy contributed by cosmological constant from such considerations.

At intermediate scales ($R \simeq 50$ Mpc) lensing has the potential of providing information about the power spectrum of fluctuations which are in the quasilinear phase. In principle the distortion of images can be inverted to obtain this information, though in practice this is extremely difficult.

At smaller scales, the "weak lensing" – leading to arcs and arclets at cluster scales – is already providing a clue to the mapping of dark matter distribution in clusters. On the other hand, direct optical and X-ray observations provide us information about the distribution of visible matter in clusters. The combination of these techniques should give us valuable information as regards the dynamical processes which separated baryons from dark matter.

At still smaller scales, galactic potentials have the capacity to produce multiple images of distant sources. The statistics of these multiple images depends crucially on the core radii of the galaxies, which in turn depends sensitively on the structure formation models. The absence of significant number of multiple images with large angular separations puts severe constraints on models for structure formation. The analytic modeling of non-linear dark matter clustering described earlier could be used to strengthen these constraints still further.

4. Scorecard for the models

The simplest models one can construct will contain a single component of dark matter, either cold or hot. Such models are ruled out by the observations. The HDM models, normalized to COBE result will have maximum power of $\Delta_m \cong 0.42h^{-2}(m/30eV)^2$ at $k = k_m = 0.11$ Mpc$^{-1}(m/30eV)$. In such a case, structures could have started forming only around $(1 + z_c) \cong (\Delta_m/1.68) \cong h_{50}^{-2}(m/30eV)^2$ or at $z_c \cong 0$. We cannot explain a host of high-z phenomena with these models. The pure CDM models face a different difficulty. These models, normalized to COBE, predict $\sigma_8 \simeq 1$, which is too high compared to the bounds from cluster abundance. When nonlinear effects are taken into account, one obtains $\xi_{gg} \propto r^{-2.2}$ for $h = 0.5$ which is too steep compared to the observed value of $\xi_{gg} \propto r^{-1.8}$. In other words, CDM models have wrong shape for $\xi(r)$ to account for the observations.

The comparison of the CDM spectrum with observations suggests that we need more power at large scales and less power at small scales. This is precisely what happens in models with both hot and cold dark matter or in models with nonzero cosmological constant. These models have been extensively studied during the last few years, and they fare well as far as large and intermediate scale observations are concerned. However, they have considerably less power at small scales compared to CDM model. As a result, they do face difficulties (Subramanian & Padmanabhan 1994) in explaining the existence of high redshift objects like quasars. For example, a model with 30% HDM and 70% CDM will have $\sigma_{0.5} \simeq 1.5$; to explain the abundance of quasars comfortably one needs $\sigma_{0.5} \simeq 3.0$. To explain the abundance of damped Lyman alpha systems one requires still larger values of about $\sigma_{0.5} \approx 4$ or so. Demanding that $\sigma(10^{12}M_\odot) > 2$ [which is equivalent to saying that $10^{12}M_\odot$ objects must have collapsed at a redshift of $z_{12} = (2/1.68) - 1 \simeq 0.2$] will completely rule out this model. Similar difficulties exist in models with cosmological constant. Notice that all models are normalized using COBE results at very large scales. Hence the severest constraints are provided by observations at smallest scales, since the "lever-arm" is longest in that case.

The comparison of models show that it is not easy to accommodate all the observations even by invoking two components to the energy density. (These models also suffer from serious problems of fine-tuning). By and large, the half-life of such quick-fix models seem to be about 2-3 years. One is forced to conclude that to make significant progress it is probably necessary to perform a careful, unprejudiced analysis of: (a) large scale observational results and possible sources of error and (b) small scale baryonic astrophysical processes.

References

Bagla, J.S. & Padmanabhan, T., 1993, IUCAA preprint 22/93

Bahcall, N. & West, M., 1992, ApJ, 270, 70

Broadhurst, T., et al., 1990, Nature, 343, 726

Carney, B.W. & Latham, D.W., 1987, Dark Matter in the Universe, eds. J. Kormendy, & G.R. Knapp (Dordrecht: Reidel) 39

Cheng, E.S., et al., 1994, ApJL, 422 L37

Dalton, G.D., et al., 1992, ApJL, 390, L1

Dekel, A., 1994, ARA&A, 32, 371

Efstathiou, G., et al., 1990, MNRAS, 247, 10p

Hamilton, A.J.S., et al., 1991, ApJL, 374, L1

Nicol, R.C., et al., 1992, MNRAS, 255, 21p

Nityananda, R. & Padmanabhan, T., 1994, MNRAS,271, 976

Padmanabhan, T., Narasimha, D., 1992, MNRAS, 259, 41P

Padmanabhan, T., 1993, Structure Formation in the Universe, (Cambridge: Cambridge University Press) chapter 4

Padmanabhan, T., et al., 1995, Princeton preprint astro-ph 9506051

Pagel, B.E.J., Phys Scr, 1991, T36, 7

Postman, M., et al., 1992, ApJ, 384, 404

Rowan-Robinson, M., et al., 1990, MNRAS, 247, 1

Saunders, W., et al., 1991, Nature, 349, 32

Schuster, J., et al., 1993, ApJL, 412, L47

Smoot, G.F., et al., 1992, ApJL, 396, L1

Subramanian, K., & Padmanabhan, T., 1994, IUCAA preprint 5/94

Vogeley, M.S., et al., 1992, ApJL, 391, L5

White, S.D.M., et al., 1993, MNRAS, 262, 1023

White, M., Scott, D., & Silk, J., 1994, ARA&A, 32, 319

TESTING COSMOGONIC MODELS
WITH GRAVITATIONAL LENSING

JOACHIM WAMBSGANSS
Astrophysikalisches Institut Potsdam
An der Sternwarte 16
14482 Potsdam, Germany

AND

RENYUE CEN, JEREMIAH P. OSTRIKER, EDWIN L. TURNER
Princeton University Observatory
Peyton Hall
Princeton, NJ 08544 USA

Abstract. Gravitational lensing provides a strict test of cosmogonic models. We use fully non-linear numerical propagation of light rays through a model universe with inhomogeneities derived from a particular cosmogonic model, i.e. three-dimensional lensing simulations, to study its lensing properties. As a first example we present results for the standard CDM scenario. The lensing test for this model predicts that we should have seen far more widely split quasar images than have been found.

1. Introduction

Gravitational lensing *directly* measures fluctuations in the gravitational potential along lines of sight to distant objects. In contrast, the conventional tools for comparing theories with observations rely on either galaxy density or velocity information, both of which unavoidably suffer from the uncertainties with regard to density or velocity bias of galaxies over the underlying mass distribution, hampering our attempts to understand the more "fundamental" questions concerning the mass evolution and distribution. Thus, gravitational lensing provides a powerful independent test of cosmogonic models (Narayan & White 1988). Each model for the de-

65

C. S. Kochanek and J. N. Hewitt (eds), Astrophysical Applications of Gravitational Lensing, 65–70.
© 1996 *IAU*

velopment of cosmogonic structure (e.g. the hot dark matter or cold dark matter [CDM] scenario) has at least one free parameter, the amplitude of the density power spectrum. But now in the light of COBE observations (Smoot et al. 1992) that parameter is fixed by the ($\pm 15\%$) determination on the $5° - 10°$ scale in the linear regime. With its amplitude fixed, a secure determination of the potential fluctuation on any scale provides a test; any single conflict between the theory and reality can falsify the former. The most leverage is obtained for tests made on scales as far as possible from the COBE measurements, since all models have an assumed power spectrum that passes through the COBE normalization point at the very large comoving scales ($\lambda \approx 1000$Mpc) fixed by that measurement. Since the slope of the power spectrum is a primary model dependent feature, the maximum variations amongst models occur at the smallest scales. Thus one looks for tests at scales as small as possible, but they should not be so small as to be greatly influenced by the difficulty in modeling the physics of the gaseous, baryonic components (≤ 10kpc). Thus critical tests are best made on scales 0.01Mpc $< r < 1$Mpc. Here we show how to use gravitational lensing from matter distributions on these scales to test cosmogonic models, with the standard CDM scenario as our first example. More detailed descriptions of the results and the method can be found in Wambsganss *et al.* (1995a,b).

2. Cosmological Model

The cosmogonic model tested here is the "standard" CDM scenario with $\Omega = 1$, $\lambda = 0$ and $H_0 = 100h = 50$ km s^{-1} Mpc^{-1}. Normalization, taken from the COBE first year results (Smoot et al. 1992), corresponds to $\sigma_8 = 1.05$. In order to allow for the existence of very large-scale waves, we first ran an $L = 400h^{-1}$Mpc size box with 500^3 cells and 250^3 particles. In order to have detailed small scale information we reran a total of 10 independent simulations with $L = 5h^{-1}$Mpc, with 500^3 cells and 250^3 particles. Knowing the distribution of overdensities on the $5h^{-1}$Mpc scale from the large simulation, we statistically convolve the small and large scale runs to produce simulated sheets or screens of matter spaced $5h^{-1}$Mpc apart between the observer at $z = 0$ and a putative galaxy or quasar in the source plane at $z = z_S$. A large number of independent runs (ten were simulated) is required so that identical structures do not repeat along a line of sight. Details and tests of the convolution method are presented in Wambsganss et al. (1995b). It is statistically reliable for describing structures in the range 30kpc $< \Delta Lh < 1.2$Mpc, which corresponds roughly to splitting angles $5" < \theta < 200"$. On these scales we expect that dark matter dominates over baryons so that a dark matter only simulation is approximately valid.

3. Lensing Method

We "fill" the universe densely with adjacent matter cubes, which are obtained from cosmological simulations as described above. Inside each cube, we project the matter onto the mid plane perpendicular to the line of sight. Then, we follow light rays through all lens planes. We speed up the calculation of deflection angles by use of an hierarchical tree code. Typically we use a few hundred (grouped) screens for each "line of sight". In a source plane at a given redshift, we then determine various lensing properties of this particular line of sight. The magnification in the *source* plane is simply given by the density of rays, relative to the unlensed case (see Figure 1a for one particular line of sight). Similarly, the magnification in the *image* plane is obtained by the differential area within a bundle of rays as compared to what it would have been had the propagation been through a universe with smoothly distributed matter (Figure 1b). Naturally we allow for crossing of ray bundles, i.e. multiple imaging. We can then use any distribution of sources (positions, sizes, shapes, redshifts) in order to determine their properties after being lensed by this three dimensional matter distribution. As an example, in Figure 1c we show the "images" of a regular grid of circular sources at $z_S = 3.0$. Deformation and change of source sizes are quite obvious in the regions that are highly magnified. In Figure 1d the "average" shear in areas of $(20 \text{ arcsec})^2$ is shown for this particular line of sight.

4. Results

For 100 different realizations, we have computed the distribution of magnifications for single and multiply-imaged point sources as a function of z_S, as well as multiplicity of images, distribution of angular splittings, rms shear and other properties. In addition, for extended sources, we have computed the expected shape distortions, frequency and properties of the giant arcs that would be seen, when the sources are lensed by intervening clusters. Figure 2a shows the probability of a splitting with separation of images greater than 5" and magnitude difference less than 1.5 mag as a function of source redshift. In fact, amplification bias will increase the probabilities over those shown in Figure 2a by a significant amount. Splittings larger than 5" should be common (when several thousand quasars have been examined), if this cosmogonic model were correct. Probably the single most revealing statistic is the distribution of image separations expected for multiple sources as shown in Figure 2b for $z_S = 1, 2$ and 3. Notice that very large splittings should be the rule. Also revealing is the distribution of expected lens redshifts as shown in Figure 2c. The lenses themselves should be close enough to be seen in almost all cases. On this issue the recent observation of a lens candidate for the double quasar QSO2345+007 (Fis-

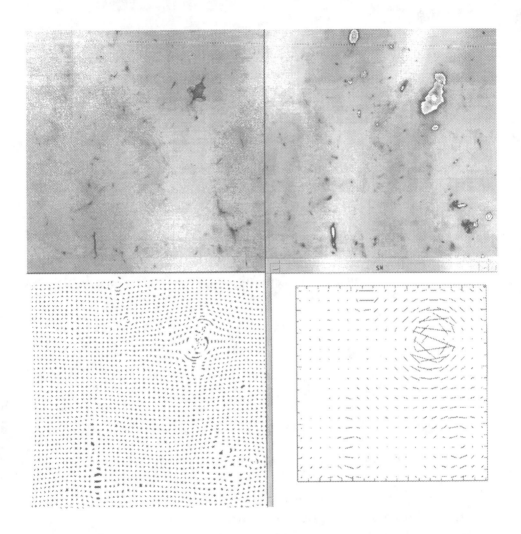

Figure 1. a) (top left) Example of the magnification due to the gravitational lens action of a CDM, $\Omega = 1$ universe for a *source* plane at $z_S = 3.0$. The size of the field is about $(5.7 \text{ arcmin})^2$. The gray scale indicates magnification, dark means high magnification. The sharp dark boundaries (caustics) indicate regions of multiple imaging. b) (top right) Magnification in the *image* plane. Again, dark indicates high magnification. The white regions just inside very dark regions indicate areas with formally negative magnification, i.e. regions that contain multiple images. The boundaries between the black and the white regions are the critical lines. c) (bottom left) "Image" of a regular grid of circular sources at a redshift of $z_S = 3.0$ for this particular line of sight. One clearly sees the effect of the lensing: variations in size indicate magnification as function of position. One can see arclets, arcs, and even multiple images (i.e., a row or column breaks up into more than one track). Interpreted as images of point sources, the relative sizes reflect the intensity ratio. d) (bottom right) Shear distribution for this line of sight: shear is averaged in square regions of about 20 arcsec at a side. The lines indicate the direction and the relative strength of the shear.

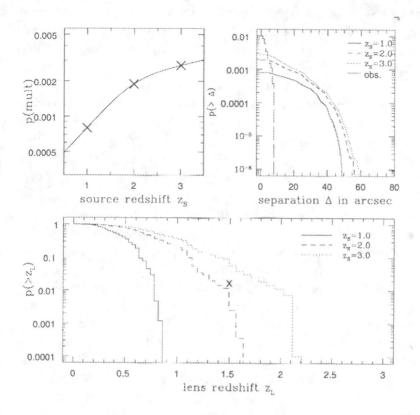

Figure 2. a) (top left) Probability of multiple imaging as a function of source redshift. Only cases with separation of images greater than 5" and less than 1.5mag difference are considered. b) (top left) Multiple-lensing probability distribution as a function of image separations for sources at $z_S = 1$, 2 and 3. Also shown as long dashed curve is the observed distribution (cf. Surdej & Soucail 1993). c) (bottom) Integrated lensing probability distribution as a function of expected lens redshift. The "X" indicates the recent observation (Fischer et al. 1994) of a lens candidate for the double quasar QSO2345+007.

cher et al. 1994) is extremely relevant. The separation of the two images is 7.06 arcsec, the quasar redshift is $z_S = 2.15$ and the putative lens is at $z_L = 1.49$. We see from Figure 2c that, although 7 arcsec separation can be produced in the CDM model, the probability that the lens is as far away as $z = 1.49$ is very small (2%) due to the relatively late formation of structure in this model. In open models structure formation occurs earlier.

It appears that all three of these results (shown in Figures 2a,b,c) are seriously in conflict with the existing observations. In particular, we find that the standard CDM model predicts that 0.0007 of all lines of sight to $z_S = 1$, 0.0014 of all lines of sight to $z_S = 2$ and 0.0020 of those to $z_S = 3$ will be multiply-imaged with angular splittings $\geq 10"$ and amplification ratios of less than 1.5 magnitudes. Various surveys and occasional serendipitous discoveries have revealed 27 confirmed or possible multiply-imaged QSO's according to a recent compilation (Surdej & Soucail 1993; an updated version of this statistics can be found in this volume). Detailed analysis of these surveys yields a lensing rate in the vicinity of a few tenths to one percent, consistent with the CDM predictions quoted above making allowance for plausible magnification biases. However, as shown in Figure 2b, all observed QSO lens systems have image splittings of less than 10", and the large majority, less than 5". This sharply contradicts and thus falsifies the model. Since the large splitting, modest brightness ratio systems predicted by the model would be typically much easier to detect and recognize than those 27 which have actually been found, no escape by appeal to observational selection seems possible. A similar conclusion for this cosmogonic model has been found by Kochanek (1995) in a semi-analytic study of the lensing properties of various cosmological models.

This failing of the model is not presented as an entirely new result, but only as a new and more robust manifestation of a previously recognized problem, namely the excessively deep potential wells produced by the dark matter component in COBE normalized standard CDM. These potential wells lead both to excess galaxy pair-wise velocity dispersions and to the predicted excessive rate of large splitting lensing events. The virtue of the lensing test is that it is independent of other tests and is not subject to the same caveats concerning "bias" of galaxies with respect to dark matter. We are in the process of testing other models (e.g. low Ω; or Λ dominated). The *directness* of gravitational lensing as a test for the growth of inhomogeneities, coupled with the rapidly increasing power of computers and numerical algorithms, makes one optimistic that calculations of the type reported on here should become a major tool for testing and discriminating among competing cosmological scenarios.

References

Fischer, P., Tyson, J.A., Bernstein, G.M., & Guhathakurta, P., 1994, ApJL, 431, L71
Kochanek, C.S., 1995, ApJ, in press
Narayan, R. & White, S.D.M., 1988, MNRAS, 231, 97p
Smoot, G.F., et al., 1992, ApJL, 396, L1
Surdej, J., & Soucail, G., 1993, in Gravitational Lenses in the Universe, eds. J. Sudej, D. Fraipont-Caro, E. Gosset, S. Refsdal, & M. Remy, (Liège: Université de Liège) 205
Wambsganss, J., Cen, R., Ostriker, J. P., & Turner, E. L., 1995a, Science, 268, 274
Wambsganss, J., Cen, R., Ostriker, J. P., 1995b, ApJ, submitted

WIDE SEPARATION LENSES

D. J. MORTLOCK AND R. L. WEBSTER
University of Melbourne,
Parkville, Victoria, 3052, Australia

AND

P. C. HEWETT
Institute of Astronomy,
Madingley Road, Cambridge, CB3 0HA, UK

There is still debate as to whether wide separation ($\Delta\theta \gtrsim 3''$) double quasars are physical binaries or gravitationally lensed sources. We proceed under the assumption that most of these objects are the result of lensing, and use maximum likelihood techniques (Kochanek 1993) to infer information about the mass distribution of deflectors that would be required to produce these wide image separations. Under the above assumptions, the most consistent explanation is the existence of a significant population of dark objects with the mass of groups of galaxies.

About half of all *candidate* lens pairs have wide separations, and, as such, cannot be the result of gravitational lensing by a galaxy. Since the Einstein radius of clusters is of the order of 20" (much larger than any of the image separations), we propose a population of group sized masses as the main deflectors. In this case, these deflectors must be very dark, as few of the quasar pairs have a visible lens between them. This picture of groups as being mostly composed of dark matter fits in well with some recent observations (*e.g.*, Mulchaey et al. 1993).

We use the ($\Lambda = 0$) cosmologies of Turner et al. (1984), and model the lenses as singular isothermal spheres (characterized by σ, their line of sight velocity dispersion). Their number density is modeled by a broken power law, with a form similar to the Schechter function for low σ, but with a shallower falloff for the higher masses. Explicitly, we take

$$\frac{dn}{d\sigma} = \begin{cases} \frac{n_*}{\sigma_*}\left(\frac{\sigma}{\sigma_*}\right)^{-4} & \text{for } \sigma < \sigma_* \\ \frac{n_*}{\sigma_*}\left(\frac{\sigma}{\sigma_*}\right)^{\beta} & \text{for } \sigma \geq \sigma_*, \end{cases}$$

C. S. Kochanek and J. N. Hewitt (eds), Astrophysical Applications of Gravitational Lensing, 71–72.
© 1996 IAU

D. J. MORTLOCK ET AL.

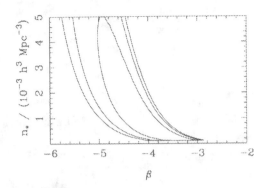

Figure 1. The maximum likelihood results shown in the β–n_* plane. The contours are at the 68%, 95% and 99% (*i.e.*, the 1, 2 and 3 standard deviation) confidence levels. The calculation implies that $\beta = -4 \pm 1$ and $n_* = (3 \pm 3) \times 10^{-3}\, h^3\, \mathrm{Mpc}^{-3}$.

where β and n_* are allowed to vary.

To constrain the parameters of the model, we used the Large Bright Quasar Survey (LBQS) (Hewett et al. 1995), a sample of 1055 quasars, with $16.0 < B_J < 18.85$. Hence this sample will not include many reddened objects, and may thus be biased against lensing events (Webster et al. 1995). The LBQS contains three lens pairs, of angular separations $1.48''$ (Hewett et al. 1994), $5.14''$ (Hewett et al. 1989) and $7.5''$ (unpublished).

Figure 1 shows the best fit to the maximum likelihood calculation as $\beta \approx -4$ and $n_* \approx 3 \times 10^{-3}\, h^3\, \mathrm{Mpc}^{-3}$. Some cosmological simulations imply the number density of *clusters* is proportional to $M^{-2} \propto \sigma^{-4}$, so our results suggest that number density may vary in this fashion from the scale of galaxies all the way up to clusters. However, the selection effects in the LBQS and the small number statistics must be more carefully considered before these conclusions can be stated with confidence; this work will be presented more fully elsewhere.

References

Hewett, P.C., Foltz, C.B. & Chaffee, F.H., 1995, AJ, 109, 1498

Hewett, P.C., Irwin, M.J., Foltz, C.B., Harding, M.E., Corrigan, R.T., Webster, R.L. & Dinshaw, N., 1994, AJ, 108, 1534

Hewett, P.C., Webster, R.L., Harding, M.E., Jedrzejewski, R.I., Foltz, C.B., Chaffee, F.H., Irwin, M.J. & Le Févre, O., 1989, ApJL, 346, L61

Kochanek, C.S., 1993, ApJ, 419, 12

Mulchaey, J.S., Davis, D.S., Mushotzky, R.F. & Burstein, D., 1993, ApJL, 404, L9

Turner, E.L., Ostriker, J.P. & Gott, J.R., 1984, Astrophysical Journal, 284, 1

Webster, R.L., Francis, P.J., Peterson, B.A., Drinkwater, M.J. & Masci, F.J., 1995, Nature, 375, 469

STATISTICS OF QUASAR LENSING
CAUSED BY CLUSTERS OF GALAXIES

K. TOMITA

Yukawa Institute for Theoretical Physics,
Kyoto University, Kyoto 606-01, Japan

1. Introduction

Cosmological lens statistics was studied first by Turner, Ostriker, & Gott (1984), and developed by Hinshaw & Krauss (1987), Gott, Park, & Lee (1989), Turner (1990), Krauss & White (1992), and Wallington & Narayan (1993). The expected number of lensed quasars in general quasar catalogs was examined by Fukugita & Turner (1991), Kochanek (1991), and Fukugita et al. (1992), and the number expected in the HST Snapshot Survey (Bahcall et al. 1992, Maoz et al. 1992, 1993) was examined by Maoz & Rix (1939) and Kochanek (1993). In most of these studies, galaxies were assumed to be the lenses, and clusters of galaxies played only auxiliary roles. As a result, the expected number of lensed quasars with separations larger than 3 arcsec is negligibly small compared to the total number of lenses. However, about 33% of observed lensed quasars have image separations larger than 3 arcsec (Maoz et al. 1993, Refsdal & Surdej 1994). This observational fact is very important.

Clusters of galaxies as lenses, on the other hand, produce conspicuous lensed images like arcs and arclets, whose sources are background galaxies. It is plausible that clusters of galaxies can similarly produce observable lensed images of quasars. In a recent paper (Tomita 1995a) we studied the statistics of lenses due to clusters of galaxies, where we assumed the clusters have a uniform spatial distribution for $z \leq z_f(= 0.5 - 0.7)$. In the next section we show the results for various cosmological models, including the super-horizon-scale inhomogeneous model introduced by (Tomita 1995b).

73

C. S. Kochanek and J. N. Hewitt (eds), Astrophysical Applications of Gravitational Lensing, 73–74.
© 1996 IAU

2. Lensing due to clusters of galaxies

For the mass distribution of clusters of galaxies we assumed the mass function derived by Bahcall & Cen (1993), which is similar in form to the Schechter function for galaxies. The bending angle was derived by regarding clusters of galaxies as isothermal gas spheres with finite cores. The parameterized dependence of the core radius and the velocity dispersion on the mass was normalized using the mean values for X-ray clusters with lensed arc images (e.g. Tyson et al. 1990, Mellier et al. 1993, 1994, Petrosian & Lynds 1992).

We derived the expected number of lenses in the HST Snapshot Survey taking into account the angular selection function and using the magnification bias model of Maoz & Rix (1993), and compared the results to the observed sample (Maoz et al. 1993). A limited number of cosmological models could satisfy the following two conditions at the same time: (1) the condition that the expected number of lensed quasars with separation angles larger than 3 arcsec is about 33% of the total number of lenses, and (2) the condition that the expected total number of lenses equals the observed number. We examined the Einstein-de Sitter model, an open low-density homogeneous model with $\Omega_0 = 0.1$ and $\lambda_0 = 0$, a flat homogeneous model with $\Omega_0 = 0.1$ and $\lambda_0 = 0.9$, and the super-horizon-scale inhomogeneous models with $\Omega_0 = 0.1$ or 0.2 in the inner homogeneous region. Of these models, only the flat model with a cosmological constant and the super-horizon-scale model with the low-density inner region satisfied the constraints. This conclusion differs significantly from earlier studies where the only lenses were galaxies formed at epochs $z > 3$.

References

Bahcall, N.A., & Cen, R., 1993, ApJL, 407, L49
Maoz, D., Bahcall, J. N., Schneider, D. P., Bahcall, N. A., Djorgovski, S., Doxsey, R., Gould, A., Kirhakos, S., Meylan, G., & Yanny, B., 1993, ApJ, 409, 28
Maoz, D., & Rix, H.-W., 1993, ApJ, 416, 425
Mellier, Y., Fort, B., & Kneib, J.-P., 1993, ApJ, 407, 33
Mellier, Y., Fort, B., Bonnet, H., & Kneib, J.-P., 1994, in Cosmological Aspects of X-Ray Clusters of Galaxies, ed. W.C. Seitter (Dordrecht: Kluwer), 219
Petrosian, V., & Lynds, R., 1992, in Gravitational Lenses, ed. R. Kayser, T. Schramm, & L. Nieser (Berlin: Springer-Verlag), 303
Refdal, S. & Surdej, J., 1994, Rep Prog Phys, 56, 117
Tomita, K., 1995a, PASJ, in press (preprint YITP/U-95-09)
Tomita, K., 1995b, ApJ, 450, 1
Turner, E.L., Ostriker, J.P., & Gott, J.R., 1984, ApJ, 284, 1
Tyson, J.A., Valdes, F., & Wenk, R.A., 1990, ApJL, 249, L1

WEAK LENSING AND THE SLOAN DIGITAL SKY SURVEY

Wide Area Weak Lensing

ALBERT STEBBINS, TIM MCKAY, AND JOSHUA A. FRIEMAN
Fermi National Accelerator Laboratory
Box 500, Batavia IL 60510, USA

Abstract. While the strategy for the first applications of weak lensing has been to "go deep" it is equally interesting to use one's telescope time to instead "go wide". The Sloan Survey (SDSS) provides a natural framework for a very wide area weak lensing survey.

Probing of the mass distribution using the distortion of galaxy image shapes by the intervening gravitational field this mass produces is a powerful new technique for probing cosmological structure (Valdes et al. 1983, Tyson et al. 1990, Miralda-Escudé 1991, Blandford et al. 1991, Kaiser 1992). The "weak lensing" technique will no doubt become one of the standard probes, on par with galaxy redshift surveys and maps of CMBR anisotropies. Except for small areas on the sky near distant rich clusters or very near galaxies, the image distortion is expected to be small and weak lensing is the appropriate technique. One must average over many galaxies to obtain a significant detection of the small image distortion; typically by measuring correlations in galaxy position-angles and thus the shear. Deep imaging is extremely useful as it allows one to get accurate estimates of the shapes of large numbers of background galaxies in the relatively small field of view of most telescopes. If one fails to go deep one can identify fewer background galaxies and, in any case, one obtains only accurate shape information for the brighter, larger galaxies. However even with moderately deep images one can, in principle, use the weak lensing technique to infer the foreground mass distributions. If the the number of galaxies per unit area for which one has accurate shape information is small then one should survey a larger area to obtain a significant signal.

The first successful applications of the weak lensing (Tyson et al. 1990, Smail et al. 1994, Fahlman et al. 1995) has naturally been to take deep images of galaxies behind rich clusters where the shear is large, and perhaps

C. S. Kochanek and J. N. Hewitt (eds), Astrophysical Applications of Gravitational Lensing, 75–80.
© 1996 IAU

more importantly, where one has a fairly good idea what one expects to find. Attempts have also been made to detect shear in the field (i.e. a direction not associated with a particular galaxy concentration), but without any definitive detection (Mould et al. 1994). When looking at the field one can expect to find contributions to image distortions from mass at various distances along the line-of-sight. While it would be useful to study the statistical properties of the shear at a given depth, one will, in the end, want to chart how the shear varies with depth. By understanding the variation with depth one can learn about the radial distribution of densities along different lines-of-sight. This can tell us something about the evolution of the density field and in particular about cosmological parameters such as Ω and Λ; as well as allow one to construct a crude map of the mass distribution. The latter application is particularly interesting as it will allow one to compare the mass distribution with the better studied *nearby* galaxy distribution (Villumsen & Gould 1994). Thus even if one had a very deep survey of galaxy image shapes one would want to study the dependence of shear with depth and in effect look at less deep surveys. The study of shear at $z \sim 0.1 - 0.4$ is interesting in its own right!

Any imaging survey of the sky is implicitly measuring the shapes of the galaxies it is able to detect. As long as the combination of depth and area of the survey are large enough to obtain a sufficient S/N one can in principle use this for weak lensing. One's calculation of depth must take into account the accuracy with which one is able measure the galaxy shapes. However it is generally true that one does not loose much by even relatively large random errors in the galaxy shapes. This is because the intrinsic non-circularness of the true projected galaxy shapes introduces random uncertainties in the inferred shear and one would have to make fairly large measurement errors to significantly add to these uncertainties. The true galactic position-angles are (assumed) random and therefore by using a sufficiently large number of galaxies one can reduce both the intrinsic and measurement uncertainties if they are random. However if uncorrected measurement errors are correlated between different galaxies one may never reach an acceptable S/N. Since the deeper the survey is the larger the signal will be, the requirement to control these systematic errors is less. It is not clear to what level one can reduce systematic errors and it is thus not clear how shallow a survey one could use for weak lensing studies.

The Sloan Digital Sky Survey (SDSS) (Kent 1994) is a prime example of a large imaging survey on which one may "piggy-back" a weak lensing program (Villumsen & Gould 1994). Perhaps the most publicized aspect of the SDSS is a redshift survey of 10^6 galaxies. To obtain the redshift targets the SDSS will image 1/4 of the sky in 5 colors, mostly around the North Galactic Cap, identifying galaxies in the North down to a nominal magnitude limit of $r' < 23.1$, and going to 25.1 in parts of the Southern survey. This will yield a catalog of $\sim 5 \times 10^7$ galaxy images. One does not really need the galaxies near the limiting magnitude to obtain a significant weak lensing signal and in this sense the SDSS can expect to do much better than a marginal detection. The multi-color photometry will be extremely useful for weak lensing as we expect to determine galaxy redshifts to $\Delta z \sim 0.04$ *photometrically* (Szalay 1995). With this redshift information one can map the shear as a function of distance. This allows one to better localize the mass distribution as a function of radius and make more of a direct comparison of the mass and galaxy distributions. Of course the SDSS redshift survey gives exactly the galaxy distribution one would want to compare to the mass distribution determined via weak lensing from the imaging survey.

In Villumsen and Gould (1994) it was estimated that nearby clusters and their extended halos would dominate the shear field measured by the SDSS. To illustrate some of the above comments in this regard let us consider the mean shear given by a model cluster with radial density profile

$$\rho(r) = \frac{3 v_1^2 r_t^2}{2\pi \, G (r^2 + r_c^2)(r^2 + r_t^2)} \qquad r_t \geq r_c \qquad (1)$$

which is a kind of truncated non-singular isothermal sphere. For this profile the image shear as a function of angle, α, from the cluster center is

$$\gamma(\alpha) = 6\pi \frac{v_1^2}{c^2} \frac{\beta}{\alpha} \frac{\alpha_t^2}{\alpha_t^2 - \alpha_c^2} \left(2 \frac{\alpha_t - \alpha_c}{\alpha} - \frac{\alpha^2 + 2\alpha_t^2}{\alpha\sqrt{\alpha^2 + \alpha_t^2}} + \frac{\alpha^2 + 2\alpha_c^2}{\alpha\sqrt{\alpha^2 + \alpha_c^2}} \right). \qquad (2)$$

Here α_c and α_t are the angles subtended by r_c and r_t at the distance of the cluster. The distances of the galaxies whose shear is measured comes into the factor β. If most of galaxies are much further away than the cluster then $\beta \approx 1$ while if most of the galaxies are at a distance comparable to or less than that of the cluster then β may be much less than unity since many of the galaxies will be in front of the cluster and not sheared at all or not far enough behind the cluster to receive the full amount of shear.

The shear around a given cluster is maximized a few core radii from the center, while the maximal shear varies roughly proportional to z until the cluster distance approaches the depth of the survey. One never finds large shear too close to the cluster center and for more nearby clusters one must look very far from the center to maximize the shear. Figure 1 illustrates that for shallow surveys one is most sensitive to nearby structures. Note that a disk radius of $5'$ does a good job of maximizing the shear over a broad range of cluster redshifts and limiting magnitudes. Wider area coverage yields a larger signal only for $z \lesssim 0.2$ clusters. Of course a large signal is of no use unless one has sufficient galaxy numbers to detect it. In Figure 2 we see

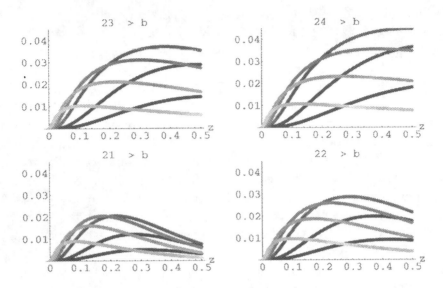

Figure 1.   Plotted is the mean tangential shear in a disk on the sky centered on a model cluster vs. the redshift of the cluster. The different curves represent different disk radii: from black to light gray the radii are $1'$, $2'$, $5'$, $10'$, $20'$, $50'$. The different plots are for different limiting b magnitudes as labeled. Here we assume a Schechter luminosity function with $\phi_* = 0.014(h/\mathrm{Mpc})^3$, $\alpha = -0.97$, and $M_* = -19.5$ corresponding to b_J magnitudes. The model cluster has 1d velocity dispersion $v_1 = 800\mathrm{km/sec}$, core radius $r_c = 250\,h^{-1}\mathrm{kpc}$, and truncation radius $r_t = 3\,h^{-1}\mathrm{Mpc}$.

that the available S/N is indeed significantly higher for deeper surveys but, with large enough area coverage, can be much larger than unity even for very shallow surveys. For extremely low redshift clusters one must survey very large areas to obtain significant signal. Yet even for $b < 21$ one can in principle obtain a significant signal from a $z = 0.03$ cluster, like Coma, if one is able to survey $\sim 1^{\circ2}$. Note however that this would require keeping systematics well below the 1% level.

Given the low tolerance for systematic errors it is important to have a good handle on how well one is determining the shear. Besides simulations and comparison with better (i.e. HST) data, one can also use an internal check of one's data. To do this, take one's measured ellipticities and rotate their position-angle by $45°$. Then use one's favorite reconstruction technique to estimate the surface density from the rotated data. The surface density one obtains should be consistent with zero up to the noise from the random galaxy orientations and known measurement errors; and from effects due to the boundary of one's sample. If not, one probably has discovered some systematic problems with one's method. The mathematics behind this is as follows. One is trying to estimate the shear tensor, γ_{ab}. Such a 2-d symmetric traceless tensor field can be decomposed into its

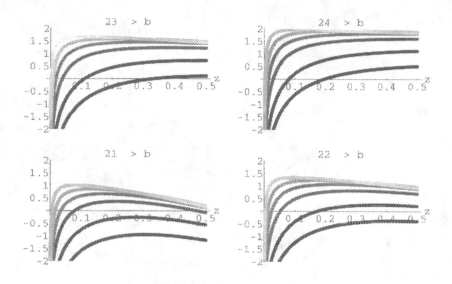

Figure 2. Plotted is $\log_{10} S/N$ for measuring the amplitude of the shear for clusters as in Figure 1. The noise is assumed dominated by the finite number of non-round galaxies in the sample assumed to have rms ellipticity 0.3.

scalar and pseudo-scalar parts

$$ \gamma_{\rm s} = \nabla^{-2} \gamma_{ab,ab} \qquad \gamma_{\rm p} = \nabla^{-2} \epsilon_{ac} \gamma_{ab,bc} \qquad \epsilon_{ab} = \begin{pmatrix} 0 & -1 \\ 1 & 0 \end{pmatrix}, \qquad (3) $$

which is analogous to decomposing a vector into its curl and a curl-free parts. For *weak* lensing $\gamma_{\rm s}$ is just proportional to the weighted surface density while for gravitationally induced shear from non-relativistic matter $\gamma_{\rm p} = 0$ since the shear is derived from a potential. Multiplying the shear tensor by ϵ_{ab} is the same as rotating the position-angle of the shear by 45°, so one obtains the above result. The two components, $\gamma_{\rm s}$ and $\gamma_{\rm p}$, are so similar that most sources of noise and error will contribute equally to both, while the true signal will contribute only to $\gamma_{\rm s}$. Thus it is probably fair to believe one's results only to the extent that, on average, $|\gamma_{\rm s}| > |\gamma_{\rm p}|$. Kaiser and Tyson report that they have used similar methods.

At this writing the SDSS telescope is not yet operational and hence it is difficult to know how it will perform in practice. To address this issue the authors have begun an observational program with a telescope at the SDSS site, the ARC 3.5m telescope, using the Fermilab Drift Scan Camera (DSC) which is similar, if much smaller, than the SDSS camera. We have not yet reduced the level of systematics to the point which would make the SDSS Northern survey useful for weak lensing, but are confident that significant improvements will be made. The SDSS collaboration is in the process of comparing DSC data in the Sloan colors and at the Sloan depth

Figure 3. By rotating the measured ellipticities (*black*) by 45° and then using these rotated ellipticities (*gray*) to reconstruct the surface density one constructs a realization of the same size as the error in the reconstructed surface density.

with deep HST WFPC-2 data. This will be extremely useful for gauging the accuracy of shear measurements that can be expected from the SDSS. If everything works well, the weak lensing data from the SDSS northern survey will be one of its major achievements. In any case we certainly do expect that the deeper SDSS southern survey will yield useful information from weak lensing studies.

Acknowledgements: This work was supported by the DOE and NASA grant # NAG-5-2788.

References

Blandford, R., Saust, A., Brainerd, T., & Villumsen, J., 1991, MNRAS, 251, 600
Fahlman, G., Kaiser, N., Squires, G., & Woods, D., 1995, ApJ, 437, 56
Kaiser, N., 1992, ApJ, 388, 272
Kent, S., 1994, Ap & Space Sci, 217, 27
Miralda-Escudé, J., 1991, ApJ, 380, 1
Mould, J., et al., 1994, MNRAS, 271, 56
Smail, I., Ellis, R., & Fitchett, M., 1994, MNRAS, 270, 245
Szalay, A., 1995, private communication
Tyson, J., Valdes, F., & Wenk, R., 1990, ApJL, 349, L1
Valdes, F., Tyson, J., & Jarvis, J., 1983, ApJ, 271, 431
Villumsen, J., & Gould, A., 1994 , ApJL, 428, L45

A 71 MEGAPIXEL MOSAIC CAMERA
FOR WEAK LENSING AT APO

A. DIERCKS, C. STUBBS, C. HOGAN AND E. ADELBERGER
Departments of Physics and Astronomy
University of Washington
Box 351580
Seattle, Washington 98195-1580

Abstract. We are developing a wide-field CCD camera system which is optimized for using weak gravitational lensing to study the distribution of dark matter in clusters of galaxies and eventually the field. The system will be used at the Apache Point Observatory (APO) 3.5 meter telescope in New Mexico.

1. Introduction

One of the significant difficulties with weak gravitational lensing studies is the need for high-resolution, wide-field imaging instruments. The camera described here is designed to meet this challenge by imaging a 30 arcminute diameter field of view with distortion free optics. The system is optimized for TDI (drift scan) operation but will also be capable of taking staring mode images. We have determined that it is most cost effective to use a simple distortion free corrector at the native plate scale of the telescope (f/10) and use a large number of CCDs to cover the entire focal plane.

The camera system contains several novel features which will help meet the demanding imaging requirements of weak gravitational lensing studies. The camera will contain dedicated, fast-readout CCDs for guiding and eventually tip-tilt correction via the secondary mirror of the 3.5m telescope as well as a tilted CCD in the focal plane to allow active focusing. The camera, corrector optics, and filters will be able to rotate with respect to each other and the telescope optics in order to study and eliminate systematic effects due to any aberrations in the optical chain.

81

C. S. Kochanek and J. N. Hewitt (eds), Astrophysical Applications of Gravitational Lensing, 81–82.

2. Observational Technique

Our observational approach is driven by the need for measurements that are designed to suppress sources of systematic error. Telescope guiding effects, pixellation effects, optical aberrations, and the effects of seeing need to be properly characterized and suppressed in order to avoid spurious coherent shape distortions in the images. This involves the use of stellar PSF's and so it is imperative that the weak lensing program have well sampled stellar PSF's. Also, large numbers of galaxies need to be observed in multiple passbands in order to gain the statistical leverage necessary to disentangle a weak lensing system.

We will take very deep images of blank and cluster fields with the redundancy that is imperative if a weak lensing signal is to be distinguished from systematic effects. The target galaxies have surface brightnesses of 27th mag (R) per square arcsec, so they are well below 1% of sky. We intend to (1) rotate the corrector plus instrument to several position angles on the sky, (2) take two full independent sets of data for each field, (3) take images in three passbands for redundancy and color discrimination, (4) rotate the CCD camera, corrector optics, filters, and telescope with respect to each other, and (5) operate in *both* drift scan and staring mode, as these two techniques have different sensitivities to a number of the sources of systematic error that concern us. We have endeavored to create a design where it is easy to perform these operations as rapidly as possible.

By demanding that the observed ellipticity distribution of the galaxies be invariant under these various operations we hope to suppress the majority of telescope and instrument artifacts that could produce a false signal, or mask a real one.

3. The APO 3.5m Telescope

The camera system is engineered to take advantage of the unique capabilities of the APO 3.5m telescope. One of these features is the ability to track at any rate and direction on the sky. Thus it is possible to "drift scan" at an arbitrary rate in an arbitrary direction, and thereby obtain long integrations. We have already employed this feature to great advantage using a smaller camera system currently available on the telescope. In drift scan mode we will be able to image a swath nearly 30 arcminutes wide, twice on a single pass. Alternatively, we will have the option of imaging the swath simultaneously in two colors.

ASSOCIATION OF DISTANT RADIO SOURCES
AND FOREGROUND GALAXIES

E. MARTÍNEZ-GONZÁLEZ & N. BENÍTEZ

Instituto de Física de Cantabria, CSIC-Univ. de Cantabria
Facultad de Ciencias, Avda. Los Castros s/n
39005 Santander, Spain

Abstract. A statistically significant (99.1%) excess of red galaxies from the APM Sky Catalogue is found around a sample of $z \sim 1$ 1Jy radio sources. The most plausible explanation for this result seems to be the magnification bias caused by the weak gravitational lensing of large scale structures at intermediate redshifts.

1. Introduction

The existence of statistically significant associations of foreground objects with high redshift AGNs, on scales ranging from a few arcseconds to $\sim 30'$, is a fact confirmed by several authors (Webster et al. 1988, Fugmann 1990, Hammer & Le Fèvre 1990, Thomas et al. 1995, Benítez et al. 1995, Seitz & Schneider 1995), and explained as a consequence of the magnification bias due to gravitational lensing by structures with sizes varying from galaxies to matter overdensities on scales at least as large as clusters of galaxies or even larger (Pei 1995, Bartelmann 1995). This effect is described by the expression $n_A(S) = n'_A(S/\mu)/\mu$, where $n_A(S)$ is the observed cumulative number density, μ is the magnification, and n'_A is the unlensed number density which is usually assumed to have the form $n'_A(S') \propto (S')^\alpha$, S' being the AGN flux in absence of magnification. Depending on the slope of the number counts of background sources, we may observe positive, insignificant or negative statistical associations of these foreground objects with background sources like distant AGNs (see Bartelmann 1995, Seitz & Schneider 1995). Radio-selected samples of high-redshift objects present a rather steep slope in their number counts-flux distribution and are thus

C. S. Kochanek and J. N. Hewitt (eds), Astrophysical Applications of Gravitational Lensing, 83–88.
© 1996 IAU

very appropriate for trying to detect excesses of foreground objects due to the magnification bias.

Here we shall report the results of a study of the distribution of galaxies taken from the APM Sky Catalogue around a sample of AGNs from the 1Jy catalogue (Stickel et al. 1994). A full-length account of this investigation can be found in (Benítez & Martínez-González 1995).

2. The Data

The 1Jy catalogue of radio-sources is described in Stickel et al. (1994). It has been often and successfully used to prove the existence of correlations between several catalogues of foreground objects and distant AGNs (Fugmann 1990, Seitz & Schneider 1995, Bartelmann & Schneider 1994).

The APM Sky Catalogue, described in Irwin & McMahon (1992) and Irwin et al. (1994), contains all the objects found by an automatic detection algorithm on the scans of the O (equivalent to $U + B$) and E (corresponding to a narrow R) Palomar Sky Survey plates.

Our AGN sample amounts to 73 QSOs and radio galaxies from the 1Jy catalogue with $0.5 < z < 1.5$ and declination $\delta > 0$, and around which there are available $90' \times 90'$ APM fields with at least 2/3 of the total surface within one POSS plate.

We first considered the APM Sky Catalogue objects contained in these fields which are classified as galaxies and have $O < 19.5, E < 21$. Our chosen magnitude limits are 0.5 mag brighter than the limiting magnitudes on each band quoted by Irwin and collaborators. Number counts-magnitude plots show that they can be reasonably considered as completion limits in the sense that fainter than these limits the number counts-magnitude relation suddenly becomes steeper and spurious detections begin to dominate the number counts.

In order to exclude as many spurious galaxies as possible, we require our objects to be detected and classified as galaxies in both the red and blue bands. This discards $\approx 40 - 50\%$ of the objects at magnitudes below our completion limits. At brighter magnitudes ($E < 18$), most of them are objects classified as galaxies in one band and as stars or merged objects in the other. This is not surprising if we remember that the classification of the objects in the APM Sky Catalogue is an automated procedure and that the pixel resolution is rather low. At fainter magnitudes we are also losing many galaxies that due to their colors cannot be detected in both bands. Nevertheless, the procedure followed is necessary in order to obtain a sufficiently homogeneous sample.

The galaxy sample is thus finally formed by the $\approx 35,000$ APM Sky Catalogue objects contained in 73 fields which are classified as galaxies in *both* the O *and* E bands, $O < 19.5$, $E < 21$.

We performed two tests in order to check for the existence of systematic gradients over the fields. First, we divided each field into four $45' \times 45'$ boxes, added the number of galaxies found in each of these boxes over all the fields and compared among them. Second, we considered the number of objects within the central part of all the fields containing half of the total $5400'' \times 5400''$ surface (excluding a central $1000'' \times 1000''$ region which contains most of the excess) and compared it with the number of objects in the outer half. In both cases there is good agreement with a $\sqrt{N}$ Poissonian dispersion, what confirms the homogeneity of the fields on average.

3. Main results and discussion

We shall look for density enhancements on a certain scale l using the following method. The fields are divided into boxes of size l^2 and the density enhancement on each box is defined as

$$q_{ij} = \frac{\sum_f n_{ij}^f}{\sum_f \frac{s_{ij}^f N^f}{S^f}}$$

where n_{ij}^f and s_{ij}^f are respectively the number of galaxies and the useful surface in the box ij of the field f, N^f is the total number of galaxies in a field and S^f is its total useful surface area.

The statistical significance of the result is estimated through the empirical probability p, $p = N_{less}/N_{total}$, where N_{less} is the number of boxes (not including the central one) with a value of q less than or equal to that of the central box, and N_{total} is the total number of boxes in the field.

We have found that p is always similar or smaller than the value of the significance obtained using the empirical rms (which is a factor of $1.2 - 1.4$ times the one deduced from Poissonian statistics (Benítez et al. 1995)) and assuming a Gaussian distribution.

Table 1 lists the results for scales of $500''$ and $1000''$ and different colors. The excess is not very prominent when we consider *all* the galaxies with $E < 19.5$ and $O < 21$, but if we select only the reddest galaxies with $O - E > 2$, the value of the density enhancement in the central box dramatically increases and reaches the highest value of all the $500'' \times 500''$ boxes in the field, $q = 1.298 \pm 0.114$ (rms error), corresponding to an empirical p of 99.1%.

This behavior agrees well with elementary considerations. Gravitational lens theory (Schneider et al. 1992) shows that background objects with redshift $z_s \approx 0.5 - 1.5$ are most efficiently magnified by foreground lenses with $z_l \approx 0.2 - 0.4$. Most types of M^* galaxies at these redshifts will have

TABLE 1. Data for the central box

$O - E$	500 arcsec	1000 arcsec
all	1.109 ± 0.078, 91.4%	1.028 ± 0.041, 71.7%
> 1	1.122 ± 0.082, 92.3%	1.039 ± 0.047, 83.9%
> 2	1.298 ± 0.114, 99.1%	1.089 ± 0.059, 91.7%

colors $O - E \gtrsim 2$ ($H_0 = 50, \Lambda = 0, \Omega = 1$). M^* and k-corrections are taken from Driver et al. (1994), and we have used the relationship $O - E \approx 2(B-V)$ quoted in Irwin & McMahon (1992). Besides, at any given redshift, early type galaxies tend to concentrate toward cluster cores. Therefore, if we constrain the galaxies to have $O - E > 2$, we are selecting the ones in the deeper potential wells and at the redshifts most efficient for lensing. This increases the excess, which is otherwise diluted in projection among the objects not contributing so strongly to the lensing effect, and makes it more easily detectable. The scale at which the density enhancement is more significant, $500''$, corresponds to $\approx 2 - 3$ Mpc at redshifts of $z = 0.2 - 0.4$.

In order to show the dependence of the excess with the angular scale we plot the AGN-galaxy cross-correlation function

$$\omega_{AG}(\theta) = \frac{\sum_f N_G^f(\theta)}{\sum_f N_{exp}^f(\theta)} - 1$$

where $N_G^f(\theta)$ is the number of galaxies found in the annular bin θ of the field f and $N_{exp}^f(\theta)$ is its expectation value based on the number of objects on the whole field normalized to the useful surface of the bin, for the galaxies with $O < 21, E < 19.5$ and $O - E > 2$ (figure 1) as a function of the bin radius in arcsec. The error bars are Poissonian $\sqrt{N}$. The previous results can be confronted with the theoretical expectations of galaxy-QSO cross-correlations $\omega_{QG}(\theta)$ produced by weak and moderate gravitational lensing. Bartelmann (1995) has calculated the expected cross-correlation for samples of galaxies with different depths and AGN catalogues simulating the 1Jy, within a CDM cosmogony. The value of the parameter q on a scale of 500 $''$ in the case of no color constraint on the APM galaxies (see table1) agrees marginally with his results. In any case, as Bartelmann expected, the amplitude is not significantly different from zero due to the small number of AGNs considered and the rather bright magnitude limit of the APM galaxies, which shows how the detectability of the magnification bias is dramatically increased by a color cutoff in the galaxy sample.

O < 21, E < 19.5, O − E > 2

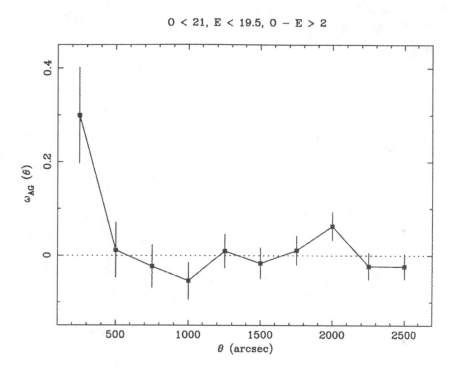

Figure 1. The cross-correlation function between 1Jy AGNs with $0.5 < z < 1.5$ and APM galaxies with $E < 19.5$, $O < 21$ and $O - E > 2$.

We have also tried to measure the effect of imposing a double flux limit by setting a cutoff on the optical magnitude of the AGNs. It seems that there is a tendency towards greater overdensities for AGNs with brighter optical magnitudes, but this constraint reduces the sample and lowers the statistical significance, which makes the results inconclusive. Larger samples are needed in order to better establish this effect.

The observed overdensity of foreground galaxies around high-redshift AGN seems to be a result gravitational lensing by large scale structure and is very difficult to reconcile with other proposed explanations such as obscuration by dust. The latter should produce the opposite effect: the objects forming the excess around high redshift AGNs would even tend to be bluer than the average, as they would be less reddened by dust.

References

Bartelmann, M. 1995, A&A, 298, 661
Bartelmann, M., & Schneider, P., 1994, A&A, 284, 1
Benítez, N. & Martínez-González, 1995, ApJL, 448, L89
Benítez, N., Martínez-González, E., González-Serrano, J.I., & Cayón L., 1995, AJ, 109,

88   E. MARTÍNEZ-GONZÁLEZ & N. BENÍTEZ

935
Driver, S.P., Phillipps, S., Davies, J.I., Morgan, I., & Disney, M.J., 1994, MNRAS, 266, 155
Fugmann, W., 1990, A&A, 240, 11
Hammer F., & Le Févre, O., 1990, ApJ, 357, 38
Irwin, M., & McMahon, R., 1992, Gemini, 37, 1
Irwin, M., Maddox, S., & McMahon, R., 1994, Spectrum, Newsletter of The Royal Observatories, 2, 14
Pei, Y., 1995, ApJ, 440, 485
Schneider, P., Ehlers, J., & Falco, E.E., 1992, Gravitational Lenses (Heidelberg: Springer)
Stickel, M., Meisenheimer, K., & Kühr, H., 1994, A&AS, 105, 211
Seitz, S., & Schneider, P. 1995, A&A, in press
Thomas, P.A., Webster, R.L., & Drinkwater, M.J., 1994, MNRAS, 273, 1069
Webster, R.L., Hewett, P.C., Harding, M.E., & Wegner, G.A., 1988, Nature, 336, 358

LIGHT PROPAGATION IN A CLUMPY UNIVERSE

LAM HUI AND UROŠ SELJAK
Department of Physics
Massachusetts Institute of Technology
Cambridge, Massachusetts 02139 USA.

The propagation of light in an inhomogeneous universe is a long standing problem. Its resolution requires, first, a realistic description of the geometry of a clumpy universe and, second, solutions to the null geodesic equations given the metric of such a universe. The Friedmann-Robertson-Walker metric has become the standard description of the large scale geometry of the universe. However, the observable universe today is manifestly inhomogeneous. The weakly perturbed Friedmann-Robertson-Walker metric is often used to describe such a universe. But its validity is only guaranteed for a weakly inhomogeneous universe, where, for instance, overdensities are small ($\delta\rho/\bar{\rho} \ll 1$), which is not true for sufficiently small scales in the universe today. It is well known, however, that the metric perturbations can still be small even if the overdensity is not small, given the right conditions and coordinates. However, spatial gradients of metric perturbations are not necessarily small any more. Here we estimate whether the second-order corrections involving them can affect significantly the expansion of the universe or the light propagation in it.

We concentrate on the energy constraint equation (or the time-time part of the Einstein's equations). The zeroth order terms of this equation constitute the Friedmann equation. The second order terms that we keep are then corrections to it. They can be viewed as the back-reaction of inhomogeneities on the expansion rate of the universe (Futamase 1988). We write the metric in the following form:

$$ds^2 = a^2(\eta)[-(1+2\phi)d\eta^2 + 2\omega_i dx^i d\eta + ((1-2\psi)\delta_{ij} + 2h_{ij})dx^i dx^j] \quad (1)$$

where ϕ, ω_i, ψ and h_{ij} are functions of time and space. h_{ij} is chosen to be traceless i.e. $\sum_k h_{kk} = 0$. We also make the slow motion approximation: peculiar velocities are small and so are the time derivatives of metric perturbations compared to their gradients. We impose the gauge conditions

C. S. Kochanek and J. N. Hewitt (eds), Astrophysical Applications of Gravitational Lensing, 89-90.
© 1996 IAU

$\omega^i{}_{,i} = 0$ and $h_{ij}{}^{,i} = 0$. By examining the rest of the Einstein equations, it can be shown that ω_i and h_{ij} and their gradients are small compared to ϕ or ψ. So, second order terms involving them are discarded. All other second order terms involving spatial gradients are kept. The time-time equation can be separated into homogeneous and inhomogeneous parts by performing a spatial average, $\langle Q \rangle = \int Q \sqrt{g^{(3)}} d^3 x / V$, where Q is any quantity that is to be averaged, $g^{(3)}$ is the determinant of the three metric and V is the volume of a finite box over which the integration is performed. Periodic boundary conditions are imposed on the box and the limit of V approaching infinity is taken. Hence, the modified Friedmann equation becomes

$$8\pi G a^2 \langle \rho \rangle = 3(\frac{a'}{a})^2 + \langle (\vec{\nabla}\phi)^2 \rangle. \tag{2}$$

The inhomogeneous part of the time-time equation is then the cosmological generalization of Poisson's equation, where the second order terms are obviously small compared to first order ones. Moreover, Poisson equation guarantees that the metric perturbation is equivalent to the Newtonian potential on small scales and remains small everywhere away from black holes. This equation is used to estimate the significance of the correction $\epsilon = \langle (\vec{\nabla}\phi)^2 \rangle / 3 H_0^2$ ($H_0 = a'/a$ is the Hubble constant today) to the Friedmann equation, given a realistic model of density fluctuations. The averaging can be rewritten in terms of the power spectrum as

$$\epsilon = \frac{4\pi}{3} \int k^3 P_\phi(k)(k/H_0)^2 d \ln k = 3\pi \int k^3 P_\delta(k)(H_0/k)^2 d \ln k. \tag{3}$$

$4\pi k^3 P_\delta(k) \approx 1$ today is at $k^{-1} \approx 10 Mpc$ from which follows $\epsilon \approx 10^{-5}$, provided that there is no divergence on very small scales. To investigate this possibility we use nonlinear evolution of realistic power spectra, using N-body simulations. The conclusion from these studies is that the logarithmic contribution to ϵ peaks at the transition from linear to nonlinear scale and is in all models of the order of 10^{-5}.

To conclude, we argue that linear metric perturbation theory in the gauge above provides an excellent description of clumpy universe today and the corrections to the Friedmann equation are found to be negligible. For light propagation, second order corrections to the geodesic equation (terms proportional to $\phi\phi_{,i}$) are necessarily small compared to the first order terms (those proportional to $\phi_{,i}$). The deflection of light ray is dominated by the linear terms and higher order terms are unimportant even in a clumpy universe, provided that the metric perturbations remain small.

References

T. Futamase, *Phys. Rev. Lett.* **61**, 2175 (1988)

EFFECTS OF LARGE-SCALE STRUCTURE
UPON THE DETERMINATION OF H$_0$ FROM TIME DELAYS

G.C. SURPI AND D.D. HARARI
Departamento de Física, FCEyN, Universidad de Buenos Aires
Ciudad Universituria - Pab. 1, 1428 Buenos Aires, Argentina

AND

J.A. FRIEMAN
NASA/Fermilab Astrophysical Center,
Fermi National Accelerator Laboratory
P.O. Box 500, Batavia, Illinois 60510, USA

Abstract. We have analyzed the effects of both large-scale inhomogeneities in the mass distribution and cosmological gravitational waves upon the time delay between two images in a gravitational lens system. We have shown that their leading order effect, which could potentially bias the determination of the Hubble parameter, is indistinguishable from a change in the relative angle between the source and the lens axis. Since the absolute angular position of the source is not directly measurable, nor does it enter the relationship between the Hubble parameter and the lens observables, the determination of H$_0$ from gravitational lens time delays follows in the usual way, as if the metric perturbations were absent.

We have considered a thin gravitational lens in the weak field approximation, embedded in a spatially-flat FRW cosmology, with small-amplitude and long-wavelength scalar and tensor metric perturbations $h_{\mu\nu}$. Our approach relied upon Fermat's principle, which is also valid in the non-stationary space-time under consideration (Kovner 1990). We extremized the time of travel within a family of zig-zag null photon paths composed of two segments, each one a geodesic of the perturbed FRW background in the absence of the deflector.

Our main result is that a lens system with the source located at an absolute angular position β (bold characters denote two-components angular

C. S. Kochanek and J. N. Hewitt (eds), Astrophysical Applications of Gravitational Lensing, 91–92.
© 1996 IAU

vectors measured with respect to the lens axis) in the presence of scalar and tensor metric perturbations is, to leading order, indistinguishable from an identical lens system in the absence of perturbations, but with a different source alignment β_{eff} given by (Frieman et al. 1994, Surpi et al. 1995),

$$\beta_{\text{eff}} \equiv \beta + \beta_{\text{pert}} \quad \text{with} \quad \beta_{\text{pert}} = \frac{1}{r_d} \int_0^{r_d} dr\, \Delta(r) - \frac{1}{r_s} \int_0^{r_s} dr\, \Delta(r) \quad , \quad (1)$$

$$\Delta^i(r) = -\frac{1}{2} \int^r dr\, (h_{00,i} + h_{33,i} - 2h_{i3,3} - 2h_{i3,0}) \quad . \tag{2}$$

Here r is the comoving coordinate distance along the lens axis, coincident with the $z-$axis. $r = 0$, r_d and r_s correspond to the observer, deflector and source respectively. The index i takes the values $i = 1, 2$. The result is valid both for scalar as well as tensor perturbations, the former described in the longitudinal gauge and the latter in the transverse-traceless gauge.

The change in the effective alignment induced by a single mode of the metric perturbations can be as large as the perturbation amplitude $|h_{\mu\nu}|$. Compatible with current limits, fixed by the cosmic microwave background anisotropy, it could be of order $10^{-5} \approx 1''$. Thus, the time delay directly induced by metric perturbations could be as large as the intrinsic delay due to the lens geometry. The alignment angle is not, however, directly observable. In particular, β_{eff} cancels out from the expression that allows determination of H_o from the time delay and other lens observables (Refsdal 1964). We thus conclude that the leading order effect of scalar and tensor metric perturbations does not compromise the program to determine H_o from gravitational lens time delays.

Our analysis is limited to the effects of small-amplitude (i.e. linear) perturbations. Non-linear structure on small scales (e.g. galaxies) could also significantly affect the distance measure, which we approximated by its FRW expression. We have also limited ourselves to long-wavelength metric perturbations, and have discarded terms $\xi/\lambda \lesssim 10^{-3}$ times smaller than the leading order terms, where ξ is the maximum transverse separation between the image paths (of order 10 kpc) and λ is the wavelength of the perturbation (larger than 10 Mpc). The neglected terms would compromise the lens reconstruction only at levels below tenths of arc seconds in the images' angular separation (Seljak 1994).

References

Frieman, J.A., Harari, D.D., & Surpi, G.C., 1994, Phys Rev D, 50, 4895
Kovner, I., 1990, ApJ, 351, 114
Refsdal, S., 1964, MNRAS, 128, 307
Surpi, G.C., Harari, D.D., & Frieman, J.A., 1995, ApJ, in press
Seljak, U., 1994, ApJ, 436, 509

EXPONENTIAL GROWTH OF DISTANCE BETWEEN NEARBY RAYS DUE TO MULTIPLE GRAVITATIONAL LENSING

T. FUKUSHIGE AND J. MAKINO

College of Arts and Science, University of Tokyo

Abstract. We have investigated multiple gravitational lensing by numerical "ray tracing" simulations. We have found that the distance between rays grows exponentially, on average, until it reaches the projected mean separation of lensing objects ($\sim RN^{-1/2}$, where R is the system size and N is the number of scattering object). This nature may affect observations of high redshift objects or the anisotropy of the cosmic background radiation.

1. Basic Concept

The explanation is summarized as follows (Fukushige and Makino 1994). When the distance between the rays is small, the rays are scattered coherently by the same object. In the single scattering event, the distance increases in proportion to the distance before the scattering, since the differential acceleration or the "tidal force" is proportional to the distance between rays. Through multiple scattering, the distance increases exponentially. This exponential growth stops when the distance becomes so large that the scattering becomes incoherent. In the community of gravitational N-body simulations, the exponential divergence of initial conditions has been well known (e.g. Goodman et al. 1993).

Previous studies neglected the effect of the exponential growth. For example, Gunn (1967) expressed differential deflection between rays through multiple lensing by a superposition of single events, under an assumption that the differential deflection through single event is small. This assumption is, however, not appropriate. The average increase of the distance between rays by a single event is proportional to its original distance. Thus,

93

© 1996 IAU

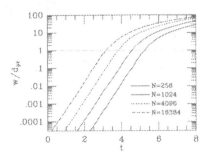

Figure 1. The distance w between rays grows exponentially until it reaches at the projected mean separation of scattering objects $(d_{pr} = RN^{-1/2})$

the distance grows exponentially through multiple scattering. The increase of the distance evidently cannot be expressed by the simple superposition.

Note that during this exponential growth the phase space volume is conserved and the "shape" of phase space volume is elongated. The exponential growth can be observed in a projection of this volume on position space or velocity space. Moreover, the exponential growth does not continues forever, since it stops at the projected mean particle separation. Note also that the exponential growth is not affected by the scattering event and that the distance converges, on average. If the scattering object exists between two rays, the two rays are scattered to approach each other. This is the more familiar case of strong gravitational lensing. However, this is very rare event when the distance between rays is small.

2. Ray Tracing Simulation

We have performed 3D "ray tracing" simulations (Fukushige et al. 1994). The lensing objects are distributed randomly in a cube, and are fixed. Figure 1 shows the distance w between rays grows exponentially until it reaches the projected mean separation $RN^{-1/2}$. We plot the median value of 1000 runs. In our units, the light velocity, the system size (R), and the gravitational constant (G) are unity, and the mean density (ρ) is 0.5. The lensing object is point a mass. We use a special-purpose computer (MD-GRAPE (Taiji et al. 1995)) for the force calculations.

References

Fukushige, T., & Makino, J., 1994, ApJ, 436, L111.
Fukushige, T., Makino, J., & Ebisuzaki, T., 1994, ApJ, 436, L107.
Gunn,, J. E., 1967, ApJ, 147, 61.
Goodman, J., Heggie, D. C., & Hut, P., 1993, ApJ, 415, 715.
Taiji, M., et al., 1995, in preparation.

EFFECT OF MULTIPLE GRAVITATIONAL LENSING ON THE ANISOTROPY OF THE COSMIC BACKGROUND RADIATION

T. FUKUSHIGE, J. MAKINO, AND T. EBISUZAKI
College of Arts and Sciences, University of Tokyo,
3-8-1 Komaba, Meguro-ku, Tokyo 153, Japan

Abstract. We investigated smoothing of the cosmic background radiation (CBR) by multiple gravitational lensing. The CBR is gravitationally scattered by galaxies, clusters of galaxies, and superclusters during the travel from the last scattering surface. Although the effect of the gravitational lensing was thought to be unimportant, we found that the multiple gravitational lensing by clusters of galaxies or by superclusters can reduced by a large factor. This result is explained by the fact that the distance between two light rays grows exponentially though multiple gravitational lensing. If such structures were formed at $z = 2 - 5$ and contain a large fraction of the mass of the universe ($\Omega_s > 0.5$), then multiple gravitational lensing can reduce the temperature anisotropy of the CBR by 40-60%, approximately up to a degree scale.

1. Introduction

The distance between rays grows exponentially, on average, until it reaches at the projected mean separation $d_{\mathrm{pr}}(= RN^{-1/2})$ of lensing objects. The basic explanation of the exponential growth is given by Fukushige and Makino (in this proceeding, hereafter FM).

The possibility that the gravitational lensing change the anisotropy of the CBR has been discussed by many authors(e.g. Cole & Efstathiou 1989, Sasaki 1989, Watanabe & Tomita 1991). The light rays sampled in a radio telescope with a finite angular width come from a large angular region of the last scattering surface by gravitational lensing. This effect influences the anisotropy correlation function. However, most authors assumed that the effect of the multiple gravitational lensing could be expressed as the

C. S. Kochanek and J. N. Hewitt (eds), Astrophysical Applications of Gravitational Lensing, 95–96.
© 1996 IAU

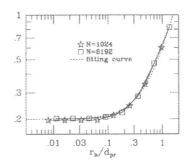

Figure 1. Non-dimensional e-folding time (τ) of the exponential growth plotted against half-mass radius (r_h) of lensing objects.

superposition of single scatterings. Thus their estimates did not take into account the exponential growth.

2. Numerical Experiment

The exponential growth does not occur if the universe is completely uniform. The scattering objects must be compact enough to be effective. We investigate the effect of the size of lensing object by 3D "ray tracing" simulations. The simulation is the same as that described in FM. The lensing object has the density profile of a King model $(W_0 = 5)$. In Figure 1, the non-dimensional e-folding time of exponential growth (τ), scaled by $1/\sqrt{G\rho}$, is plotted as a function of the half-mass radius of lensing objects (r_h). The exponential growth is suppressed if r_h is larger than d_{pr}.

We estimated the growth factor α_{LS} of the angle between two photons that come from the last scattering surface by solving differential equations for the angle θ and the distance w: $c\theta = -dw/dt$ and $dw/dt = -c\theta_0 - w/(\tau/\sqrt{G\rho})$. The growth factor is estimated to be $\alpha_{LS} = 4 - 5$ if the superclusters were formed at $z = 2$ and the universe consists only of the superclusters. The actual effect on the measured anisotropy depends on the power spectrum of the anisotropy, but it is likely to be significant. More detailed discussion is presented by Fukushige et al. (1994,1995).

References

Cole, S., & Efstathiou, G., 1989, MNRAS, 239, 195.
Fukushige, T., Makino, J., & Ebisuzaki, T., 1994, ApJ, 436, L107.
Fukushige, T., Makino, J., Nishimura, O., & Ebisuzaki, T., 1995, PASJ, in press.
Sasaki, M., 1989, MNRAS, 240, 415.
Watanabe, K., & Tomita, K., 1991, ApJ, 370, 481.

INFLUENCE OF GRAVITATIONAL LENSING ON ESTIMATES OF Ω IN NEUTRAL HYDROGEN

MATTHIAS BARTELMANN[1,2] AND ABRAHAM LOEB[1]

[1] *Harvard-Smithsonian Center for Astrophysics*
60 Garden Street, Cambridge, MA 02138, USA, and
[2] *Max-Planck-Institut für Astrophysik*
Postfach 1523, 85740 Garching, Germany

1. Introduction

A wealth of observational data supports the commonly held view that damped Lyman-α (Lyα) absorption in QSO spectra is associated with neutral-hydrogen (HI) disks in spiral galaxies. Most of the HI probed by QSO absorption lines is traced by damped Lyα lines because of their high column densities, $N > 10^{20} \, \mathrm{cm}^{-2}$. The spiral galaxies hosting the HI disks can act as gravitational lenses on the QSOs. If the HI column density increases towards the center of the disks, as suggested by observations of local galaxies, the magnification bias preferentially selects for high column-density systems. The estimates of HI in damped Lyα systems can then systematically be distorted by gravitational lensing.

2. A Model for Lensing by Damped Lyα Absorbers

We use a simple model to quantify this influence of lensing on the observed cosmological density parameter in HI, Ω_{HI} (Bartelmann & Loeb 1995). Its ingredients are:

- The lenses are modeled as singular isothermal spheres with a mean velocity dispersion of $\sigma_v \sim 160 \, \mathrm{km \, s}^{-1}$.
- Their spatial number density is assumed to be constant in comoving coordinates.
- The HI column densities in the absorbers are modeled by exponential disks adapted to observational data.
- The disks are randomly inclined relative to the line-of-sight.

97

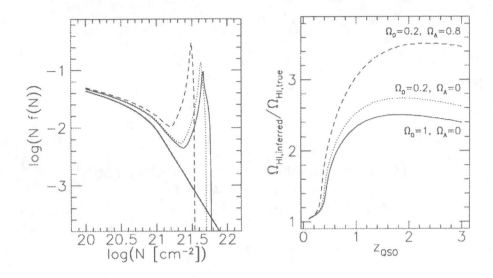

Figure 1. $f(N)$ and $\Omega_{\rm HI}$ are influenced by lensing; see the text for an explanation.

We investigate the influence of gravitational lensing on the column-density distribution, $f(N)$, of the damped Lyα absorption lines, and on $\Omega_{\rm HI}$. In terms of $f(N)$, $\Omega_{\rm HI}$ is determined by $\Omega_{\rm HI} \propto \int {\rm d}N\, N\, f(N)$.

3. Results

The left frame in figure 1 shows $N\, f(N)$ for a QSO redshift of $z_{\rm QSO} = 3$ and absorber redshifts $0.4 \leq z_{\rm abs} \leq 1$. The heavy line shows the result neglecting lensing for comparison. The different lines represent the results for three different choices for the cosmological parameters, as indicated in the right frame. The most prominent feature in $f(N)$ is the peak followed by an abrupt cut-off. The peak arises because the magnification bias exaggerates the number of high column-density systems. The cutoff is due to the fact that the bright image of a QSO can not come arbitrarily close to the center of the absorber because light rays are bent by the lens. Although the fainter image can approach the absorber center, the equivalent width of the absorption line and therefore the inferred column density are dominated by the spectrum of the bright image. The effects on $\Omega_{\rm HI}$ are seen in the right frame, where the absorber redshift varies between zero and $z_{\rm QSO}$. $\Omega_{\rm HI}$ is overestimated by factors of a few, and the redshift dependence of lensing effects can produce a substantial spurious evolution of $\Omega_{\rm HI}$.

References

Bartelmann, M., & Loeb, A., 1995, ApJ, submitted

GRAVITATIONAL LENSES AND DAMPED LY-α SYSTEMS

A. SMETTE

Kapteyn Astronomical Institute
Postbus 800, NL-9700 AV Groningen, The Netherlands

J.-F. CLAESKENS

E.S.O. (Chile) and Aspirant au FNRS (Belgium)
Institut d'Astrophysique, Université de Liège
5 Avenue de Cointe, B-4000 Liège, Belgium

AND

J. SURDEJ

Directeur de Recherches au FNRS (Belgium),
STScI, 3700 San Martin drive, Baltimore, MD 21218, USA

Abstract. We study the influence of gravitational lensing on determinationing the number density and column density distributions of damped Ly α systems.

1. Results

Systems showing damped Ly-α absorption lines seen in quasar spectra (hereafter, DLAs) are thought to arise in the progenitors of present day disk galaxies (Wolfe 1988) and contain at least a large fraction if not all of the neutral hydrogen content of the Universe at high-z (Lanzetta, Wolfe, Turnshek 1995). Statistics of DLAs are based on the assumption that the lines-of-sight towards background quasars are distributed uniformly over the sky and thus are unaffected by the DLAs themselves.

However, gravitational lensing (GL) provides three effects that undermine this assumption (Smette 1995):

— the 'by-pass' effect: a random line-of-sight has its effective impact parameter increased relatively to the case when no lensing is taking place;

C. S. Kochanek and J. N. Hewitt (eds), Astrophysical Applications of Gravitational Lensing, 99–100.
© 1996 IAU

- amplification bias (due to the galaxy associated with an observed DLA as a whole): a set of quasars selected on the basis of their (bright) apparent magnitude is likely to contain a significant fraction of quasars whose apparent brightness has been boosted by GL amplification;
- micro-lensing amplification bias: similar, but due to stars in the DLA.

We studied the two first effects and performed statistical tests devised to check whether existing surveys of DLAs are affected by GL. We assumed that the DLAs are the only GL agents, that they arise in spiral galaxies, immersed in dark matter halos, and can be described by a simple model.

The by-pass effect (function of the background QSO redshift z_q) may decrease by more than 20% the effective cross-section of the galaxies for DLA absorption; furthermore, their central part is avoided. Because of the amplification bias (function of z_q and QSO magnitude b_q), DLAs corresponding to an impact parameter equal to twice the Einstein radius of the host galaxy are preferentially observed in DLA surveys. These two effects may lead to severely over-estimate the number of DLAs with high-column densities at low-z (say in the range 0.3-0.6) in front of bright ($b_q < 17$), $z_q > 1$ quasars (Smette, Claeskens & Surdej 1995).

However, the existing DLA surveys have characteristics, specially at high-z, that may preclude the detection of strong GL effects:

- DLAs can be detected even in relatively low S/N spectra: thus faint QSOs can be used, for which the amplification bias does not work well;
- for high-z_q QSOs, the redshift range in which DLAs are searched for is limited to a redshift domain just somewhat smaller than z_q.

Indeed, we find that the high-z ($z > 1.6$) survey (Lanzetta et al. 1991) is not significantly affected by GL. However, if the IUE ($z < 1.6$) survey (Lanzetta et al. 1995) only detects 3 DLAs, they all lie in regions of the (z_{DLA}, b_q, z_q) domain for which the strongest GL effects are expected. No definite conclusion can be drawn due to the paucity of data, but HST direct imagery of the 3 background QSOs may reveal the signature of strong GL effects in the form of multiple QSO images and should thus allow to set unique constraints on the mass of the DLAs. Bartelmann & Loeb (these proceedings) have independently presented a similar work.

References

Lanzetta, K.M., Wolfe, A.M., Turnshek, D.A., et al., 1991, ApJS, 77, 1
Lanzetta, K.M., Wolfe, A.M., & Turnshek, D.A., 1995, ApJ 440, 435
Smette, A., Claeskens, J.-F., & Surdej, J., 1995, submitted
Smette, A., 1995, in the ESO Workshop on QSO Absorption Lines, eds G. Meylan et al., in press
Wolfe, A.M., 1988, in QSO Absorption Lines: Probing the Universe, eds. J.C. Blades, D.A. Turnshek & C.A. Norman (Cambridge: Cambridge Univ. Press) 297

THE GRAVITATIONAL LENS CANDIDATE HE 1104–1805

AND THE SIZE OF ABSORPTION SYSTEMS

A. SMETTE
Kapteyn Astronomical Institute
Postbus 800, NL-9700 AV Groningen, The Netherlands

J.G. ROBERTSON
School of Physics, University of Sydney
Sydney, NSW 2006, Australia

P.A. SHAVER
European Southern Observatory
Karl-Schwarzschild-Strasse 2, D-85748 Garching, Germany

AND

D. REIMERS, L. WISOTZKI AND TH. KÖHLER
Hamburger Sternwarte, Universität Hamburg
Gojenbergsweg 112, D-21029 Hamburg, Germany

Abstract. We obtained 1.2 Å resolution spectra over the range 3175 - 7575 Å for the two components of the gravitational lens candidate HE 1104–1805 ($z = 2.31$, $m_B = 16.7$ and 18.6, separation = 3.0 arcsec; *cf.* Wisotzki et al. 1993), with the aim of setting limits on the sizes of the clouds producing the Ly-α, C IV, and Mg II absorption systems. We refer to Smette et al. (1995) for a detailed account of this study.

1. Results

The Ly-α absorption lines are strongly correlated in equivalent width, suggesting that the two lines of sight pass through the same clouds in all cases, and there are no significant differences in velocities to within $\simeq 10$ km s^{-1} between corresponding pairs of lines (the separation of the Ly-α clouds ranges from 0 to $6 - 25\ h_{50}^{-1}$ kpc, depending on the lens redshift). From the 72 Lyα lines with $W_{\text{rest}} > 0.085$ Å (and $\lambda > 3395$ Å) detected at 5σ in A

C. S. Kochanek and J. N. Hewitt (eds), Astrophysical Applications of Gravitational Lensing, 101–102.
© 1996 IAU

and 3σ in B we statistically derive a 2σ lower limit of $100\ h_{50}^{-1}$ kpc for the diameter of spherical Ly-α clouds, assuming that the lens redshift $z_L > 1$, $H_0 = 50\ h_{50}$ km s^{-1} Mpc^{-1}, $q_0 = 0.5$, and $\Lambda = 0$. Similarly, 2σ lower limits of 100, 60 and 20 h_{50}^{-1} kpc are obtained for the region inside the clouds giving rise to lines with $W_{\text{rest}} > 0.17,\ 0.32$ and 0.60 Å respectively. The inferred sizes are within an order of magnitude of the maximum size that avoids overlap of the Ly-α clouds. These values and the strong correlation between Ly-α cloud line equivalent widths are hardly compatible with the spherical mini-halo model.

For the systems causing the C IV and Mg II absorption lines, we find 2σ lower limits of 28 and 22 h_{50}^{-1} kpc respectively for $W_{\text{rest}} > 0.3$ Å. Again the equivalent widths of corresponding lines are correlated, although some lines are not seen in both spectra. The redshift differences between corresponding pairs of lines are small.

A damped Ly-α system is present in one spectrum (A) but not in the other (B), indicating a typical size on the order of the separation of the two lines of sight at that redshift, $\sim 8 - 25 h_{50}^{-1}$ kpc at $z_{\text{abs}} = 1.6616$, depending on the actual geometry of the system. Some high-ionization lines in this system have similar characteristics in both spectra, while all low-ionization lines are much weaker in the spectrum of B. This suggests that, while the region causing the low-ionization lines may be similar in size to the damped Ly-α region, the region causing some of the high-ionization lines must be considerably larger.

These spectra present further evidence for the gravitationally lensed nature of HE 1104–1805. An investigation of the continuum in the two images suggests the possibility that the quasar continuum source in A is microlensed, in which case its size generally increases with wavelength, but is relatively constant between 970 and 1215 Å (rest wavelengths). The equivalent widths of the C IV absorption lines in A and in B are similar near the quasar redshift, but differ for smaller redshifts, as expected in a gravitational lens geometry. The Ly-α cloud size that we derive assuming that HE 1104–1805 is not a lens greatly exceeds the separation between the lines-of-sight for the quasar pair UM 680/681 (Shaver & Robertson 1983) for which no significant correlation was observed.

References

Shaver, P.A., & Robertson, J.G., 1983, ApJL, 268, L57

Smette, A., Robertson, J.G., Shaver, P.A., Reimers, D., Wisotzki, L., & Köhler, T., 1995, A&AS, in press

Wisotzki, L., Köhler, T., Kayser, R., & Reimers, D., 1993, A&A 278, L15

A COMMON HIGH-COLUMN DENSITY LY-α LINE
IN THE SPECTRA OF Q 1429–008 A & B

A. SMETTE
Kapteyn Astronomical Institute
Postbus 800, NL-9700 AV Groningen, The Netherlands

G.M. WILLIGER
MPI für Astronomie
Königstuhl 17, D-69117 Heidelberg, Germany

J.G. ROBERTSON
School of Physics, University of Sydney
Sydney, NSW 2006, Australia

AND

P.A. SHAVER
European Southern Observatory
Karl-Schwarzschild-Strasse 2, D-85748 Garching, Germany

Abstract. We observed a common high-column density Ly-α absorption line in the spectra of both Q 1429–008 A & B, but with different equivalent widths.

1. Common high-column density Ly-α absorption at $z = 1.662$

Q 1429–008 is a probable gravitational lens candidate (Hewett et al. 1989). The two quasar images, A & B, are separated by $5.14''$, have the same redshift (2.076), and have R-band magnitudes of $m_R = 17.7$ and 20.8. No lensing galaxy is observed.

We obtained 2 Å resolution spectra of the Ly-α forest for A & B with the CTIO 4m telescope, and additional high-resolution ($R \approx 33000$) spectra for A with UCLES on the AAT.

We observed a common high-column density Ly-α absorption line at $z = 1.662$ in the spectra of both Q 1429–008 A & B, but with different rest-frame

C. S. Kochanek and J. N. Hewitt (eds), Astrophysical Applications of Gravitational Lensing, 103–104.
© 1996 IAU

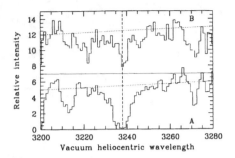

Figure 1. The high N_{HI} Ly-α absorption line in A (lower panel) and in B (upper panel). The continua used to measure the equivalent widths are drawn as dotted lines.

equivalent widths W_{rest}. We obtain $W_{\mathrm{rest}} = 10.6$ Å for the line in the A spectrum, which corresponds to $N_{\mathrm{HI}} = 3 \times 10^{20}$ cm^{-2}. The measurement of W_{rest} for the corresponding line in the B spectrum critically depends on the definition of the continuum, a delicate operation for low S/N spectra such as the one of the B component. A preliminary estimate of the continuum leads to $W_{\mathrm{rest}} = 2.5 \pm 1$ Å so that $N_{\mathrm{HI}} \simeq 10^{19}$ cm^{-2}.

We also observe a C IV doublet at $z = 1.42$ (i.e. on top of the Ly-α emission line), with a velocity difference of 580 km s^{-1} between the lines in A and in B. This suggests the presence of a cluster of galaxies, which could be the main lensing agent. If we assume that this velocity difference is a good approximation of the one-dimensional velocity dispersion of a Singular Isothermal Sphere cluster, then we only need an additional L_* elliptical galaxy to account for the separation of the two images (5.14″). Therefore, we propose that the lens redshift is 1.42.

In this model, the linear separation between the two lines-of-sight at the redshift of the $z = 1.662$ system is 23 h_{50}^{-1} kpc ($H_0 = 50\ h_{50}$ km s^{-1} Mpc^{-1}, $q_0 = 1/2$, $\Lambda = 0$). Similar column densities are seen for the $z = 1.6616$ Ly-α absorption line seen in the A and B components of HE 1104-1805 (Smette et al. 1995, Smette et al. in these proceedings), but the linear separation is less well determined in this system. These values are compatible with the current view of damped Ly-α systems as the progenitors of present-day spiral galaxy disks with diameters of $\sim$ 40-60 h_{50}^{-1} kpc, embedded in the $\sim$ 160 h_{50}^{-1} kpc diameter halos that give rise to the Lyman limit systems.

References

Hewett, P.C., Webster, R.L., Harding, M.E., Jedrzejewski, R.I., Foltz, C.B., Chaffee, F.H., Irwin, M.J., & Le Fèvre, O., 1989, ApJ, 346, L61

Smette, A., Robertson, J.G., Shaver, P.A., Reimers, D., Wisotzki, L., & Köhler, T., 1995, A&AS, in press

THE CLUSTERING EVIDENCE OF LYMAN α FOREST

WU YINGEN AND HUANG KELIANG

Department of Physics, Nanjing Normal University, China

Abstract. Using high resolution spectral data of the Ly α forest, we found evidence for clustering of Ly α absorbers on scales of $8\text{-}10h_0^{-1}$ Mpc.

1. Introduction, Sample, & Results

The absorption lines of quasars provide us chances to study the early universe, and they are a very important tool for understanding the formation and early evolution of galaxies, the early evolution of element abundances, and the large scale structure of the universe. Ly α forest lines are absorption lines in the spectra of QSOs produced by primeval intervening clouds. Analyses of low or medium resolution spectral data found no clustering of the clouds. However, blending in crowded regions of Ly α forest lines, means that spectral data of 1-2 A resolution can only be used to study clustering on scales greater than 300 km s^{-1}. To study the clustering on smaller scales, higher resolution spectral data are necessary.

We obtained echelle spectra of QSO 1225+317, using the echelle spectrograph on the 4 m Mayall telescope at KPNO (Huang et al. 1995). The spectrograph was equipped with a UV camera that transmits to the atmospheric cutoff near 3100 A. We used the echelle grating to obtain spectra of orders 96 through 65, or 3130 A to 4500 A. The spectrograph output was coupled through the camera to the intensified CCD detector. The detector consists of an RCA two-stage magnetically focused image tube with a 38 mm cathode, lens coupled to the T13 CCD. The system gain is about 20. The total exposure time was about 10 hours, which yielded an average $S/N = 20$. Each object exposure was preceded and followed by TH-A arc lamp images for wavelength calibration. The average FWHM measured in single Th-A arc exposures was 18 km s^{-1}. Data were reduced with stan-

C. S. Kochanek and J. N. Hewitt (eds), Astrophysical Applications of Gravitational Lensing, 105–106.
© 1996 IAU

dard IRAF routines. The spectra were plotted and examined to identify lines stronger than 5σ.

We found 35 Ly α forest lines between Ly α emission and Ly β emission, that can be used to understand the Ly α forest. We combined our sample with high resolution data for the QSOs 2000−330 (Carswell et al. 1987), 2126−158 (Giallongo et al. 1993), 0420−388 (Atwood et al. 1985), 0014+813 (Rauch et al. 1992), 0055−269 (Cristiani et al. 1994), 1033−0327 (Williger et al. 1994), and 1331+170 (Kulkarni et al. 1995). We use the method of Liu & Liu (1992) (i.e. calculating the free path of a photon passing through two adjacent absorbers at $z = 0$ and comparing the distribution of free path with one when absorbers are distributed at random). The advantage of this method is that all available data can be used and the sample is much larger than one obtained from one quasar. We need to calculate the probability that the free path is less than $l_0 : P(< l_0) = 1 - Ae^{-\frac{l_0}{B}}$, here A and B are constants. If $A = 1$, it means the distribution of absorber is random. Fitting the data, we find $A = 0.97$ and $B = 17.31$. If deleting some points for which free path is in the range of $4.5 - 10h_0^{-1}(h_0 = H_0/100)$, the fitted probability distribution would be consistent with random distribution. It means that the absorber may be clustered on scale of $4.5\text{-}10h_0^{-1}$ Mpc.

We also checked the clustering by use of the traditional two point correlation function. We calculated the correlation function of all the data instead of the data for each separate quasar (Mo et al. 1992). The result shows that correlation function has an obvious excess greater than 3σ at $8\text{-}10h_0^{-1}$ Mpc in agreement with the previous result. Next we checked a high column density subsample with $\log N_0 > 13.5$, and found similar results. Thus our results show a possible clustering of Ly α absorbers on scales of $8\text{-}10h_0^{-1}$ Mpc.

Acknowledgements: This work was supported by the National Climbing Programme on Fundamental Researches and the National Natural Science Foundation of China.

References

Atwood, B., Baldwin, J.A. & Carswell, R.F., 1985, ApJ, 292, 58.
Carswell, R.F., Webb, J.K., Baldwin, J.A. & Atwood, B., 1987, ApJ, 319, 709.
Cristiani, S., D'Odorico,S., Fontana, A., Giallongo, E., & Savalio, S., 1995, MNRAS, in press.
Giallongo, E., Cristiani,S., Fontana, A., & Trevese, D., 1993, ApJ, 416, 137.
Huang, K.L., et al., 1995, in preparation.
Kulkarni, V.P., Huang, K.L., Green, R.F., Bechtold,J., et al., 1995, MNRAS, in press.
Liu, Y.Z., & Liu, Y., 1992, AAp, 264, L17.
Mo, H.J., Xia, X.Y., Deng, Z.G., Borner, G., & Fang, L.Z., 1992, AAp, 256, L23.
Rauch, M., Carswell, R.F., Chaffee, F.H., et al., 1992, ApJ, 390, 387.
Williger, G.M., Baldwin, J.A., Carswell, R.F., Cooke, A.J., et al., 1994, ApJ, 428, 574.

MAPPING DARK MATTER NEAR GALAXY CLUSTERS

J.A. TYSON
AT&T Bell Labs
Murray Hill, NJ 07974 USA

Abstract. Weak gravitational lensing can provide a direct measure of mass overdensity on scales of kpc to several Mpc. The total mass and light distribution in a survey of 32 clusters of galaxies is reviewed. The mass is derived from apodized inversion of thousands of weak lensing arclets in deep CCD shift-and-stare exposures to uniform faint surface brightness in two colors. Rest frame V band mass-to-light ratios of several hundred h solar are found.

1. Introduction

The total mass and light distribution in deep survey data of 32 clusters of galaxies is reviewed. The mass is derived from thousands of weak lensing arclets found in deep multicolor CCD shift-and-stare exposures to uniform faint surface brightness. Masses of individual cluster galaxies may also be obtained via their additional local microlensing effects. Realistic simulations of the entire source-lens-atmosphere-detector process were performed, including multiple background galaxy redshift shells, masses for individual cluster galaxies, clumped dark matter, atmospheric seeing, and pixel sampling and sky shot noise. "Blank" field HST WFC-2 Medium Deep Survey data, together with seeing deconvolved ground-based data, are used to derive the source galaxy angular scales. Given accurate seeing and source galaxy angular scales, we find that the mass scale may be calibrated from these weak lensing data alone. Strong lensing (long arcs at the Einstein critical radius) then forms an independent check on this weak lensing mass scale calibration.

There are several source requirements for mass mapping by statistical gravitational lens inversion. Sources must (1) have redshifts large compared

C. S. Kochanek and J. N. Hewitt (eds), Astrophysical Applications of Gravitational Lensing, 107–112.

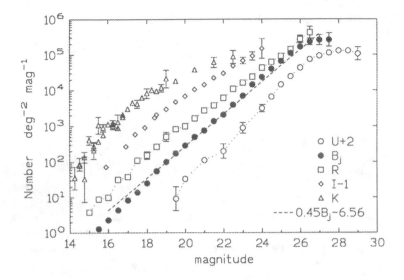

Figure 1. Faint galaxy differential number counts dN(m) as a function of magnitude in five color bands are plotted to their completeness limits. Note the decrease in slope with wavelength.

with the lens, (2) have a number density on the sky sufficient to sample the lens on relevant scales, (3) have an intrinsic angular diameter larger than the ratio of seeing FWHM to the magnification of the lens, and (4) have other properties (blue color and very low surface brightness) enabling efficient separation of the sources from lens and other foreground objects.

2. The sources

Figure 1 summarizes faint galaxy count data for five wavelength bands. The U counts are from several surveys; the B_j, R, and I counts are from our most recent data on 15 fields; and the K counts are from Gardner et al. (1993). Number-magnitude counts which rise with magnitude like dex (0.4 mag) or steeper continue to add to the integrated light at the faintest magnitudes, a type of Olbers' paradox. This corresponds to a number-flux ($N \sim S^{-n}$) slope of $n = 1$ near 600 nm wavelength. Lens-induced source number density enhancement is zero for $n = 1$.

2.1. COLORS, REDSHIFTS, AND SIZES OF FAINT GALAXIES

While the mean B_j-R color of zero redshift galaxies is about 1.5–2 mag at 21 B_j mag, most galaxies at 26–28 B_j mag have $B_j - R < -0.2$. These extreme blue colors suggest one is seeing starburst galaxies at redshifts of 1–2, so that the UV excess is redshifted into the B_j passband. In a redshift-

magnitude plot of current surveys the trend to redshift ≈ 1 at 25th B_j
magnitude is clear. Most of these faint galaxies appear to have redshift less
than three. A typical galaxy seen at z=1 may be a 0.1 $L^\star$ galaxy and a
survey at 25th magnitude would cover a wide range in redshifts extending
from 1–3. For arclet inversion of z < 0.3 lenses, the lack of redshift data for
these sources produces less than a 10% mass scale error.

Galaxies fainter than 26 B_j mag are found to have average exponential
scale lengths of 0.2–0.5 arcsec and typical half-light diameters of 1 arcsec.
Most faint galaxies imaged at $>$ 29 mag arcsec^{-2} surface brightness have
apparent angular diameters larger than the minimum required for statistical
lensing studies of foreground mass. Since angular sizes of galaxies at each
magnitude show a big range, IIST has been useful in resolving the small
size end of this distribution (see Im et al. (1995) for a review and model
comparisons). This complements the lower surface brightness studies using
large ground-based telescopes.

3. Mass density profile and maps from arclet inversion

The mass contrast in rich clusters of galaxies distorts all the faint background galaxies within several arcminutes of the cluster. Foreground galaxy
clusters at redshifts 0.2-0.5 with radial velocity dispersions above 700 km
s^{-1} have sufficient mass density to significantly distort background galaxies of redshift greater than 0.4-1. Lensing preserves the surface brightness
and spectrum of the source, so that most arcs have the very faint surface
brightness and blue color of the faint blue galaxies.

To construct a rough map of the gravitational lens projected mass distribution, the distortion statistic $T_g(x,y) = (a^2 - b^2)/(a^2 + b^2)$ is computed
over a grid of positions as candidate lens centers (Tyson et al. 1990). Here
a(x,y) and b(x,y) are the (r,θ) principal-axis transformed second moments
of the background galaxy image (arclet) g. At any point $\vec{r}$ in the image
plane we can sum over the tangential alignment of all arclets about that
point, creating a continuous scalar distortion statistic

$$D(\vec{r}) = \int K(\vec{u})\, T(\vec{r} - \vec{u})\, d\vec{u}, \qquad (1)$$

where the apodization kernel $K(\vec{u})$ weights source images at large radius
less, and is generally of the form $K(\vec{u}) = (u^2 + u_0^2)^{-1}$ (Kaiser and Squires
1993). Our kernel also is a function of the distance to the image edges,
compensating for edge effects. The distortion map D(x,y) uniquely locates
the lens mass, and gives its morphological shape on the sky.

J.A. TYSON

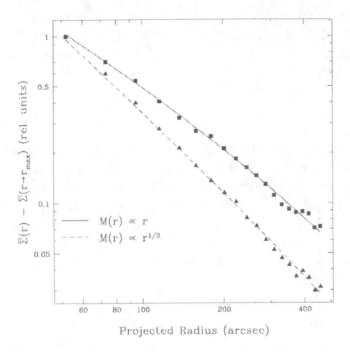

Figure 2. A radial plot of the projected mass density contrast obtained by inversion of the arclets in two 29 B mag arcsec^{-2} noisy simulations with $\sigma_v = 1000$ km sec^{-1}. The solid lines are the input mass density.

Rearranging one of Jordi Miralda's equations, the average projected mass density interior to radius r is given by

$$\bar{\Sigma}(r) = \Sigma_c C B(r) \int_r^{r_{max}} T(r) \, dlnr + \bar{\Sigma}(r, r_{max}), \qquad (2)$$

where C is the seeing and efficiency correction found in Monte Carlo tests of the data, $B(r) = (1 - r^2/r_{max}^2)^{-1}$, $\bar{\Sigma}(r, r_{max})$ is the average density in the annulus between r and r_{max}, and Σ_c is the critical density. For a sufficiently large field $\bar{\Sigma}(r, r_{max})$ is small compared to the peak density.

Automated detection, sub-image splitting and photometry reduce the CCD images to a catalog of galaxy image properties. Faint blue galaxies have their positions, total magnitudes, and apodized second moments passed to the lens arclet inversion software. Finally, the lens projected mass overdensity (Eqn. 2) as a function of radius from the centroid of the mass distribution may be compared to the lens cluster light (baryonic) distribution. A radial plot of the projected surface mass density found in this way is shown in Figure 2 for $z_l = 0.3$, $\sigma_v = 1000$ km sec^{-1}.

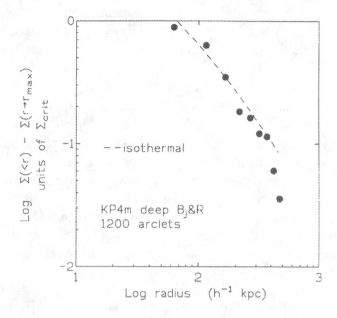

Figure 3. A radial plot of the projected mass density contrast for the $z = 0.39$ cluster 0024+17. The dashed line is an isothermal profile.

4. Results

Selecting 50 clusters with a wide range of optical richness and X-ray luminosity, we have imaged 32 clusters over the past seven years to a uniform surface brightness limit of 29 B and 28 R mag arcsec^{-2}: A1285, A1146, A545, A2218, A1689, A665, 2130-428, MS0839+29, A778, A2163, A732, A963, A2397, Zw3146, A2390, A1942, A2645, A1525, A1352, A528, A483, S506, MS2137-23, 2244-02, S67, 0957+561, 0024+17, 0940+46, 0303+17, 3C295, MS0451-03 and 1602+42, in order of increasing redshift, plus fifteen blank fields. Although some of these fields must be repeated with larger CCDs, several clusters are mapped beyond 1 h^{-1} Mpc from the center. From the inversion of 6000 arclets surrounding the rich $z = 0.18$ cluster A1689, we find a steeper than isothermal profile (Tyson & Fischer 1995) beyond 300 h^{-1} kpc radius. Some clusters at high redshift are nearly as compact in mass. Figure 3 shows the mass profile for CL 0024+17. The projected mass density contrast was obtained from 1200 arclets, excluding the bright long arcs, in the 40 arcmin^2 field of the 4-meter.

In rich compact clusters we find that mass traces the cluster red light, on scales $> 100 \ h^{-1}$ kpc, with rest-frame V band mass-to-light ratios of a few hundred h in solar units. The mass core radius is significantly smaller than some observed X-ray core radii in nearby clusters, suggesting that the X-ray gas may be less relaxed dynamically than the dark matter. At least for rich clusters these lens studies appear to confirm the large mass which was implied by virial calculations using velocity dispersions. For the optically compact clusters we find the dark matter peak densities are about 10^5 times the critical density for closure, with core radii up to 70 kpc. The relation of the faint diffuse optical nebulosity near the cores of many of these clusters to the dark matter or X-ray is intriguing, and merits further study; it is not correlated with either bright galaxies or the mass map.

The scale of the sub-clumping of the dark matter within clusters of galaxies is also interesting. The dark matter cannot all be clumped on galaxy mass scales, otherwise the long arcs seen in several relaxed cluster lenses would be broken up on several arcsecond scales and arcs or distorted "ringlets" would form around cluster galaxies. These appear in simulations where masses in excess of the Faber-Jackson mass or non-truncated isothermal mass distributions are assigned to cluster galaxies, but are not seen in the observations. On scales < 100 kpc, dark matter does not clump with individual cluster galaxies other than the central cD galaxy. Most cluster galaxies appear to be no more massive than field galaxies, so the dark matter in clusters is primarily in the diffuse component. We have found several cases of multiple mass clumps on 300 h^{-1} kpc scales.

Ultimately, comparing the mass morphologies found near clusters with high resolution N-body simulations for various cosmogonies will further narrow the candidates for dark matter. Weak gravitational lensing arclet inversion provides a direct measure of mass overdensity on scales of kpc to several Mpc. Larger scale applications of this dark matter mapping may eventually find clumped dark matter unrelated to galaxies or clusters of galaxies. Mosaics of CCDs make such a large scale search for coherent alignment in the distant faint galaxies particularly attractive, and dark matter on angular scales up to degrees can in principle be studied in this way. Preliminary measurements of the arclet orientation correlation function in random fields are below the CDM prediction.

Acknowledgements: My collaborators in this research are Gary Bernstein, Phil Fischer, Raja Guhathakurta, George Rhee, Jordi Miralda-Escudé, Ed Turner, Wes Colley and Rick Wenk.

References

Gardner, J.P., Cowie, L.L., & Wainscoat, R.J., 1993, ApJL, 415, L9
Im, M., Casertano, S., Griffiths, R.E., Ratnatunga, K.U. & Tyson, J.A., 1995, ApJ, 441, 494
Kaiser, N. & Squires, G., 1993, ApJ, 404, 441
Tyson, J.A. & Fischer, P., 1995, ApJ, 96, 1
Tyson, J.A., Valdes, F., & Wenk, R.A., 1990, ApJL, 349, L1

THE CORES OF CLUSTER LENSES

Mass Distribution & Background Galaxies Properties

J.P. KNEIB
Institute of Astronomy
Madingley Road, Cambridge CB3 0HA, U.K.

AND

G. SOUCAIL
Observatoire Midi-Pyrénées
14 Av. E. Belin, 31400 Toulouse, France

Abstract. The gravitational lensing of faint background galaxies by rich clusters is emerging as a very efficient method both to constrain the mass distribution of clusters of galaxies and to probe the statistical properties of faint background galaxies. We review results concerning the cores of cluster lenses, where recent Hubble Space Telescope (HST) images are giving new insights into the gravitational distortion phenomena.

1. Introduction

The confirmation by Soucail et al. (1988) that the nature of the giant arc in A370 is due to gravitational lensing (GL) has brought a new very interesting class of lenses in the GL field: clusters of galaxies. This discovery clearly demonstrated that the core of cluster lenses are very massive and that the mass distribution is peaked, and therefore allows us to see background galaxies distorted but magnified (Fort & Mellier 1994). Modeling of multiple arcs (Mellier et al. 1993, Kneib et al. 1993) have shown that the mass distribution in the very center follows the geometrical distribution (ellipticity and orientation) of the stars in the halo of the central cluster galaxy. Total mass estimates derived from X-ray analysis and lensing analysis apparently differ (Miralda & Babul 1995), however recent analyses show a more complex picture that will lead to better understanding of physical processes in the core of clusters. Knowing the mass distribution of these lenses, we can use them as gravitational telescopes to study the

113

C. S. Kochanek and J. N. Hewitt (eds), Astrophysical Applications of Gravitational Lensing, 113–118.
© 1996 IAU

J.P. KNEIB AND G. SOUCAIL

TABLE 1. Cluster-lenses with giant arcs

Cluster	Type	z_c	z_{arc}	S	r_{arc} (kpc)	$M(< r_{arc})$ (10^{14} $M_{\odot}$)	M
PKS0745	(gE)	0.103	0.433	*	45.9	0.25	e
MS0955	(gE)	0.145	0.800		38.5	0.22	c
A2104	(gE)	0.155	0.800		31.1	0.10	c
A1689	(g)	0.170	0.800		161.3	2.40	c
A2218a	(cD+gE)	0.175	0.702	*	78.5	0.50	m
A2218b	(cD+gE)	0.175	1.034	*	236.5	2.38	m
MS0440	(g)	0.190	0.530	*	90.0	0.93	c
A2163	(gE)	0.201	0.728	*	66.1	0.75	c
A963	(gE)	0.206	0.771	*	51.7	0.35	c
A2219N	(gE)	0.225	1.000		100.0	1.06	m
A1942	(gE)	0.226	0.800		37.2	0.18	c
A2390	(gE)	0.231	0.913	*	177.0	1.60	m
A2397	(gE)	0.240	0.800		69.8	0.45	c
S295	(gE)	0.299	0.930	*	32.9	0.14	c
AC114	(gE)	0.310	1.860	*	56.0	0.25	c
MS2137	(gE)	0.313	1.000		87.4	0.63	e
Cl0500	(gE)	0.316	0.800		124.7	1.50	c
GHO2154	(gE)	0.320	0.721	*	34.2	0.20	c
Cl2244	(g)	0.331	2.236	*	46.5	0.20	c
A370	(2gE)	0.374	0.724	*	140.0	1.00	m
Cl0024	(g)	0.398	1.300		166.7	2.00	e
Cl0302	(gE+E)	0.424	0.800		122.3	1.60	c
Cl2236	(gE+E)	0.560	1.116	*	87.6	0.30	m
MS2053	(gE)	0.583	1.000		116.8	1.50	c

properties of background galaxies and derive a likely distance estimate of the faint background galaxies. We use $H_0 = 50h$ km s^{-1} Mpc^{-1} and $\Omega_0 = 1$ throughout this contribution.

2. Known Cluster Lenses

In Table 1 we give a non-exhaustive list of clusters in which giant arcs have been discovered (and in a few cases confirmed by spectroscopy). There are 23 such clusters, 11 with known multiple images and 12 with at least one known arc redshift. About 66% of the clusters have a single bright giant elliptical (cD or gE) galaxy at their center, ~17% with a dense group of galaxies (g) at the center and ~17% show bimodal (2gE) or complex morphologies (e.g. cD+gE). For most of them, we give the total mass within the radius of the giant arc. This value comes either from precise models using multiples images (m, $\star$), a simple elliptical model (e,$\square$) or a circular model (c,$\downarrow$). Furthermore, if no information exists on the distance of the giant arc, we assume $z_{arc} = 0.8$.

For a circular model we can write simply: $M(< r_{arc}) = \pi \Sigma_{crit} r_{arc}^2$ with $\Sigma_{crit} = c^2/4G \, D_S D_L/D_{LS}$. We should stress here that this result is independent of the mass profile only if the mass distribution is circular and centered on the brightest cluster galaxy. However, in most cases the circular model gives an upper mass estimate because of the structure of the mass distribution (elliptical or more complex distribution). Modeling the mass distribution with an elliptical mass distribution generally decreases the mass estimate by 10 to 20%. Figure 1 shows the mass enclosed in the radius of the arc versus the distance to the central cluster galaxy. The lines correspond to isothermal mass distribution with a core of $50h^{-1}$kpc and a velocity dispersion of respectively 1000km/s and 1500km/s. This plot nicely delimits the possible mass profiles, although it does not give a perfect picture because there is some scatter in the total mass from cluster to cluster. Moreover it clearly contradicts the virial masses estimated from cluster velocity dispersions higher than 1500 km/s. This demonstrates that many clusters are still dynamically young and not virialized with ongoing mergers.

3. Mass Distribution

3.1. MULTIPLE IMAGE MODELING

The circular model, although robust in the mass estimate, does not give any clues about the exact mass distribution. The thinness of the arc is the product of the intrinsic size of the arc and the slope of the mass profile at the location of the arc, and therefore must be analyzed carefully. The curvature radius of the arc is sensitive to the "core-radius" and the ellipticity of the mass distribution. In the limit of $r_{curv} \sim r_{arc}$ this gives the constraint $r_{core} < r_{arc}$. Translated to our sample this gives $r_{core} < 100h^{-1}$kpc.

In the case of giant arcs, they are very often multiple-image systems, and a more detailed analysis can be done (e.g. Mellier et al. 1993, Kneib et al. 1993). Three classes of large arcs are examined:
- fold arc (2 images): either one counter-image (in the case of naked cusp which is likely the case of very elliptical mass distribution) or three counter-images (but one de-amplified) otherwise [e.g. MS2137, Cl2244, A2218, A2219, AC114, Cl0302?, GHO2154?, MS0440?],
- cusp arc (3 images): no-counter images in the naked cusp case or two counter-images (but one de-amplified) otherwise [e.g. A370, Cl0024, S295],
- radial arc (2 images): generally only one counter-image [e.g. MS2137, A370, A2390, AC118].

Finding the multiple images in a cluster is usually an iterative process. High resolution (either with seeing< 0.7″ for ground telescope or the use of the HST) or color matching allows to identify multiple images systems. Once a multiple image system is determined and a mass model can explain it, it is easy to check if other images within the multiple image area have counter-image or not and to use the predictive power of the model.

From the multiple-image system models it is now convincing that the total mass distribution follows the orientation and ellipticity of the halo of the brightest cluster galaxies, and that the core radius of the mass distribution is $\sim 50h^{-1}$kpc. The compactness is true in the case of a cluster with a giant elliptical in the center but also in the case where a dense group of galaxies sits in the center (Cl0024: Smail et al. 1995).

More recently, thanks to the high resolution and stability of the HST, a plethora of arcs and arclets have been revealed in the core of some cluster-lenses (A2218, Cl2244, A2390, AC118, AC114, MS0440, MS2137, Cl0024). In the case of A2218, at least 7 sets of multiple images are detected (Kneib et al. 1995) which allow a precise model of the mass distribution. Furthermore, the HST images suggest that the mass is concentrated around the brightest elliptical galaxies.

3.2. COMPARISON WITH ROSAT/HRI X-RAY MAP

The first comparison of the X-ray maps and lens models in A370 (Mellier et al. 1994) revealed identical morphologies, namely two clumps of mass centered on the two brightest ellipticals. However it was difficult to be more precise as the number of counts detected in the X-ray was quite low. It is likely however that A370 corresponds to two clusters separated by $\sim$ 30Mpc as the relative velocity of the two giant ellipticals suggests. Miralda-Escudé & Babul (1995) noticed that the total mass estimate in A2218 and A1689 differ by a factor of 2 between lensing and X-ray, although A2163 gives better agreement. Allen et al. (1995) made a more precise study of the X-ray gas distribution using both ASCA & ROSAT data and demonstrated with a multi-phase model for the cooling flow cluster PKS0745 that the total mass estimate can agree within a few percent inside the arc radius. Deeper ROSAT/HRI images can however tell us more on the substructure of the mass distribution. For example Pierre et al. (1995) made a very deep exposure with ROSAT of A2390 and were able to relate a substructure in the X-ray to an enhancement of the total mass as revealed by the presence of the straight arc and an almost "straight" shear field. Further studies will put better constraints on the total mass profile in the central part and probably improve the understanding of physical processes in the core of clusters.

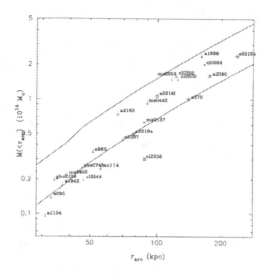

Figure 1. Mass within the Einstein radius for clusters of Table 1.

4. Background Galaxies Properties

4.1. MORPHOLOGY, SIZE, CONTENTS, DYNAMICS OF HIGHLY MAGNIFIED GALAXIES

Smail et al. in this proceedings show that the intrinsic sizes of the giant arcs with known redshift are smaller than their present counterparts. The morphology viewed from HST of the highly magnified arcs display generally complex structure with many knots, but there is at least one case (A2218 #359) where the arc is an elliptical red galaxy. Pelló et al. (1992) first measured a velocity gradient in the straight arc of A2390. Other velocity gradients were detected in Cl2236 (Melnick et al. 1993) and PKS0745 (Allen et al. 1995). Velocity gradients probably exist in other arcs (as suggested by their morphology) and are easiest to measure in the most amplified images. Argus/2D spectroscopy mode (specially when used with VLT telescopes) coupled with HST images of giant arcs will allow us to have a better understanding of the dynamics of some very distant galaxies.

4.2. REDSHIFT DISTRIBUTION OF THE FAINT GALAXIES

Using the mass model derived from the multiple images with spectroscopically known redshift (therefore absolutely calibrated), it is possible to invert the lens equation for each object in the central part and assign it a most likely redshift. Kneib et al. (1994) have first used this method for the arclets in A370, and derived a median redshift $z_{med} \sim 0.8$ for galaxies with intrinsic magnitudes $24 < B_j < 27$. However the largest limitation of this study was probably due to error measurements of the ellipticity of the background galaxies.

The refurbished HST will bring a large improvement for this method. Kneib et al. (1995) have already applied this inversion to the shallow exposure of A2218 (3 orbits) and were able not only to reconstruct the mass of the core of the cluster but to constrain the redshift distribution of the background galaxies. Their main result is that the no-evolution prediction (down to R=26) for the background galaxies redshift distribution fits well the observed distortion in A2218.

5. Conclusion

The future prospects concerning the core of cluster lenses are:
• the determination of the precise mass distribution down to the galaxy size (i.e. 5–$400h^{-1}$ kpc) especially with the HST images.
• comparison analysis between lensing and X-ray in the "low" redshift clusters $(0.15$–$0.3)$ for which ROSAT and ASCA data are easily available.
• a better understanding of the physical formation processes of cluster core.

• a better view of high redshift galaxies (morphology, size, content).

• a better constraint on the redshift distribution of faint galaxies by stacking the results from different clusters & check with direct spectroscopy of the brightest arclets.

Acknowledgments: We thank all our collaborators in this long-term lensing program, namely B. Fort, Y. Mellier, R. Pelló, R. Ellis, A. Fabian, A. Edge, S. Allen, I. Smail, J. Miralda-Escudé, M. Pierre, H. Börhinger, W. Couch, R. Sharples as well as T. Brainerd, P. Schneider, R. Blandford, M. Rees, for fruitful discussions.

References

Allen, S.W., Fabian, A.F., & Kneib, J.-P., 1995, MNRAS, in press
Fort, B., & Mellier, Y., 1994, Astron. Astr. Rev., 5, 239
Kneib, J.-P., Mellier, Y., Fort, B., & Mathez, G., 1993, A&A, 273, 367
Kneib, J.-P., Mellier, Y., Fort, B., Soucail, G., & Longaretti, P.Y., 1994, A&A, 286, 701
Kneib, J.-P., Mellier, Y., Pelló, R., Miralda-Escudé, J., Le Borgne, J.-F., Börhinger, H., & Picat, J.-P., 1995, A&A, in press
Kneib, J.-P., Ellis R.S., Smail, I.R., Couch W., & Sharples, R., 1995, ApJ, preprint
Melnick, J., Altieri, B., Gopal-Krishna, & Giraud, E., 1993, A&A, 271, L8
Mellier, Y., Fort, B., & Kneib, J.-P., 1993, ApJ, 407, 33
Mellier, Y., Fort, B., Bonnet, H., & Kneib, J.-P., 1994, in Cosmological Aspects of X-ray Clusters of Galaxies" NATO ASI series C 441, ed. W. Seitter, in press
Miralda-Escudé, J., & Babul, A., 1995, ApJ, in press
Pierre, M., Le Borgne, J.-F., Soucail, G., & Kneib, J.-P., 1995, A&A, preprint
Smail, I., Dressler, A., Kneib, J.-P., Ellis, R.S., Couch, W.J., & Sharples, R.M., Oemler, A., 1995, ApJ, in press

HST OBSERVATIONS OF GIANT ARCS

IAN SMAIL AND ALAN DRESSLER

The Observatories of the Carnegie Institution of Washington, 813 Santa Barbara St., Pasadena, CA 91101-1292

JEAN-PAUL KNEIB AND RICHARD S. ELLIS

Institute of Astronomy, Madingley Rd, Cambridge CB3 0HA, UK

WARRICK J. COUCH

School of Physics, University of NSW, Sydney 2052, NSW Australia

RAY M. SHARPLES

Dept. of Physics, University of Durham, South Rd, Durham DH1 3LE, UK

AND

AUGUSTUS OEMLER JR.

Yale Observatory, New Haven, CT 06511

Abstract. We discuss HST imaging of eight spectroscopically-confirmed giant arcs, pairs and arclets. Although our HST observations include both pre- and post-refurbishment images, the depth of the exposures guarantees that the majority of the arcs are detected with diffraction-limited resolution. We present the size information on these distant field galaxies in the light of HST studies of lower redshift samples. We suggest that the dominant population of star-forming galaxies at $z \sim 1$ is a factor of 1.5–2 times smaller in size than the equivalent population in the local field. This implies either a considerable evolution in the sizes of star-forming galaxies within the last ~ 10 Gyrs or a shift in the relative space densities of massive and dwarf star-forming systems over the same time scale.

C. S. Kochanek and J. N. Hewitt (eds), Astrophysical Applications of Gravitational Lensing, 119–124.
© 1996 IAU

1. Introduction

The key to understanding the nature and evolution of the faint field galaxy population may come through morphological studies. Using WFC-1 HST Medium Deep Survey (MDS) data Mutz et al. (1994) studied the scale sizes of galaxy images as a function of redshift. They find only weak evidence for size evolution of spiral disks over the last ~ 3 Gyrs. Unfortunately, the Mutz et al. (1994) sample only probes to $I \sim 21$, corresponding to a median depth of $z \sim 0.2$, and while deeper morphological samples are available (e.g. Glazebrook et al. (1995) and Driver et al. (1995)) these have only statistical redshift information.

Ideally we would like to study the distributions of morphology and size at earlier times to observe the changes in the nature of the faint field population responsible for the steep optical counts. Such a search is hampered, however, by the faintness of these distant galaxies which makes redshift identification extremely time consuming. Thus at high redshift we are constrained to study only the intrinsically brightest and hence possibly least representative objects. Gravitationally lensed features, in particular giant arcs, may offer an alternative approach to study a small sample of "normal" distant galaxies.

If we wish to use the giant arcs to study the scale sizes of distant galaxies we must also discuss the arc detection as a function of intrinsic source size. For the giant arcs used here the high tangential amplification means there should be no bias against finding thick arcs formed from intrinsically large background sources (e.g. Wu & Hammer 1993). However, a bias may exist against identifying the strongly lensed images of compact sources, which appear as multiple, discrete images and are therefore difficult to identify. Hence, we might expect compact sources to be under represented in the current giant arc sample, allowing us to determine only a firm upper limit on the scale size distribution of distant galaxies from the study of the widths of giant arcs.

Using HST we have acquired deep optical imaging at 0.1 arcsec resolution (spatial scales of $\sim 0.5h^{-1}$ kpc) of eight spectroscopically confirmed, high redshift field galaxies which appear as arcs, pairs or arclets in the fields of moderate redshift clusters.

2. Observations and Reduction

Four of the five clusters presented (A370, AC114, Cl0024+16, A2218) were observed with WFC-1 for studies of galaxy evolution in distant clusters (Dressler et al. 1994, Couch et al. 1994). The presence of giant arcs in these fields is purely serendipitous and in some respects reflects the prevalence of such features in these rich, moderate redshift clusters. The fifth cluster

Cl2244−02 was observed in Cycle-2 to study the arc present in the cluster and these images were retrieved from the STScI archive.

To measure half-light radii (r_{hl}) for the arcs we have chosen to correct for the extended PSF of the WFC-1 images by deconvolving our frames. The frames were therefore processed by deconvolving for 40 iterations of the Lucy-Richardson algorithm using model PSF's created with TINYTIM (Krist 1992). This procedure has been extensively tested by Windhorst et al. (1994), who report that it is adequate for restoring a galaxy's average light profile. To measure the half-light radii of the arcs we extract profiles through the arcs, orthogonal to the local shear direction indicated by the arc's shape. These profiles are effectively one dimensional slices through the source and they must be corrected, by radially weighting the profile to reflect the two dimensional geometry of the source, to give standard half-light radii, with typical measurement errors of ~ 0.1 arcsec. Figure 1 gives half-light radii for the arcs measured from our HST frames. It should be noted that for all arcs, except A5 in A370, we detect the source out to $\gtrsim 3 \times$ the measured half-light radius. To estimate our resolution limit we have analyzed similarly processed stars from our images, showing that the resolution limit of HST is approximately $r_{hl} \sim 0.1$–0.2 arcsec. Thus HST well resolves all but one arc, Cl2244−02, in our sample.

3. Discussion and Conclusions

The main aim of this study is to gain a first view of the sizes of typical star-forming field galaxies at cosmologically interesting look-back times, back to $z \sim 1$–2 or from 8 to 11 Gyrs ago (for $h = 0.5$). To do this we must address the possibility that lensing amplification will change the observed source profile. The radial amplification, A_{rad}, of a giant arc depends upon the compactness of the lensing potential (Wu & Hammer 1993). For a lens with an isothermal mass profile the amplification is $A_{rad} = 1$. The majority of giant arcs which have been modeled in detail show that the gross properties of the very central regions of the cluster potential can be characterized by a nearly singular isothermal mass distribution with core radii of $\lesssim 25h^{-1}$ kpc (e.g. Kneib et al. 1993, Smail et al. 1995). With such a small core radius we expect the radial magnification to be $A_{rad} \sim 1$ within 10–20%. Hence we can adopt the radial light profile of the arc as an unmagnified one-dimensional slice through the source.

The half-light radii for our sample are shown in Figure 1. We have chosen to compare our sample with spirals, the dominant local population of star-forming galaxies. We use, therefore, the median size of $r_{hl} \sim 4.4h^{-1}$ kpc taken from the local spiral sample of Mathewson et al. (1992) and a more distant sample analyzed from WFC-1 MDS images by Mutz et al. (1994).

We plot the variation in apparent angular size of the typical Mathewson disk as a function of cosmology in Figure 1. The observed arc widths follow a smooth progression from the scale lengths observed in the lower redshift samples. However, the distribution of arc widths is apparently incompatible with a non-evolving population of disks with intrinsic half-light radii $r_{hl} \sim 4.4h^{-1}$ kpc, irrespective of the adopted geometry. The mean size of the arc sources is $< r_{hl} >= (2.3 \pm 0.8)h^{-1}$ kpc. Thus the average arc source is a factor of ~ 1.5–2 times smaller than would have been expected from extrapolation of the sizes of local bright star-forming systems, in standard geometries.

To indicate the strength of evolution required to connect the arc source sizes to local spirals we plot on Figure 1 a model, $r_{hl} \propto (1 + z)^{-1}$, which Mutz et al. (1994) state is consistent with their observations (also see Im et al. 1995). A more detailed comparison with the Mutz et al. (1994) sample is hampered by the small samples, but what is apparent, however, is the absence of the larger sources ($r_{hl} \gtrsim 5h^{-1}$ kpc) relative to smaller sources in the arc catalogue. As we discussed earlier this cannot be explained by selection effects in the original sample and we must therefore accept that either: 1) large galaxies with very extended star-forming regions are rarer at $z \sim 1$–2 than they are today or 2) the relative proportions of large and small galaxies has changed between $z \sim 1$ and today. The small number of giant arcs precludes us distinguishing between these alternatives. However, the latter possibility is discussed at length in Driver et al. (1995) and Glazebrook et al. (1995).

These conclusions are similar to those reached by Smail et al. (1993) who used K imaging of a sample of giant arcs, which considerably overlaps that used here, to measure the rest-frame near-infrared luminosities of the sources. They found that the arc sources had sub-K^* luminosities compared to the local field. One explanation for this difference is a lack of massive, luminous galaxies at $z \gtrsim 1$. Alternatively, this result could arise from a change in the relative abundance of dwarf galaxies compared to more massive systems. Thus the distributions of both K luminosities and sizes of $z \gtrsim 1$ field galaxies indicate that dominance of large, massive galaxies is a relatively recent feature of the field population.

The power of using arcs and arclets for studying the evolution of the sizes of faint galaxies comes from the large redshift range probed. As Kneib et al. (1995) show, it is possible, using HST observations of a well-constrained lensing cluster, to study the redshift distribution of very faint galaxies, $R \lesssim 26$–27, from the shear induced in the galaxy images by the cluster. In principle with deep enough data this technique can be extended to study the distribution of scale sizes as a function of redshift for large samples of very faint galaxies, well beyond the reach of conventional spectroscopy.

4. Acknowledgements

We thank Roger Blandford, Bernard Fort, David Hogg, Yannick Mellier and Jordi Miralda-Escudé for useful discussions. We also wish to thank Dr. S. Mutz for providing the data from his HST study. Support via a Carnegie Fellowship, Space Telescope Grant GO–05352 and an IAU Travel Grant are gratefully acknowledged. The observations presented here were obtained with the NASA/ESA Hubble Space Telescope, which is operated by the Association of Universities for Research in Astronomy, Inc., under NASA contract NAS5-26555.

References

Couch, W.J., Ellis, R.S., Sharples, R.M. & Smail, I., 1994, ApJ, 430, 121.

Dressler, A., Oemler, A., Butcher, H. & Gunn, J.E., 1994, ApJ, 430, 107.

Driver, S.P., Windhorst, R.A., Ostrander, E.J., Keel, W.C., Griffiths, R.E. & Ratnatunga, K.U., 1995, ApJL, submitted

Glazebrook, K., Ellis, R.S., Santiago, B. & Griffiths, R.E., 1995, MNRAS, in press

Im, M., Casertano, S., Griffiths, R.E. & Ratnatunga, K.U., 1995, ApJ, in press

Kneib, J.-P., Mellier, Y., Fort, B. & Mathez, G., 1993, A&A, 273, 367

Kneib, J.-P., Ellis, R.S., Smail, I., Couch, W.J. & Sharples, R.M., 1995, ApJ, submitted

Krist, J., 1992, TINYTIM User's Manual, STScI, Baltimore

Mathewson, D.S., Ford, V.L. & Buchhorn, M., 1992, ApJS, 81, 413

Mutz, S.B., Windhorst, R.A., Schmidtke, P.C., Pascarelle, S.M., Griffiths, R.E., Ratnatunga, K.U., Casertano, S., Im, M., Ellis, R.S., Glazebrook, K., Green, R.F. & Sarajedini, V.L., 1994, ApJL, 434, L55

Smail, I., Ellis, R.S., Aragòn-Salamanca, A., Soucail, G., Mellier, Y. & Giraud, E., 1993, MNRAS, 263, 628

Smail, I., Hogg, D.W., Blandford, R.D., Cohen, J.G., Edge, A.C. & Djorgovski, S.G., 1995, MNRAS, in press

Windhorst, R.A., Schmidtke, P.C., Pascarelle, S.M., Gordon, J.M., Griffiths, R.E., Ratnatunga, K.U., Neuschaefer, L.W., Ellis, R.S., Gilmore, G., et al., 1994, AJ, 107, 930

Wu, X.P. & Hammer, F., 1993, MNRAS, 262, 187

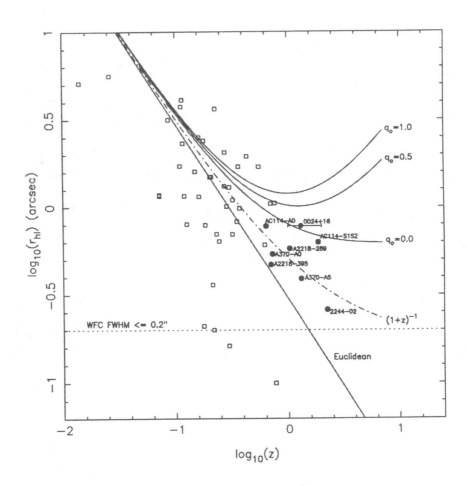

Figure 1. The distribution of observed half-light radii versus redshift for the arc sample
(•). We assume no radial magnification by the lensing process. We plot for comparison the
half-light radii of a moderate redshift sample (□, Mutz et al. 1994). Finally, we show the
observed size of a typical local spiral galaxy from Mathewson et al. (1992), $r_{hl} = 4.4h^{-1}$
kpc, as a function of redshift in different cosmologies and the effect of a scale evolution of
the form $r_{hl} \propto (1+z)^{-1}$ in a $q_o = 0.5$ cosmology. The relative dearth of large sources over
smaller systems, compared to that expected from a non-evolving population of sources
is readily apparent. This plot is adapted from Mutz et al. (1994).

A LUMINOUS ARC IN A Z=0.042 CLUSTER OF GALAXIES

The nearest gravitational arc known?

LUIS E. CAMPUSANO
Departamento de Astronomía, Universidad de Chile
Casilla 36-D, Santiago, Chile

AND

EDUARDO HARDY
Departamento de Astronomía, Universidad de Chile and
Département de Physique, Université Laval, Québec, Canada

Abstract. We report the discovery of a bright, arc-like feature with a redshift of z= 0.073 associated with the nearby cluster of galaxies ACO3408, at z = 0.042. The redshift, position, and geometry of the arc with respect to the central cD galaxy of the cluster strongly suggest that this is by far the nearest case known of gravitational lensing of a faint background galaxy.

1. A Nearby Arc-like Feature

The serendipitous discovery took place on 6/7 February 1995 during a run with the CTIO 0.9m telescope, while performing direct imaging of clusters of galaxies for a program conducted in collaboration with Giovanelli and Haynes (Cornell). On 1 April 1995, the spectroscopy of the arc was obtained with the CTIO 4m telescope, revealing that its redshift was nearly twice that of the galaxy cluster.

The observed photometric properties of the arc are as follows: R=18.60, R-I = 0.42, and B-R =0.79. The mean BRI surface brightnesses are 24.23, 23.44, and 23.02 mag/sq.arcsec, respectively. Its length is $\sim$ 11 arcsec, corresponding to a linear size of 22 h_{50}^{-1} kpc These parameters are consistent with, for example, a lensed image of a faint BCD galaxy at z=0.07. Could the arc be instead an edge-on unlensed spiral? Its size is compatible with a medium-size one. But the object exhibits emission lines all along its body (within our resolution of $\sim$1.5"). There is some velocity structure along the arc, but with a small amplitude, more reminiscent of an irregular galaxy

C. S. Kochanek and J. N. Hewitt (eds), Astrophysical Applications of Gravitational Lensing, 125–126.
© 1996 IAU

Figure 1.   CTIO B-image showing the arc and the central cD galaxy of ACO3048

of the Magellanic type. The size however is too large by far for the arc to be an unlensed Irregular. Notice that the direct calculation of the absolute magnitude of the arc gives $M_B = -18.9$, a value expected for a small spiral such as M33, or for an irregular, such as the LMC, which have sizes of only a few kpc.

The "gravitational radius" of the arc is $\sim$ 47" (i.e.$\sim$54 h_{50}^{-1} kpc), although this value remains somewhat uncertain for no counter-arc is observed. Under the hypothesis of a singular isothermal sphere (SIS; see Blandford & Narayan 1992 and Fort & Mellier 1994) we derive the following cluster parameters: Velocity dispersion $\sigma_v \sim$ 2000 km/s, a mass interior to the arc $\sim$2 x $10^{14} h_{50}^{-1}$ solar masses, and a (M/L) ratio $\sim$ 740h_{50}. Although these parameters might be overestimated due to simplified hypotheses (SIS), their very large values invite caution. The crucial test to confirm the lensing nature of the arc presented here would be the observation of shear in the fainter images of far-away background galaxies near the central cD galaxy of ACO3048. This test is ideally suited for the Hubble Space Telescope.

Acknowledgments: LEC and EH acknowledge the financial support of Fundación Andes(Chile).

References

Blandford, R.D. & Narayan, R., 1992, ARA&A, 30, 311
Fort, B. & Mellier, Y., 1994, A&A Review, 5, 239

WEAK LENSING BY THE CLUSTER MS0302+1658

DAVID FISHER, KONRAD KUIJKEN AND MARIJN FRANX
Kapteyn Institute
PO Box 800, 9700 AV, Groningen, The Netherlands

Abstract. We present preliminary results of a weak lensing investigation of the $z = 0.43$ cluster MS0302+1658.

We have started a program in Groningen to investigate the mass distributions in clusters of galaxies using weak gravitational lensing of distant background galaxies. Here we report preliminary results for the Einstein cluster MS0302+1658 (redshift 0.43), based on data obtained with the William Hershell Telescope. Total exposure times for the images analyzed were 7000s in R, and 6000s in V, and seeing was about 1".

This cluster contains two large central galaxies, with a 'straight arc' (Mathez et al. 1992) in between them, and a second arc in the envelope of the brightest cluster galaxy (Fig. 1a). Mathez et al. (1992) found the cluster potential to be poorly constrained by the arcs, since the central galaxies are a strong influence on them. We have therefore measured the weak lensing of background galaxies at larger radii (out to 2') by the cluster in order to provide a stronger measurement. At these radii the gravitational potential is dominated by the cluster as a whole, rather than by individual galaxies.

The data were analyzed following the methods of Kaiser, Squires and Broadhurst (1995). The V-R color-magnitude diagram indicates that most of the faint resolved objects are indeed blue galaxies, and probably lie behind the cluster. Averaging the complex second moments of the faint resolved objects and applying a (rather large and still preliminary) correction for circularization by seeing results in a 'polarization map': it gives the distortion of an intrinsically circular object at the mean source distance. Significant tangential distortion consistent with weak lensing is detected.

Comparison of these polarizations with a singular isothermal model results in a best-fit velocity dispersion of 700 km s^{-1}, assuming the sources are at infinite distance and are all being lensed (Fig. 1b). Given the high cluster

C. S. Kochanek and J. N. Hewitt (eds), Astrophysical Applications of Gravitational Lensing, 127–128.

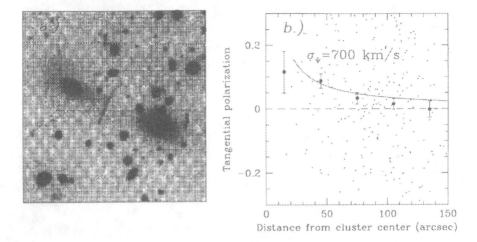

Figure 1. a.) The central 1′ of the cluster, containing the two large galaxies and giant arcs. b.) Elongations (small dots) of background galaxies and binned mean tangential polarizations (error bars) observed as a function of distance from the cluster center. The line shows the expected polarization of infinite-distance background sources by a 700 km s^{-1} singular isothermal sphere.

redshift, most sources are likely to be at most a factor of two further away than the cluster, reducing the efficiency of lensing. Therefore, the true velocity dispersion of the cluster potential is likely a factor 1.5–2 higher than the number quoted above. *If the cluster potential is isothermal, we estimate a velocity dispersion of $\sigma_\Psi = 1200$ km s^{-1},* but more careful study of the corrections for the seeing and finite source distances is necessary.

The galaxies in the cluster have a radial velocity dispersion of 920 km s^{-1} (Fabricant et al. 1994), and a deprojected radial number density $n_g \propto r^{-2.5}$. Thus, this cluster may provide an illustration of *velocity bias:* the galaxies are more centrally concentrated than the dark matter, and therefore show a smaller velocity dispersion. In a simple constant-velocity dispersion spherical cluster model with a power-law density and a singular isothermal potential, n_g varies with radius as a power $-2(\sigma_\Psi^2/\sigma_r^2 + \beta)$, where $\beta = 1 - \sigma_t^2/\sigma_r^2$ is the galaxy orbits' radial anisotropy. The line-of-sight velocity dispersion in such a model is $\sigma_{RV}^2 = (1 - 12\beta/5)\sigma_r^2$, so the observed velocity dispersion and number count slope may imply a slight radial anisotropy ($\beta = 0.15$) of the galaxy orbits. A detailed error analysis as well as results for other cluster data in hand will be forthcoming.

References

Mathez, G., Fort, B., Mellier, Y., Picat, J.-P. & Soucail, G. 1992, A&A, 256, 343
Kaiser, N., Squires, G. & Broadhurst, T. 1995, ApJ, 449, 460
Fabricant, D.G., Bautz, M.W. & McClintock, J.E. 1994, AJ, 107, 8

RECONSTRUCTION OF CLUSTER MASS DISTRIBUTIONS:
APPLICATION AND RESULTS FOR CL0939+4713

CAROLIN SEITZ

Max-Planck-Institut für Astrophysik
Postfach 1523, D-85740 Garching, Germany

Abstract. We show the reconstructed mass distribution of the cluster CL0939+4713 ($z_d = 0.4$) using data obtained with the WFPC2 instrument on HST. We find a remarkable correlation with the light distribution of the bright galaxies, which are probably cluster member galaxies. A bootstrapping analysis confirms this correlation, whereas an anti-correlation between the mass distribution and the faint galaxies - used for the mass reconstruction - is found. We give lower bounds on the mass inside the data field, and confirm that this cluster is indeed quite massive.

Because of the limited space, we do not describe the data analysis and the method of reconstruction, but refer the reader to the paper by Seitz et al. (1995, hereafter Paper 1). We point out that due to the high redshift of the cluster ($z_d = 0.4$) the redshift distribution of the faint galaxies has to be accounted for in the reconstruction. In Figure 1 we show the reconstructed dimensionless mass density κ for a source at 'infinity', assuming a mean redshift of $\langle z \rangle = 0.8$ for the faint galaxies and using the invariance transformation derived in Paper I such that the minimum of the resulting κ-map is roughly zero. Comparing the mass distribution with the light distribution of the bright galaxies shown in Figure 2, we find that the position of the maximum mass density corresponds well with the position of the cluster center suggested by Dressler & Gunn (1992). The secondary maximum corresponds to a group of bright cluster galaxies, and the minimum to a region where very few cluster galaxies are observed.

We performed a error bootstrap analysis, i.e., we generate a number of synthetic data sets by drawing N galaxies at a time with replacement from the original data set consisting of $N = 230$ galaxies with $m \in (24, 25.5)$. For each of the synthetic data sets we obtain the mass-reconstruction and

C. S. Kochanek and J. N. Hewitt (eds), Astrophysical Applications of Gravitational Lensing, 129–130.

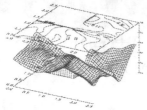
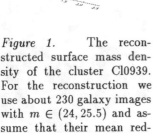
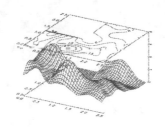
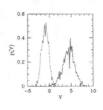

Figure 1. The reconstructed surface mass density of the cluster Cl0939. For the reconstruction we use about 230 galaxy images with $m \in (24, 25.5)$ and assume that their mean redshift is $\langle z \rangle = 0.8$. The field of view is $2'.5$.

Figure 2. The gaussian-smoothed light distribution of all bright galaxies with $m \in (18, 22)$. We use a smoothing length of $s = 0'.3$. Dressler & Gunn (1992) propose the cluster center to be in the upper left corner of the lower left CCD

Figure 3. The distribution $p(V)$ calculated from reconstructed mass distributions for 900 bootstrapping data sets (see text). The solid line shows $p(V)$ for the bright galaxies ($m \in (18, 22)$), the dotted one for the faint galaxies ($m \in (24, 25.5)$)

calculate the number $V := \sum_{\text{galaxies}} \kappa_\infty(\vec{x}_{\text{galaxy}}) - \langle \kappa_\infty \rangle$, both for the faint galaxies used for the reconstruction and for the bright galaxies, which are mostly cluster member galaxies. From 900 simulations we find the distributions $p(V)$ shown in Figure 3. Clearly, a correlation between the bright galaxies and the reconstructed mass distribution shows up (solid line), whereas there is an anti-correlation for the faint galaxies (dotted line). This anti-correlation is most likely an effect of the magnification (anti)bias, and is expected from the slope of the observed number counts (see Paper I). Furthermore, we find that the main features of the reconstructed mass distribution shown in Figure 1, (i.e., the maximum of the mass distribution, the decrease to the right and the prominent minimum), are stable.

We confirm that this rich cluster is also quite massive; as a lower bound on the total mass inside the data field (side length about $1h_{50}^{-1}$ Mpc we find for a mean redshift $\langle z \rangle = 0.6 \ (0.8, 1.0, 1.5)$ about $M \geq 5.3 \times 10^{14} h_{50}^{-1} M_\odot$ (3.6, 3.0, 2.3). These limits depend only slightly on the actual form of the assumed redshift distribution as shown in Paper I.

Acknowledgements: This work was supported by the "Sonderforschungsbereich 375-95 für Astro–Teilchenphysik" der Deutschen Forschungsgemeinschaft.

References

Dressler, A. & Gunn, J.E., 1992, ApJS, 78, 1
Seitz, C., Kneib, J.P., Schneider, P. & Seitz, S., 1995, in preparation (Paper I)

THE MASS DISTRIBUTION IN CLUSTERS OF GALAXIES FROM WEAK AND STRONG LENSING

JORDI MIRALDA-ESCUDÉ
Institute for Advanced Study
Olden Lane, Princeton NJ 08540, USA
Email: jordi@sns.ias.edu

Abstract. This paper is intended as an introduction to the theory of weak lensing. A review of the inversion formula introduced by Kaiser and Squires is presented. We then prove the formula of the aperture densitometry method in a simple way that allows a clear understanding of where the various terms come from. This is particularly useful to measure quantitatively masses in any region of a lens. We then summarize what has been learned from observations of strong lensing about the dark matter distribution; weak lensing should provide similar information on larger scales in clusters of galaxies.

1. The Weak Lensing Inversion

Gravitational lensing can be described as a mapping from the image plane (with coordinates x_I, y_I) to the source plane (with coordinates x_S, y_S), which is given by the lens equation:

$$(x_S, y_S) = (x_I, y_I) - \nabla \phi(x_I, y_I) , \qquad (1)$$

where $\nabla \phi$ is the deflection angle (in the single screen approximation), the projected gravitational potential ϕ is related to the surface density Σ by $\nabla^2 \phi = 2(\Sigma / \Sigma_{crit})$, and Σ_{crit} is the critical surface density (see Schneider, Ehlers, & Falco 1992). The distortion of a small image is given by the magnification matrix, which is the Jacobian of this mapping:

$$A^{-1} \equiv \frac{\partial(x_S, y_S)}{\partial(x_I, y_I)} \equiv \begin{pmatrix} 1 - \kappa - \lambda & -\mu \\ -\mu & 1 - \kappa + \lambda \end{pmatrix} . \qquad (2)$$

C. S. Kochanek and J. N. Hewitt (eds), Astrophysical Applications of Gravitational Lensing, 131–136.
© 1996 IAU

Using the lens equation, the convergence can be expressed as $\kappa = 1/2\,\nabla^2\phi = \Sigma/\Sigma_{crit}$, while the two components of the shear are

$$\lambda = \frac{1}{2}\left(\frac{\partial^2\phi}{\partial^2 x_I} - \frac{\partial^2\phi}{\partial^2 y_I}\right)\,, \qquad\qquad \mu = \frac{\partial^2\phi}{\partial x_I\,\partial y_I}\,. \qquad (3)$$

The magnification matrix can be diagonalized by rotating the coordinates by some angle β, to the form

$$A^{-1} = \begin{pmatrix} 1-\kappa-\gamma & 0 \\ 0 & 1-\kappa+\gamma \end{pmatrix}\,, \qquad (4)$$

where $\lambda = \gamma\cos(2\beta)$, and $\mu = \gamma\sin(2\beta)$. The axes in this rotated frame are called the "principal axes of the shear", because the images are "stretched" along these axes. The "stretching factor" q, or the axis ratio of the image of a circular source, is the ratio of the two eigenvalues $q = (1-g)/(1+g)$, where $g = \gamma/(1-\kappa)$. Thus, the change in the ellipticities of background sources depends only on the quantity g. In the limit of weak lensing, we assume $\gamma \ll 1$ and $\kappa \ll 1$, so $q \simeq 2\gamma$ and the observed quantity from the galaxy ellipticities is the shear. The problem of weak lensing is then reduced to obtaining the surface density given the shear.

To solve this problem, we first see how the shear is expressed in terms of the surface density. The projected potential is a linear superposition of the potentials caused by every element of mass in the lens:

$$\phi(\mathbf{r}_I) = \int d\mathbf{r}'_I \, \frac{\Sigma(\mathbf{r}'_I)}{\pi\Sigma_{crit}} \, \log\|\mathbf{r}_I - \mathbf{r}'_I\|\,, \qquad (5)$$

where $\mathbf{r}_I$ is the vector (x_I, y_I) (this results from the fact that $\nabla^2\log\|\mathbf{r}_I - \mathbf{r}'_I\| = 2\pi\,\delta^2(\mathbf{r}_I - \mathbf{r}'_I)$). Using equation (3), the shear is given by

$$[\lambda(\mathbf{r}_I), \mu(\mathbf{r}_I)] = \int d\mathbf{r}'_I \, \frac{\kappa(\mathbf{r}'_I)}{\pi} \, \frac{-[\cos 2\eta, \sin 2\eta]}{\|\mathbf{r}_I - \mathbf{r}'_I\|^2}\,. \qquad (6)$$

Here, the quantities written inside square parentheses are "polars" which, like the shear, rotate by twice the angle by which the coordinates are rotated. We have defined η as the angle between the direction from $\mathbf{r}_I$ to $\mathbf{r}'_I$ and the x-axis. With the minus sign, the orientation of the polar inside the integral rotates to the tangential direction which the shear at (x_I, y_I) produced by a mass element at (x'_I, y'_I) should have.

The inversion of the operator in equation (6) is most easily found by considering the Fourier transforms of the convergence, the shear and the potential, which we denote as κ_k, $[\lambda_k, \mu_k]$, and ϕ_k, and will be functions of the two Fourier coordinates k_x, k_y. The second derivatives giving the convergence and the shear in terms of the potential are converted into multiplications with k_x and k_y in Fourier space, so equation (6) is simply given by:

$$[\lambda_k, \mu_k] = \left[\frac{k_x^2 - k_y^2}{k_x^2 + k_y^2}, \frac{2k_x k_y}{k_x^2 + k_y^2} \right] \kappa_k . \tag{7}$$

It is most easily seen that the scalar product of the operator in equation (7) with itself is equal to unity. In other words, this operator is its own inverse. If true in Fourier space, this must also be true in real space, where the operator has the form in equation (6). Thus, the surface density is expressed in terms of the shear as

$$\kappa(\mathbf{r}_I) = \int d\mathbf{r}'_I, \frac{[\lambda(\mathbf{r}'_I), \mu(\mathbf{r}'_I)]}{\pi} \cdot \frac{-[\cos 2\eta, \sin 2\eta]}{\|\mathbf{r}_I - \mathbf{r}'_I\|^2} . \tag{8}$$

This equation was found by Kaiser & Squires (1993). As they explained, it is obviously not directly applicable since the surface density can only be recovered with finite resolution. Several methods to obtain convolved maps, solving the problem of the finite size of the fields, and extending the analysis beyond the weak lensing limit have been extensively discussed (Kaiser 1995; Schneider & Seitz 1995; Seitz & Schneider 1995; Bartelmann 1995). Here, we shall give an alternative proof of the inversion equation which gives us directly the surface density convolved with a particularly useful window function, in the aperture densitometry method.

The main operation in equation (8) is to calculate an average of the tangential component of the shear (which is obtained from the scalar product of the shear with the polar $[\cos 2\eta, \sin 2\eta]$) around every point, as proposed originally by Tyson, Valdes, & Wenk (1990). We can then simply do this average over a circle for a point mass lens, which can only depend on the distance from the point mass to the center of the circle. Since any arbitrary lens can be decomposed into small mass elements that can be treated as point masses, this will give us an inversion formula for the surface density smoothed with a certain filter. Using the symbols defined in Figure 1, the tangential component of the shear at a point along the circle is $\lambda_t(\phi) = (b/D)^2 \cos 2\beta$, where b is the critical radius of the point mass, and using simple trigonometry, the average along the circle λ_t^C is

$$\lambda_t^C = \frac{b^2}{R^2} \int_0^{2\pi} \frac{d\phi}{2\pi} \frac{(1 - a \cos \phi)^2 - (a \sin \phi)^2}{(1 - a \cos \phi)^2 + (a \sin \phi)^2} . \tag{9}$$

where $a = d/R$. The remarkable result is that this integral is equal to 1 for $a < 1$, and equal to zero for $a > 1$. Thus, *any mass outside the circle has no effect on λ_t^C, while any mass which is inside has an effect which is independent of its position within the circle.* In fact, b^2/R^2 is the average

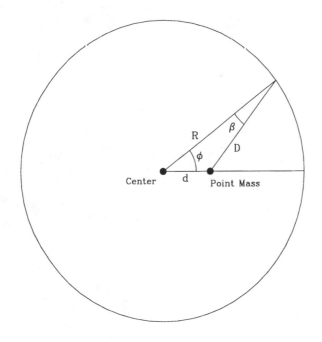

Figure 1.

convergence within the circle, $\bar{\kappa}$, due to the point mass (since, by definition, the average surface density within the critical radius is equal to the critical surface density). Thus, for any arbitrary lens, the contribution from the mass inside the circle is $\lambda_t^C = \bar{\kappa}$. We still need to include the contribution from the mass *on* the circle. We first notice that, from symmetry considerations, the contribution from any element of mass cannot depend on its azimuthal angle, and therefore the contribution from the mass on the circle can only depend on the surface density averaged over the circle, κ^C. Since we also know that the shear is zero when the surface density is uniform, the contribution from the mass on the circle must be $-\kappa^C$, so the result for any arbitrary lens is

$$\lambda_t^C = \bar{\kappa} - \kappa^C \ . \tag{10}$$

This equation was known for a circularly symmetric lens (where $\lambda_t^C = \gamma(r)$ is simply the total shear), but its general validity was not realized until recently. The integration of this equation over an interval in $\log R$, where R is the radius around a fixed center, yields (using $\bar{\kappa}(R) = 2/R^2 \int dR' \, R' \, \kappa^C(R')$ and exchanging the order of the integrals)

$$2 \int_{R_1}^{R_2} d\log R \ \lambda_t^C(R) = \bar{\kappa}(R_1) - \bar{\kappa}(R_2) \ , \tag{11}$$

which gives the surface density convolved with a compensated top-hat (see Fahlman et al. 1994, Kaiser 1995). This formula is useful not only to measure mass differences on the largest scale of an observed frame, but also to obtain maps of the small-scale variations in surface density (which could indicate the presence of clumpiness in the dark matter) by applying the formula around all possible centers.

2. Observations of Strong Lensing and Future Prospects

After weak lensing was first detected by Tyson et al. (1990), it is only recently that a large number of observational results are being reported (Bonnet, Mellier, & Fort 1994; Fahlman et al. 1994; Smail, Ellis, & Fitchett 1994; Smail et al. 1995; Tyson & Fisher 1995). Weak lensing by individual galaxies, first searched by Tyson et al. (1984), is now also being found (Brainerd, Blandford, & Smail 1995), and the average shear in blank fields (the weakest lensing in the sky) is also being searched for (Mould et al. 1994; Villumsen 1995). The weak shear is only sensitive to differences in surface density, as is clearly seen from the inversion formulae extended only over finite fields (10), (11). Recent work by Broadhurst, Taylor, & Peacock (1995) has highlighted the use of the magnification, measured from the fluctuations in the number counts of galaxies in different colors, which has the advantage of probing directly the surface density although it may be more severely subject to systematic errors due to the clustering of the lensed sources.

Meanwhile, much progress has been done on strong lensing, the observations of the highly distorted "arcs" and multiple images. After the initial realization that this ruled out a flat core larger than $\sim 50\,h^{-1}$ Kpc for the dark matter, it has been found that lensing always occurs around bright central cluster galaxies or around highly compact galaxy clumps, and the mass-to-light ratios are typically $M/L_B \sim 300\,h$ within the radius of the arcs (Fort & Mellier 1994). Models reproducing the multiply imaged sources require the presence of substructure in the form of dark matter clumps around the brightest galaxies; furthermore, these clumps are found to have similar ellipticities as the outer stellar isophotes of the bright galaxies (see Kneib et al. 1993, 1995; Mellier et al. 1993). This is not surprising, since the outer stellar isophotes can be observed to radii similar to those of the observed arcs, and therefore they must be in orbits in the dark matter clumps yielding the lensing deflection angles; it also implies that the stellar velocity dispersions of the bright galaxies must rapidly rise within the radius of the arcs (Miralda-Escudé 1995). These observations suggest that clusters of galaxies are typically not in a static equilibrium, but they are constantly undergoing mergers on a wide range of scales, down to the scales of the observed arcs and the stellar halos of the central galaxies. This picture seems consistent with hierarchical theories of formation of clusters and should offer new clues on the formation of central cluster galaxies (Merritt 1985 and references therein). The determination of masses can also provide new information on the hot intracluster gas (Loeb & Mao 1994, Miralda-Escudé & Babul 1995, Waxman & Miralda-Escudé 1995, Allen, Fabian, & Kneib 1995, Squires et al. 1995). The present rapid growth of weak lensing observations will probably lead to new discoveries on similar aspects of clusters of galaxies at larger radii: the mass-to-light ratios and the density profiles, substructure in the dark matter and its relation to the galaxy distribution, and the physical state of the intracluster gas.

I thank Nick Kaiser for many illuminating discussions on this subject. I gratefully acknowledge support from the W. M. Keck Foundation.

References

Allen, S. W., Fabian, A. C, & Kneib, J. P., 1995, MNRAS, in press
Bartelmann, M., 1995, A&A, in press
Bonnet, H., Mellier, Y., & Fort, B., 1994, ApJ, 427, L83
Brainerd, T. G., Blandford, R. D., & Smail, I., 1995, ApJ, submitted
Broadhurst, T. J., Taylor, A. N., & Peacock, J. A., 1995, ApJ, 438, 49
Fahlman, G., Kaiser, N., Squires, G., & Woods, D., 1994, ApJ, 437, 56
Fort, B., & Mellier, Y., 1994, A&AR, 5, 239
Kaiser, N., 1995, ApJ, 493, L1
Kaiser, N., & Squires, G., 1993, ApJ, 404, 441
Kneib, J. P., Mellier, Y., Fort, B., & Mathez, G., 1993, A&A, 273, 367
Kneib, J. P., Mellier, Y., Pelló, R., Miralda-Escudé, J., Le Borgne, J. F., Böhringer, H., & Picat, J. P., 1995, A&A, in press
Loeb, A., & Mao, S., 1994, ApJ, 435, 109
Mellier, Y., Fort, B., & Kneib, J. P., 1993, ApJ, 407, 33
Merritt, D., 1985, ApJ, 289, 18
Miralda-Escudé, J., 1995, ApJ, 438, 514
Miralda-Escudé, J., & Babul, A., 1995, ApJ, 449, 18
Mould, J., Blandford, R., Villumsen, J., Brainerd, T., Smail, I., Small, T., & Kells, W., 1994, MNRAS, 271, 31
Schneider, P., Ehlers, J., & Falco, E. E., 1992, Gravitational Lenses, (Berlin: Springer-Verlag)
Schneider, P., & Seitz, C., 1995, A&A, 294, 411
Seitz, C., & Schneider, P., 1995, A&A, 297, 287
Smail, I, Ellis, R. S., & Fitchett, M. J., 1994, MNRAS, 270, 245
Smail, I, Ellis, R. S., Fitchett, M. J., & Edge, A. C., 1995, MNRAS, 273, 277
Squires, G., Kaiser, N., Babul, A., Fahlman, G., Woods, D., Neumann, D. M., & Böhringer, H., 1995, ApJ, submitted
Tyson, J. A., & Fisher, P., 1995, ApJLet, in press
Tyson, J. A., Valdes, F., & Wenk, R. A., 1990, ApJLet, 349, L1
Tyson, J. A., Valdes, F., Jarvies, J. F., & Mills, A., 1984, ApJ, 281, L59
Villumsen, J. V., 1995, MNRAS, submitted
Waxman, E., & Miralda-Escudé, J., 1995, ApJ, in press

THE RECONSTRUCTION OF CLUSTER MASS PROFILES FROM IMAGE DISTORTIONS

PETER SCHNEIDER

Max-Planck-Institut für Astrophysik
Postfach 1523, D-85740 Garching, Germany

Abstract. I will review the basic principles and some of the most recent developments of parameter-free reconstructions of cluster mass profiles from the distortions of background galaxy images.

1. Introduction

The discovery of cluster lensing in the form of giant luminous arcs was immediately recognized as a new tool for determining the mass in the inner part of the lensing clusters. Less dramatic image distortions as seen in arclets (Fort et al. 1988) and in the coherent image alignments (Tyson et al. 1990) have increased the angular extent over which mass estimates can be obtained – out to about 1 Mpc in some cases (see, e.g., Bonnet et al. 1994). A detailed account of the observation, theory and modeling of arc clusters is given in Fort & Mellier (1994). In their pioneering paper, Kaiser & Squires (1993; hereafter KS) have shown that the observed image ellipticities can be directly translated into an estimate for the surface mass distribution of the cluster lens, without the need for fitting parametrized mass models. The underlying idea is that the image distortions are caused by the tidal gravitational field, which in turn is directly related to the mass distribution. The KS inversion method, described in the next section, has already been applied to a number of clusters (Fahlman et al. 1995; Smail et al. 1995), and further applications can be found elsewhere in these proceedings (see contributions by N. Kaiser, C. Seitz); at least in one case a surprisingly large mass-to-light ratio was found (Fahlman et al. 1995). The faint level to which background galaxies can now be imaged, with the corresponding high number density of objects, together with the imaging capability of

C. S. Kochanek and J. N. Hewitt (eds), Astrophysical Applications of Gravitational Lensing, 137–142.

© 1996 IAU

the HST, the improvements in ground-based imaging, the increase of the field of view, and the development of improved inversion techniques, render the mass inversion from image distortion a unique tool for studying the mass distribution in clusters of galaxies. Here I shall review some of the latest developments of techniques for the reconstruction of the cluster mass distribution.

2. The "classical" Kaiser & Squires method

The shear (or tidal field) is described by a traceless symmetric 2×2 tensor, or, equivalently, by a two-component quantity, for which we shall use the complex shear γ, which is related to the dimensionless surface mass density κ through $\gamma(\boldsymbol{\theta}) = \frac{1}{\pi} \int_{\mathbf{R}^2} d^2\theta' \, \mathcal{D}(\boldsymbol{\theta} - \boldsymbol{\theta}') \, \kappa(\boldsymbol{\theta}')$, with $\mathcal{D}(\mathbf{z}) = 1/(z_1 - iz_2)^2$. The inversion of this convolution-type integral can be easily obtained in Fourier space and yields (KS)

$$\kappa(\boldsymbol{\theta}) = \frac{1}{\pi} \int_{\mathbf{R}^2} d^2\theta' \, \mathcal{R}e[\mathcal{D}^*(\boldsymbol{\theta} - \boldsymbol{\theta}') \, \gamma(\boldsymbol{\theta}')] + \kappa_0, \tag{1}$$

where κ_0 is an arbitrary constant, unconstrained from the $k = 0$ mode of the Fourier transform, and an asterisk denotes complex conjugation. Hence, (1) yields a parameter-free estimate of the surface mass distribution, up to an overall additive constant, which, however, can be constrained due to the non-negativity of the mass. Several modifications of this "classical" KS method have been recently suggested in the literature: (a) In general, the shear γ is not an observable; however, in the case of weak distortions ($\kappa \ll 1$, $|\gamma| \ll 1$) the observable is directly related to γ (see Sect.3). The generalization to the strong lensing regime is described briefly in Sect.4. (b) The integral in (1) extends over the 'whole sky', whereas actual data fields are finite. The leads to a bias of (1) at the boundaries of the data field; the generalization to unbiased estimators are described in Sect.5. (c) The integration constant in (1) describes an invariance transformation, which leaves image ellipticities unchanged; this invariance can be broken by invoking magnification effects, which will be briefly outlined in Sect.6.

3. The observable: distortion

Unless stated otherwise, we shall, for simplicity, consider in the following the case that the distance ratio D_{ds}/D_s is the same for all background sources, which is a good approximation for $z_d \lesssim 0.2$. From the tensor Q_{ij} of second brightness moments one defines the complex ellipticity $\chi = (Q_{11} - Q_{22} + 2iQ_{12})/\mathrm{tr}Q$. The locally linearized lens equation yields a transformation between the source and image ellipticity, $\chi^{(s)} = \chi^{(s)}(\chi, g)$, where $g = \gamma/(1 - \kappa)$ is the 'reduced shear'. With the fundamental assumption, underlying all these reconstruction methods, that the intrinsic orientation of the sources are random, $\langle \chi^{(s)} \rangle = 0$, this translates to an equation determining g: $\langle \chi^{(s)}(\chi, g) \rangle = 0$, where here the angular brackets denote a local average over images, i.e., one introduces a smoothing length. As it turns out, this equation always has two solutions – if g is a solution, so is $1/g^*$. Hence, the observable is the distortion (Kochanek 1990, Miralda-Escudé 1991, Schneider & Seitz 1995) $\delta = \frac{2g}{1+|g|^2} = \frac{2\gamma(1-\kappa)}{(1-\kappa)^2+|\gamma|^2}$. In the case of weak lensing ($\kappa \ll 1$, $|\gamma| \ll 1$), $\gamma \approx \delta/2 \approx -\langle \chi \rangle /2$ can be 'observed'. The resulting value for δ is independent of the intrinsic ellipticity distribution, but the accuracy of the estimate decreases for broader intrinsic distributions. If the intrinsic ellipticity distribution is known, a maximum likelihood estimate yields more precise values for δ, except if the former is a Gaussian.

4. Non-linear generalization of KS

The KS inversion (1) can now be written in a form which accounts also for the strong lensing regime, by setting $\gamma = (1 - \kappa)g$:

$$\kappa(\boldsymbol{\theta}) = \frac{1}{\pi} \int_{\mathbf{R}^2} \mathrm{d}^2\theta' \, [1 - \kappa(\boldsymbol{\theta}')]\mathcal{R}e[\mathcal{D}^*(\boldsymbol{\theta} - \boldsymbol{\theta}') \, g(\boldsymbol{\theta}')] + \kappa_0, \qquad (2)$$

with $g = \delta \left(1 - \mathrm{sign}(\det A)\sqrt{1 - |\delta|^2}\right) / |\delta|^2$, and A is the Jacobi matrix of the lens equation. The integral equation (2) for κ can be solved iteratively (Seitz & Schneider 1995a). The resulting mass distribution $\kappa(\boldsymbol{\theta})$ is then determined up to a global invariance transformation (Schneider & Seitz 1995)

$$\kappa(\boldsymbol{\theta}) \to \lambda\kappa(\boldsymbol{\theta}) + (1 - \lambda), \qquad (3)$$

which is the mass sheet degeneracy pointed out by Gorenstein, Falco & Shapiro (1988) and which leaves the observable δ invariant. In the linear (weak lensing) case, this reduces to the additive constant in (1).[1]

5. Unbiased finite-field inversions

The second modification mentioned in Sect.2 was to account for the finiteness of the data field $\mathcal{U} \in \mathbb{R}^2$. There are two quick fixes to deal with the

---

[1] If the redshift distribution of the sources becomes important, i.e., if D_{ds}/D_s is not nearly the same for all sources, this degeneracy is broken, but unless the cluster is nearly critical, an approximate invariance transformation still holds (Seitz & Schneider 1995c). For critical clusters, the breaking of invariance is stronger and may be used, at least in principle, to fix the mass density uniquely.

$\mathbb{R}^2$-integral in (1): one is to set $\gamma = 0$ for $\theta \notin \mathcal{U}$ (used in the presently published reconstructions), or, arguably more reasonable, to extrapolate the shear γ outside $\mathcal{U}$, as suggested in Seitz & Schneider (1995a) and quantitatively tested in Bartelmann (1995). The success of these fixes depends on the mass distribution and the CCD field $\mathcal{U}$, as discussed below. The road for an unbiased finite-field inversion was paved by Kaiser (1995) who pointed out that the gradient of κ is related to a combination of derivatives of the shear components. He showed that

$$\nabla\kappa = \mathbf{U} \quad \text{and} \quad \nabla K = \mathbf{u}, \tag{4}$$

where the vector field $\mathbf{U}$ contains first derivatives of the components of the shear γ, or, alternatively, the gradient of $K = \ln(1 - \kappa)$ is related to the reduced shear g and its first derivatives, which is combined to the vector field $\mathbf{u}$. In the linear case, where γ is an observable, the first of eq.(4) shows that κ can be determined only up to an additive constant, whereas in the general case, K can be determined from the observable g only up to an additive constant, so that we reobtain the mass sheet degeneracy mentioned in the previous section. Either of the two equations (4) can be integrated by path integration – selecting the first for illustration, that becomes $\kappa(\theta) = \int_{\theta_0}^{\theta} \mathrm{d}\mathbf{l} \cdot \mathbf{U} + \kappa(\theta_0)$, where $\theta_0 \in \mathcal{U}$ is a reference point, and $\mathbf{l}$ denotes an integration curve. Averaging over a set $\{\theta_0\}$ of starting points, a generalization reads $\kappa(\theta) = \sum_{\{\theta_0\}} w(\theta_0) \int_{\theta_0}^{\theta} \mathrm{d}\mathbf{l} \cdot \mathbf{U} / \sum_{\{\theta_0\}} w(\theta_0) + \text{const.}$, where the $w(\theta_0)$ are arbitrary weight factors; this yields a set of unbiased finite-field inversions. There remains a huge freedom in selecting the set $\{\theta_0\}$, the weights $w(\theta_0)$ and the curves connecting θ_0 and θ, and different choices have been suggested by Schneider (1995) and Kaiser et al. (1995). The reason why different choices are not equivalent is that the vector field $\mathbf{U}$ is obtained from observations and thus is not a gradient field in general – if it were a gradient field, even a single line integration would be sufficient. The 'noise' of $\mathbf{U}$ is handled differently by different choices. However, the rotational component of $\mathbf{U}$ is a noise contribution which is readily identified as such, and one can construct a finite-field inversion formula which is designed to filter out this rotational component,

$$\kappa(\theta) = \int_{\mathcal{U}} \mathrm{d}^2\theta' \, \mathbf{H}(\theta'; \theta) \cdot \mathbf{U}(\theta') + \bar{\kappa}, \tag{5}$$

where the kernel function $\mathbf{H}$ was explicitly constructed in Seitz & Schneider (1995b); in that paper we have also performed extensive simulations on synthetic data to show that (5) is the best unbiased finite-field method yet published. The application of (5) is not more complicated than that of (1), so *there is no reason* not *to apply this unbiased estimator*. [2]

6. Breaking the degeneracy: Magnification effects

Under the transformation (3), the magnification transforms as $\mu \rightarrow \mu/\lambda^2$, i.e. magnification can be used to break the invariance. Two methods have been suggested: Broadhurst et al. (1995; see also contribution by A. Taylor) point out that the local number density $n(> S)$ of background sources with flux $> S$ is changed according to $n(> S) = n_0(> S/\mu)/\mu$ (n_0: unlensed counts). Assuming locally a power law $n_0 \propto S^{-\alpha}$ one has $n/n_0 = \mu^{(\alpha-1)}$. [3] Whereas density fluctuations of the background source population may be a problem for local determination of the magnification, the effect averaged over the whole data field should be able to break the mass sheet degeneracy.

A second method was proposed by Bartelmann & Narayan (1995, see also contribution by M. Bartelmann), using the fact that the surface brightness I of images is unchanged by lensing. A source of angular size R_0 thus retains its surface brightness, and its image attains the size $R = \sqrt{\mu}R_0$. If $\langle R_0 \rangle (I)$ denotes the mean size of sources with surface brightness I, an estimate of the local magnification is $\mu = (\langle R \rangle (I)/ \langle R_0 \rangle (I))^2$, where $\langle R \rangle (I)$ is the local average of image sizes with surface brightness I. [4]

---

[2] The importance of the modifications of the KS method discussed in this and the previous section depends on the data set and the mass distribution. For example, if one has a fairly large CCD field centered on an isolated cluster, the finite-field corrections will be moderate, and except for the central part of the cluster, the resulting mass distribution will only be weakly affected by the non-linear corrections. However, if there is a significant mass component close to the edge, or outside of the data field, or if the data field is small, or if one is interested in the distribution in the central part of the cluster, these corrections no longer provide 'minor modifications', but are essential. The WFPC2 images of cluster centers can be used to construct their mass profile, as demonstrated in the contribution by C. Seitz for the cluster 0939+4713 (Seitz et al. 1995), and one should not even think about attempting this reconstruction without the two modifications discussed here (of course we did to see the – huge – differences). Note that the aperture densiometry (Kaiser 1995), also discussed in the contribution by J. Miralda-Escudé, yields unbiased lower limits of the mass inside circular apertures.

[3] Whereas the faint blue galaxies have $\alpha \approx 1$ and are thus unusable for this effect, the red galaxies appear to have flatter counts, $\alpha \sim 0.4$, so that magnification depletes the number counts locally. In fact, this depletion has been seen in the inner part of A1689 (Broadhurst 1995) and in Cl0939 (Seitz et al. 1995).

[4] The intrinsic size $\langle R_0 \rangle (I)$ can be obtained from images in empty fields. The accuracy of the local determination of μ depends of course on the intrinsic width of the size distribution at fixed surface brightness; in the galaxy model used in Bartelmann & Narayan (1995), this comes out to be $\Delta \ln R \sim 0.5$. With such a distribution, the size effect alone can provide a reasonable estimate of the surface mass distribution. But even if the width of the intrinsic size distribution turns out to be significantly broader, the size effect averaged over the data field allows breaking of the degeneracy.

7. Conclusions

Unbiased, non-linear cluster reconstruction methods, generalizing the original KS method, have been developed; they are quantitatively tested on synthetic data, are easy to apply, and therefore should be applied! In cases where the center of a cluster should be reconstructed from HST images, the modifications discussed here are essential. Currently we (with S. Seitz, M. Bartelmann & R. Narayan) are developing an improved inversion technique which (1) avoids the introduction of an arbitrary smoothing length needed for performing the local averages, (2) accounts for both the distortion and the magnification effects in a single step, and (3) yields an objective measure for the quality of the reconstruction. The method maximizes the likelihood for the observed image ellipticities and sizes (and/or local number densities) with respect to the surface mass density described on a grid. In order to avoid overfitting the data, the likelihood function is regularized. See the contribution by S. Seitz for some details. Whereas this method is significantly more complicated than the direct inversion (1) or (5), the effort is still negligible compared to obtaining the data, and it most likely will increase the accuracy of the inversion substantially.

Acknowledgements: I thank C. and S. Seitz for helpful comments on this manuscript. This work was supported by the "Sonderforschungsbereich 375-95 für Astro–Teilchenphysik" der Deutschen Forschungsgemeinschaft.

References

Bartelmann, M., 1995, A&A, in press
Bartelmann, M. & Narayan, R., 1995, ApJ, in press
Bonnet, H., Mellier, Y. & Fort, B., 1994, ApJL, 427, L83
Broadhurst, T.J., 1995, preprint
Broadhurst, T.J., Taylor, A.N. & Peacock, J.A., 1995, ApJ, 438, 49
Fahlman, G., Kaiser, N., Squires, G. & Woods, D., 1994, ApJ, 437, 56
Fort, B. & Mellier, Y., 1994, A&AR, 5, 239
Fort, B., Prieur, J.L., Mathez, G., Mellier, Y. & Soucail, G., 1988, A&A, 200, L17
Gorenstein, M.V., Falco, E.E. & Shapiro, I.I., 1988, ApJ, 327, 693
Kaiser, N., 1995, ApJL, 439, L1
Kaiser, N. & Squires, G., 1993, ApJ, 404, 441
Kaiser, N., Squires, G., Fahlman, G., Woods, D.& Broadhurst, T., 1995, preprint
Kochanek, C.S., 1990, MNRAS, 247, 135
Miralda-Escude, J., 1991, ApJ, 370, 1
Schneider, P., 1995, A&A, in press
Schneider, P. & Seitz, C., 1995, A&A, 294, 411
Seitz, C. & Schneider, P., 1995a, A&A, 297, 287
Seitz, S. & Schneider, P., 1995b, A&A, in press
Seitz, C. & Schneider, P., 1995c, in preparation
Seitz, C., Kneib, J.-P., Schneider, P. & Seitz, S., 1995, in preparation
Smail, I., Ellis, R.S., Fitchett, M.J. & Edge, A.C., 1995, MNRAS, 273, 277
Tyson, J.A., Valdes, F. & Wenk, R.A., 1990, ApJ, 349, L1

REDSHIFTS OF FAINT BLUE GALAXIES FROM GRAVITATIONAL LENSING

MATTHIAS BARTELMANN [1,2] AND RAMESH NARAYAN [1]
[1] *Harvard-Smithsonian Center for Astrophysics*
60 Garden Street, Cambridge, MA 02138, USA, and
[2] *Max-Planck-Institut für Astrophysik*
Postfach 1523, 85740 Garching, Germany

1. Introduction

The possibility to reconstruct cluster mass distributions from their gravitational lensing effects on background sources is now firmly established and is discussed in a number of other contributions to these proceedings. Here, we want to suggest a new method, the *lens parallax method* (Bartelmann & Narayan 1995), to infer the redshifts of the background sources, which are in general too faint to allow spectroscopy. The method is based on gravitational lensing, and apart from arclet redshifts it yields calibrated, accurate cluster reconstructions. The key idea is simple and rests on three facts: (1) The strength of lensing effects increases with source redshift, if the lens redshift is kept fixed. (2) The surface brightness of the arclet sources steeply decreases with redshift z. Neglecting spectral effects, the surface brightness is $\propto (1 + z)^{-4}$. (3) The surface brightness of the sources is invariant to lensing. In combination, these facts imply that lensing effects become stronger with decreasing surface brightness, so that relative distances of arclet sources can be inferred from the variation of lensing effects with the surface brightness of lensed sources. If the redshift of the brightest sources can be measured spectroscopically, the relative distances can be calibrated and the average redshift as a function of surface brightness can be found.

Conversely, knowledge of the source redshifts removes an uncertainty from cluster reconstructions, which usually assume the sources to be all at one redshift. All reconstructions based on image distortions alone suffer from the so-called mass-sheet degeneracy (Falco et al. 1985; Gorenstein

C. S. Kochanek and J. N. Hewitt (eds), Astrophysical Applications of Gravitational Lensing, 143–148.
© 1996 IAU

et al. 1988; Kaiser & Squires 1993; Schneider & Seitz 1995), which states that the surface-mass density of the lens can only be reconstructed up to a one-parameter family of transformations. Cluster masses can be reliably determined only if this degeneracy is broken. We suggest including the magnification of the arclet sources in the reconstruction algorithm to eliminate the mass-sheet degeneracy and obtain accurate cluster masses (see also Broadhurst et al. 1995).

The results of numerical simulations designed to test the algorithm we suggest are promising. Using a sample of 12 numerically simulated clusters and a realistic model for the intrinsic source properties, we show that source redshifts can be inferred with an accuracy of $\Delta z \sim 0.1$, and that the masses of the lensing clusters can be reconstructed to within $\pm 5\%$.

2. Outline of the Method

We require the following observational data, which have to be obtained in a sample of cluster fields and a number of empty control fields: The ellipticities ε of the sources, their surface brightnesses S, and their sizes R. In the following, primed symbols refer to quantities obtained in control fields. The surface brightnesses are sorted into bins S_i. The data in the cluster fields depend on position and thus have to be sorted into spatial bins $\vec{x}_j$, while the control data should be spatially homogeneous. We thus start from the data sets

- $\varepsilon'(S_i)$ and $R'(S_i)$, the ellipticity and the size of the sources as a function of their surface brightness in the control fields; and
- $\varepsilon(\vec{x}_j, S_i)$ and $R(\vec{x}_j, S_i)$, the same properties of the arclets in the cluster fields.

The $\varepsilon'(S_i)$ are used to find the intrinsic spread of ellipticities to calibrate a major source of noise in the cluster reconstructions. The sizes are combined to form the quantity

$$r_{ij} \equiv r(\vec{x}_j, S_i) \equiv \left(\frac{R'(S_i)}{R(\vec{x}_j, S_i)} \right)^2 , \tag{1}$$

which estimates the inverse magnification at position $\vec{x}_j$ for sources with surface brightness S_i.

For linear lenses which satisfy $\kappa \ll 1$ and $|\gamma| \ll 1$, the ellipticities estimate the shear and the sizes estimate the magnification,

$$\begin{aligned}
\varepsilon_{ij} &\equiv \varepsilon(\vec{x}_j, S_i) \approx \gamma(\vec{x}_j, S_i) \equiv \gamma_{ij} \\
r_{ij} &\approx 1 - 2\kappa(\vec{x}_j, S_i) \equiv 1 - 2\kappa_{ij} .
\end{aligned} \tag{2}$$

Their redshift dependence is expressed by their dependence on surface brightness. We describe it by distance factors d_i to relate the data in the surface-brightness bins S_i, $i \geq 2$, to the (brightest) bin S_1,

$$(\kappa, \gamma)_{ij} = (\kappa, \gamma)_{1j} \cdot d_i . \tag{3}$$

The d_i constitute a "distance ladder" for the sources which depends on the adopted cosmological model. Specifying Ω and Λ, the d_i can be converted to redshifts once the redshift corresponding to S_1 is measured. From each spatial bin, we obtain two independent estimates for the d_i,

$$d_i \approx \frac{\varepsilon_{ij}}{\varepsilon_{1j}} \approx \frac{1 - r_{ij}}{1 - r_{1j}} , \tag{4}$$

which can be weighted by their respective estimated accuracies and combined. Conversely, the ellipticities and sizes in each surface-brightness bin provide estimates for convergence and shear according to (2),

$$\gamma_{1j} \approx \frac{\varepsilon_{ij}}{d_i} \quad , \quad \kappa_{1j} \approx \frac{1 - r_{ij}}{2 d_i} , \tag{5}$$

which can equally well be combined.

If the cluster lens is nonlinear, the algorithm becomes somewhat more complicated. Then, ellipticities and sizes measure combinations of convergence and shear rather than each quantity separately,

$$\varepsilon_{ij} = \frac{\gamma_{ij}}{1 - \kappa_{ij}} \quad , \quad r_{ij} = (1 - \kappa_{ij})^2 - \gamma_{ij}^2 ; \tag{6}$$

(cf. Kochanek 1990; Miralda-Escudé 1991; Schneider & Seitz 1995; Kaiser 1995). Since both κ and γ are initially unknown, the algorithm has to proceed iteratively, but the principle remains as sketched above. The data, ε_{ij} and r_{ij}, are used to iteratively reconstruct the distance factors d_i and to build up estimates for γ and κ of the clusters involved.

In the linear as in the nonlinear case, the reconstructed γ can be transformed into κ, however not uniquely because of the mass-sheet degeneracy. The direct estimate of κ from the image magnifications is fairly inaccurate because of the large variance of intrinsic source sizes with given surface brightness (cf. the lower right frame of figure 1). The convergence reconstructed from γ, however, can be calibrated by the direct estimate of κ obtained from the image sizes. This breaks the mass-sheet degeneracy and allows to unambiguously select that cluster model which fits both ellipticities and sizes best, given the data and their errors.

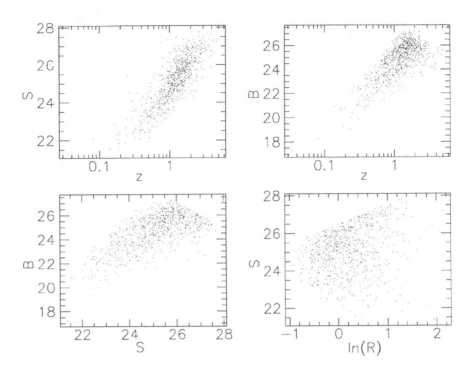

Figure 1. Scatter plots showing the distribution of "observed" properties of our simulated sources. *Upper left frame:* Surface brightness S in mag arcsec^{-2} vs. redshift z; *Upper right frame:* Blue magnitude B vs. z; *Lower left frame:* B vs. S; *Lower right frame:* S vs. size R. Sharp boundaries in the plots reflect the detection criterion.

3. Numerical Model

To test the feasibility of the proposed algorithm, we use a sample of 12 cluster models simulated in the CDM cosmogony, with $\Omega_0 = 1$ and $\Lambda = 0$; see Bartelmann et al. (1995) for a detailed description. The clusters are all at about the same redshift, $z_d \sim 0.35$, and they span a broad mass range. Their velocity dispersions σ_v fall within $[950, 1550]\,\mathrm{km\,s^{-1}}$.

Our model for the sources is guided by the observation that there seems to be little galaxy evolution in red light, although the galaxy number counts evolve significantly in blue light (e.g. Broadhurst et al. 1988; Colless et al. 1990; Colless et al. 1993; Lilly et al. 1991; Lilly 1993; Crampton et al. 1994; Lilly et al. 1995; Koo & Kron 1992) We adopt the view that the blue light comes from star-forming regions in otherwise normal galaxies. Consequently, we assume a constant comoving number density and a Schechter luminosity function with constant parameters. We further choose their surface-brightnesses, their colors, and their ellipticities from distribution functions which well resemble observational data. The sources are subject to a detection criterion which depends on both surface brightness

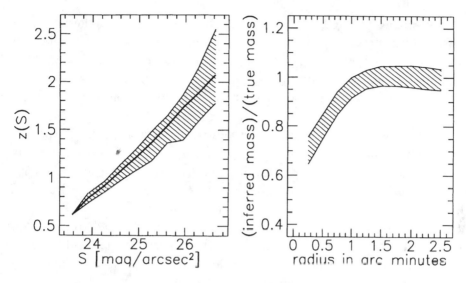

Figure 2.   Left frame: Recovered source redshift as a function of surface brightness. The hatched region is centered on the average redshift, and its width corresponds to the 1–σ error obtained from 12 cluster and control fields. The heavy line shows the intrinsic source redshift. *Right frame:* Cumulative reconstructed cluster masses relative to the true masses. While smoothing of the data broadens cluster peaks close to the center and causes the mass underestimate, the reconstructed mass agrees with the true mass to within ±5% at the field boundary.

and magnitude. The "observed" properties of the simulated sources closely match the observations (cf. figure 1).

4. Results

The simulated sources are "observed" in the 12 cluster fields, where they are lensed, and in an equal number of control fields. The simulated data are processed with the iterative algorithm described before. We show the results for the recovered source redshifts and for the reconstructed cluster masses in figure 2.

The left frame shows the source redshifts as a function of surface brightness. The hatched region is centered on the average redshift recovered and its width is the 1–σ error from the 12 cluster fields. The heavy line represents the intrinsic redshift distribution. The plot shows that the average source redshift is well recovered with an error of $\Delta z \sim 0.1$ at intermediate redshifts. The right frame of the figure displays the cumulative reconstructed mass fraction of the cluster lenses, i.e. the ratio between the reconstructed and the true cluster mass within a given radius. Smoothing of the simulated data broadens the central cluster peaks and causes the mass underestimate within $\sim 1'$, but towards the boundary of the field the reconstructed mass equals the true mass to within ±5% accuracy.

Summarizing, we find that our simulations indicate that the proposed iterative algorithm succeeds in reconstructing arclet source redshifts and cluster mass distributions to very satisfactory accuracy. The inclusion of the source magnifications breaks the mass-sheet degeneracy inherent in all cluster reconstructions based on image distortions alone and allows for accurate, unambiguous cluster mass determinations.

The *lens parallax method* can be extended in several ways. First, additional information about the sources can incorporated. The most obvious choice are their colors which are also invariant to lensing. Second, the conversion of distance factors to redshifts depends on the cosmological parameters chosen. Possible inconsistencies between the source redshifts derived from cluster samples with different lens redshifts can be used to constrain especially the cosmological constant Λ to which lensing is particularly sensitive (Turner 1990; Kochanek 1993). Third, it can be tested whether a single lens is responsible for the image distortions. If several lenses at different redshifts contribute, the principal axis of the shear in a given spatial bin should change its direction with redshift and thus with the surface brightness of the sources.

This work was supported in part by NSF grant AST 9423209 and by the Sonderforschungsbereich SFB 375-95 of the Deutsche Forschungsgemeinschaft.

References

Bartelmann, M., & Narayan, R., 1995, ApJ, in press (Sep. 20 issue)
Koo, D. C., & Kron, R. G. 1992, ARA&A, 30, 613
Bartelmann, M., Steinmetz, M., & Weiss, A., 1995, A&A, 297, 1
Broadhurst, T. J., Ellis, R. S., & Shanks, T., 1988, MNRAS, 235, 827
Broadhurst, T. J., Taylor, A. N., & Peacock, J. A., 1995, ApJ, 438, 49
Colless, M. M., Ellis, R. S., Taylor, K., & Hook, R. N., 1990, MNRAS, 244, 408
Colless, M. M., Ellis, R. S., Broadhurst, T. J., Taylor, K., & Peterson, B. A., 1993, MNRAS, 261, 19
Crampton, D., Morbey, C. L., Le Fevre, O., Hammer, F., Tresse, L., Lilly, S. J., & Schade, D. J., 1994, SISSA preprint astro-ph/9410041
Falco, E. E., Gorenstein, M. V., & Shapiro, I. I., 1985, ApJ, 289, L1
Gorenstein, M. V., Falco, E. E., & Shapiro, I. I., 1988, ApJ, 327, 693
Kaiser, N., 1995, ApJ, 493, L1
Kaiser, N., & Squires, G., 1993, ApJ, 404, 441
Kochanek, C. S., 1990, MNRAS, 247, 135
Kochanek, C. S., 1993, ApJ, 419, 12
Koo, D. C., & Kron, R. G. 1992, ARA&A, 30, 613
Lilly, S. J., 1993, ApJ, 411, 501
Lilly, S. J., Cowie, L. L., & Gardner, J. P., 1991, ApJ, 369, 79
Lilly, S. J., Tresse, L., Hammer, F., Crampton, D., & Le Fevre, O., 1995, ApJ, submitted
Miralda-Escudé, J., 1991, ApJ, 370, 1
Schneider, P., & Seitz, C., 1995, A&A, 294, 411
Turner, E. L., 1990, ApJ, 365, 43

GRAVITATIONAL MAGNIFICATION
AND CLUSTER MASSES

A.N. TAYLOR
Institute for Astronomy, University of Edinburgh,
Royal Observatory, Blackford Hill, Edinburgh, UK

1. Introduction

The shear distortion of galaxies behind a large cluster provides us with only a relative measure of the mass of the cluster (Tyson et al. 1990, Kaiser & Squires 1993) due the invariance of shear to a sheet of matter. Thus we can at best only place lower limits on cluster masses, with the constraint that the surface density, $\kappa(\theta) = \Sigma(\theta)/\Sigma_c$, is non-negative. However, an absolute measurement of the mass can be obtained via the magnification of background galaxies (Broadhurst, Taylor & Peacock 1995, hereafter BTP). Here we describe the magnification effect, its observational consequences and mass reconstruction in the linear and nonlinear regimes. We also describe a number of applications in progress.

2. Lens Magnification

In general, gravitational lensing induces a change in the size and shape of an image galaxy. This distortion in the image is characterized by the image magnification, $A(z)$, for a galaxy at redshift z, such that the ratio of source to image areas is

$$A = |(1 - \kappa)^2 - \gamma^2|^{-1} \simeq 1 + 2\kappa, \tag{1}$$

where γ is the shear and the approximation holds in the weak lensing regime. Surface brightness is a conserved quantity, so the change in image area produces a shift in apparent magnitude, $\Delta m = 2.5 \log_{10} A(z)$. As a result the effective flux limit decreases. BTP have shown that the lensed redshift distribution is related to the unlensed distribution by

$$N'(z) = A(z)^{\beta(z)-1} N(z) \tag{2}$$

149

where $\beta(z) \equiv -d \ln N(z)/d \ln L_m$, and L_m is the limiting survey luminosity. Hence, one can hope to measure the magnification by comparing a lensed background source distribution with that of an unlensed field. In the linear regime, this will provide us with an absolute estimate of the surface mass density. In the nonlinear regime, shear information must also be included to solve equation (1) for κ. However, there is now a four-fold degeneracy of solutions. We can overcome this by assuming continuity of the surface density field and integrating our solutions from the linear regime inwards.

BTP have shown that with redshift information the uncertainty in surface density is only shot–noise limited when estimated from distortions in the background magnitude distribution, but has an additional uncertainty due to intrinsic clustering when estimated from redshift distributions. These uncertainties are similar ($\simeq 20\%$) when the mass resolution is about $\theta_R \simeq 10'n^{-1/2}$, where n is the mean number density of background galaxies per square arcminute. In the absence of redshifts it is still possible to measure a magnification by looking at galaxy counts behind the lensing cluster (BTP). However while this increases the number of available sources we are again subject to an increased uncertainty due to intrinsic density perturbations as well as contamination by foreground galaxies.

We are currently undertaking three interelated applications. Broadhurst et al. (1995) have examined the number counts behind A1689 (z=0.18) and shown that a measurable magnification can be seen from the suppression of red band galaxies (Broadhurst 1995). We are also awaiting Hubble Space Telescope imaging of A1689 to measure both the magnification and shear fields. When this is available we expect to place strong constraints on the absolute mass and cluster profile in both the linear and nonlinear regimes. Finally, we have obtained narrow band imaging of A1689 and A2218. Using a succession of narrow band filters we can estimate galaxy redshifts and so reduce the effects of foreground contamination as well as providing valuable redshift information for estimating the magnification.

3. Conclusions

We have shown that lens magnification is a measurable effect and can be used to estimate the absolute cluster mass in both linear and nonlinear regimes. We are currently applying these methods to A1689.

References

Broadhurst, T.J., Taylor, A.N. & Peacock, J.A., 1995, ApJ, 438, 49 (BTP)
Broadhurst, T.J., 1995, in Dark Matter, eds. S. Holt, & C. Bennett, 320
Kaiser, N. & Squires, G., 1993, ApJ, 404, 441
Tyson, J.A., Valdes, F., & Wenk, R.A., 1990, ApJL, 349, L1

OPTIMIZED CLUSTER RECONSTRUCTION

STELLA SEITZ

Max-Planck-Institut für Astrophysik
Postfach 1523, D-85740 Garching, Germany

Abstract. We outline the noise-filtering finite-field inversion kernel and the regularized maximum likelihood methods for cluster reconstruction.

Unbiased finite-field mass reconstructions start from the relation $\nabla K = \mathbf{u}$, where $K = \ln(1 - \kappa)$, and the vector field $\mathbf{u}$ is a combination of the observable reduced shear $g = \gamma/(1 - \kappa)$ and derivatives of it. Estimates of K are thus obtained up to an arbitrary constant by averaging line integrals of observables. Noise degrades $\mathbf{u}$ to a combination of a gradient and a rotational field $\mathbf{u} = \nabla \tilde{K} + \mathbf{R}$, where only the rotational ('noise') component is treated differently by different line averaging methods. Requiring that the contribution from $\mathbf{R}$ is filtered out in the reconstruction is therefore a sensible requirement. For this, the above decomposition has to be specified uniquely by a boundary condition for $\mathbf{R}$: for ideal data, where $\mathbf{u}$ is indeed a gradient field, $\mathbf{R}$ has to vanish identically; this implies that in this case the scalar field from which $\mathbf{R}$ can be derived has to be constant everywhere, in particular on the boundary. This boundary condition is consistent with the assumption that in the general case there should not be a systematic rotational component in the observed region $\mathcal{U}$: $< \mathbf{R} >_{\mathcal{U}} = 0$. Since K and $\mathbf{u}$ are linearly related, we can make the ansatz that $K(\boldsymbol{\theta}) - \bar{K} = \int_{\mathcal{U}} \mathrm{d}^2\theta' \, \mathbf{H}(\boldsymbol{\theta}', \boldsymbol{\theta}) \cdot \mathbf{u}(\boldsymbol{\theta}')$, where the additive constant $\bar{K}$ is chosen to be the average of K over $\mathcal{U}$. We now replace $\mathbf{u}$ by its decomposition in gradient and rotational field and integrate by parts. Boundary terms vanish if we require that $\mathbf{H} \cdot \mathbf{n} = 0$ on the boundary $\partial \mathcal{U}$, and the rotational component is 'filtered out' if $\mathbf{H}$ is a gradient field, $\mathbf{H} = \nabla \mathcal{L}$. With these requirements, the ansatz holds iff $\Delta \mathcal{L}(\boldsymbol{\theta}', \boldsymbol{\theta}) = \nabla \cdot \mathbf{H}(\boldsymbol{\theta}', \boldsymbol{\theta}) = -\delta(\boldsymbol{\theta}' - \boldsymbol{\theta}) + \frac{1}{A}$, where A is the area enclosed by $\partial \mathcal{U}$. Together with $\mathbf{H} \cdot \mathbf{n} = 0$ this differential equation defines a Neumann boundary problem for $\mathcal{L}$, which implies an unique solution for $\mathbf{H}$. For a rectangular data field we have compared (see

C. S. Kochanek and J. N. Hewitt (eds), Astrophysical Applications of Gravitational Lensing, 151–152.

© 1996 IAU

Seitz & Schneider 1995 and references therein) reconstructions using the new kernel with earlier finite-field reconstructions of Schneider and Kaiser et al. or Bartelmann, hereafter named KSFWB/B, and the classical KS-reconstruction. The ranking of the unbiased finite-field reconstructions in terms of the power spectrum of the error of the reconstructed K-field is always the following: the noise filtering reconstruction is the best, followed by the Schneider and KSFWB/B reconstructions. A clear peak of the power spectrum at large wavelengths reflects the artifacts introduced by the KS-reconstruction.

I now outline the maximum likelihood reconstruction technique developed with Bartelmann, Schneider and Narayan: this method uses shape and magnification information (and possibly any other information) with the same weight simultaneously, it does not require the choice of a smoothing length and finally yields an objective measurement for the quality of the reconstructed mass profile. We define a model by choosing the deflection potential on a grid and calculate the mass density, shear and magnification, and finally the expectation value of the observables at every galaxy position and compare it to the 'measured' observables. The likelihood or, in the case of gaussian probability distributions (which we consider for simplicity), the χ^2 per galaxy, measures the quality of the model considered. One now can minimize the χ^2 to obtain a model which 'best' fits the data; this has a χ^2 per galaxy considerably below unity, which illustrates that the data are 'overfitted'; the corresponding surface mass density shows strong short scale fluctuations, to fit the noise as closely as possible. To avoid this, we add to the χ^2 a regularization term proportional to the sum of squares of derivatives of κ at every grid point and minimize the sum. These derivatives can be of first or higher order, or combinations of these. The factor of proportionality must be adjusted such, that at the minimum, the χ^2 for the shape and size and/or number-density distribution is equal to one. Any regularization contains a prejudice or prior knowledge about the mass distribution generating the observables. A regularization which is not compatible with the lens reveals itself immediately, since then the χ^2 for the shape and magnification distribution can not become equal to one at the same time, or the χ^2 locally becomes much larger then 1 (e.g. non-resolved substructure).

Acknowledgements: This work was supported by the "Sonderforschungs-bereich 375-95 für Astro–Teilchenphysik" der Deutschen Forschungsgemeinschaft.

References

Seitz, S. & Schneider, P. 1995, A&A, in press.

BELTRAMI EQUATION AND CLUSTER LENSING

Characteristic Equations & Applications

T. SCHRAMM
Technical University of Hamburg-Harburg
Computer Center – D 21071 Hamburg
and
Hamburg Observatory
Gojenbergsweg 112 – D 21029 Hamburg

Abstract. Arclets in clusters of galaxies can be used to determine the lens mapping and not only to constrain the mass density of the cluster. Multiply imaged arclets are therefore easily identified without further modeling.

1. The Beltrami equation

In Schramm & Kayser (1995) we introduced the complex **Beltrami Equation** as an appropriate framework for the analysis of arclets in cluster lensing. Corresponding real formalisms have been developed by Kaiser and Schneider & Seitz (this volume, compare also the references in Schramm & Kayser 1995). Here, we show how the solutions of the Beltrami *differential* equation can be used to identify multiply imaged arclets. The Beltrami Equation

$$\frac{\partial w}{\partial \bar{z}} = \mu \frac{\partial w}{\partial z} \tag{1}$$

states that a small ellipse in the deflector ($z = x + iy$) plane given by μ is mapped locally by w onto a circle in the source ($w = u + iv$) plane. The axial ratio ϵ of the ellipse and the direction angle φ are given by $\epsilon = (1 - |\mu|)/(1 + |\mu|)$ and $2\varphi = \pi + \arg(\mu)$, respectively.

2. Known source sizes

The **Jacobian** is also easily found

$$J = \left| \frac{\partial w}{\partial z} \right|^2 - \left| \frac{\partial w}{\partial \bar{z}} \right|^2 \tag{2}$$

153

C. S. Kochanek and J. N. Hewitt (eds), Astrophysical Applications of Gravitational Lensing, 153–154.
© 1996 IAU

Trivially, the mass density σ is uniquely determined if the Jacobian and the Beltrami parameter can be measured at the location of an arclet. Since $\partial w/\partial z = 1 - \sigma$ we can solve the Beltrami Equation and the Jacobian for the mass density:

$$(1 - \sigma)^2 = \frac{J}{1 - |\mu|^2} \tag{3}$$

3. Characteristic equations

Normally we are not so lucky to have the Jacobian but we assume we can measure the μ-field with some accuracy. However, the formulation as **differential equation** yields some insights. For the measurable Beltrami parameter we find for (one-plane) lens mappings

$$\mu = \mu_r + i\mu_i = \frac{\frac{\partial u}{\partial x} - \frac{\partial v}{\partial y} + 2i\frac{\partial v}{\partial x}}{\frac{\partial u}{\partial x} + \frac{\partial v}{\partial y}} \tag{4}$$

which results in two decoupled linear, homogeneous partial differential equations

$$\mu_i \frac{\partial u}{\partial x} - (1 + \mu_r)\frac{\partial u}{\partial y} = 0 \quad , (1 - \mu_r)\frac{\partial v}{\partial x} - \mu_i \frac{\partial v}{\partial y} = 0 \tag{5}$$

The **characteristics** of these equations are given by:

$$x'(t) = \mu_i, \quad y'(t) = -(1 + \mu_r) \Longrightarrow \frac{dy}{dx} = -\frac{1 + \mu_r}{\mu_i} \tag{6}$$

$$x'(t) = (1 - \mu_r), \quad y'(t) = -\mu_i \Longrightarrow \frac{dy}{dx} = -\frac{\mu_i}{1 - \mu_r} \tag{7}$$

where the solutions of these equations are the curves u, v=constant. The lens equation is therefore uniquely determined if the values are known on curves (not identical to characteristics). Even without this knowledge, characteristics are of interest: each two (possibly) multiply-**intersecting** characteristics of u and v map onto a cross-hair in the source plane so that **multiply imaged arclets** can be identified.

REFERENCES

Schramm, T. & Kayser, R. 1995, A&A, 299, 1

EFFECT OF SUB-STRUCTURE IN CLUSTERS ON THE LOCAL WEAK-SHEAR FIELD

PRIYAMVADA NATARAJAN
Institute of Astronomy, University of Cambridge, Cambridge U.K.

AND

JEAN-PAUL KNEIB
Institute of Astronomy, University of Cambridge, Cambridge U.K.

Abstract. Weak shear maps of the outer regions of clusters have been successfully used to map the distribution of mass at large radii. The effects of substructure in clusters on such reconstructions of the total mass have not been systematically studied. We propose a new method to study the effect of perturbers (bright cluster galaxies or sub-groups within the cluster) on the weak shear field. We present some analytic results below.

1. Analysis of the local weak shear field

Working in the coordinate frame of the perturber, we model the total cluster potential as the sum of a global smooth piece (assumed to be circular for illustration) and a perturbing piece which in general has both an elliptical part and a circular part.

$$\phi_{\text{tot}} = \phi_{\text{cluster}} + \phi_{\text{perturber}}$$

$$\phi_{\text{c}} = \phi_{\text{oc}} \left(|\vec{r} + \vec{r}_{\text{c}}| \right)$$

$$\phi_{\text{p}} = \phi_{\text{op}} r \left(1 + \frac{\epsilon}{6} \cos 2(\theta - \theta_0) \right)$$

where ϵ and θ_0 are respectively the intrinsic ellipticity and orientation of the perturber. We calculate the shear in the weak limit at a given point as the sum of contributions from these two sources. In order to obtain a quantity

155

C. S. Kochanek and J. N. Hewitt (eds), Astrophysical Applications of Gravitational Lensing, 155–156.
© 1996 IAU

that can be compared to observations we define an averaging procedure that involves integration over an annulus of finite radius which gives,

$$< g_x > = \int_0^{2\pi} \int_{r_1}^{r_2} g_x(r, \theta) \ r \ dr \ d\theta$$

$$< g_y > = \int_0^{2\pi} \int_{r_1}^{r_2} g_y(r, \theta) \ r \ dr \ d\theta$$

$$< g_x >_c = -\frac{\phi_{oc}}{2r} \ ; \ < g_y >_c = 0$$

$$< g_x >_p = -\frac{\phi_{op}}{12(r_1 + r_2)} \epsilon \cos 2\theta_0$$

$$< g_y >_p = -\frac{\phi_{op}}{12(r_1 + r_2)} \epsilon \sin 2\theta_0$$

The aim is to extract the parameters of the perturber (ϕ_{op} and ϵ) independently, so we convolve with an appropriate window function $\hat{W}(\theta)$ that maximizes the signal. An optimum choice is the following window,

$$\hat{W}_x(\theta) = \cos 2\theta \ ; \ \hat{W}_y(\theta) = \sin 2\theta$$

Some interesting features which are primarily due to the particular choice of averaging procedure are that for a circular perturber and an elliptical perturber oriented at $\frac{\pi}{4}$ with respect to the cluster center the contribution to the local weak shear field as defined above vanishes.

2. Conclusions

Applications of this formalism provide us with a probe of the structure of cluster galaxies. With high resolution, wide field data we can put limits on halo sizes and masses of cluster galaxies, which are relevant for understanding the details of the process by which clusters assemble. We shall also be able to quantify the errors in mass estimates from lensing due to the presence of substructure in clusters.

References

Kaiser, N., & Squires, G., 1993, ApJ, 404, 441
Blandford, R., & Narayan, R., 1987, ApJ, 321, 658
Kaiser, N., 1992, ApJ, 388, 272
Kneib, J., 1993, PhD. thesis
Kneib, J., Mellier, Y., Fort, B., & Mathez, G., 1993, A&A, 273, 367
Natarajan, P., Kneib, J.P., & Rees, M., 1995, in preparation

A MULTI-WAVELENGTH ANALYSIS OF THE MATTER DISTRIBUTION IN CLUSTERS OF GALAXIES

HAIDA LIANG

CEA/DSM/DAPNIA Service d'Astrophysique
F-91191 Gif sur Yvette, Cedex, France

Abstract. We combine the X-ray data and the Sunyaev-Zel'dovich effect with the constraints from gravitational lensing for an intermediate redshift cluster to check the consistency of various mass estimates and the validity of the assumption of hydrostatic equilibrium, and to deduce the properties of the intra-cluster gas, the profile of the cluster potential and H_o.

1. Introduction

Recent developments in the field of gravitational lensing have made detailed and independent analyses of the mass distribution of clusters of galaxies possible. The traditional tracers of the cluster potential are the cluster galaxies and the intracluster gas. By assuming that the cluster is in virial equilibrium, the observed velocity dispersion of the member galaxies can be used to deduce the cluster mass. The intracluster gas is observed either through their thermal X-ray emission or the Sunyaev-Zel'dovich effect (SZ effect; Syunyeav and Zel'dovich 1981). The cluster potential can then be deduced from the X-ray surface brightness and temperature and/or the Sunyaev-Zel'dovich effect by assuming that the gas is in hydrostatic equilibrium. Unlike these two methods, gravitational lensing effects are independent of the dynamical state of the cluster since the distortions of the background galaxies only depends on the gravitational field of the cluster. It has been shown that the position of a giant arc constrains very well the mass interior to the radius of the arc (Kochanek 1992), and at the same time the weak distortion of the background galaxies provides a good estimate of the shape of the cluster potential in the outer parts of the cluster (Kaiser and Squires 1993). In this paper we will combine the various in-

C. S. Kochanek and J. N. Hewitt (eds), Astrophysical Applications of Gravitational Lensing, 157–162.
© 1996 IAU

dependent methods of cluster mass analysis for a medium redshift cluster. We will check the validity of the assumption of hydrostatic equilibrium and then combine the lensing constraints with the X-ray data and the SZ effect to constrain the cluster potential and deduce H_0.

2. A Multi-wavelength Analysis of A2218

A2218 is an Abell richness class 4 cluster at a redshift of z=0.175 with a velocity dispersion of 1370^{+160}_{-120} kms^{-1} (Le Borgne et al. 1992). There has been a large amount of data collected for the cluster in the optical, X-ray and radio wavelengths by various groups. It has a X-ray luminosity of $L_x = 1.01 \times 10^{45}$ ergs/s in the 2-10 keV energy range and a temperature of 8×10^7K measured by the *Ginga* satellite[1] (McHardy et al. 1990, David et al. 1993). There are gravitational lensing arcs found in this cluster, both around the central cD and the bright galaxy of a smaller clump (Pello et al. 1992). Furthermore, A2218 is one of the few clusters where the Sunyaev-Zel'dovich decrement has been successfully measured (Jones et al. 1993; Figure 3). Given the wealth of data available on this cluster, we can obtain a better understanding of the dynamical state and/or properties of the intra-cluster medium, as well as the matter distribution and H_0 than most other clusters available.

Both the X-ray surface brightness distribution and the SZ effect can be linked to the cluster gravitational potential via the equation of hydrostatic equilibrium. If the intracluster gas is in hydrostatic equilibrium then the gas density is given by $\rho_g/\rho_0 = \exp(-\frac{\mu m_p}{kT_g}[\phi(r) - \phi_0])$, where we have also assumed that the gas is isothermal for simplicity. Thus the X-ray surface brightness is given by

$$S_x(R) \propto \int_R^{r_t} \exp(-\frac{2\mu m_p}{kT_g}[\phi(r) - \phi_0])\Lambda(T_g)\frac{rdr}{\sqrt{r^2 - R^2}} \qquad (1)$$

and the SZ effect is given by

$$\Delta T_r(R) \propto \int_R^{r_t} \exp(-\frac{\mu m_p}{kT_g}[\phi(r) - \phi_0])T_g\frac{rdr}{\sqrt{r^2 - R^2}} \qquad (2)$$

where ϕ is the cluster potential, T_g is the gas temperature, $\Lambda(T_g)$ is the temperature dependent emissivity. The cluster potential ϕ is constrained by the amount of shear produced on the background galaxies by the cluster. In the case where there is a giant arc at the center of the cluster, the projected mass within the arc is well determined and thus the central parts

---

[1]Preliminary results from ASCA also show that the gas temperature is ~ 6.8keV and that the gas clearly extends to 4 Mpc (M. Bautz 1994)

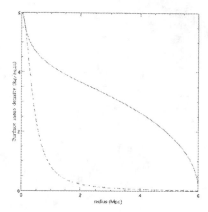

Figure 1.   The cluster surface mass density in units of kg/m^2. The solid curve gives the surface mass density corresponding to the potential that fits all the X-ray and SZ effect data. The dashed curve gives the surface density profile corresponding to $[1 + (r/r_c)^2]^{-1}$ but normalized at the arc radius to fit the constraint imposed by the arc. This functional form was found to fit the galaxy number density distribution (Dressler 1976).

of the potential is constrained through Poisson's equation. In a spherically symmetric cluster, the position of the giant arc corresponds to the locus of the critical circle. The mean surface density inside the critical circle of radius b is given by the critical surface density (Blandford & Kochanek 1988), i.e.

$$\Sigma(b) = \Sigma_{crit} = \frac{c^2}{4\pi G}\frac{D_s}{D_l D_{ls}} \tag{3}$$

where D_s, D_l and D_{ls} are the distance to the source, to the lens and the distance between the lens and the source. In the case of A2218, one of the large arcs (object number 359 in Pello et al. 1992; see also Kneib et al. this proceeding) has a spectroscopic redshift of $z = 0.702$. As pointed out in Miralda-Escudé & Babul (1995; MB here after), there is another object in the field (object number 328 in Le Borgne *et al.* 1992) which is very likely another image of the arc (number 359). The critical line should thus intersect the arc which puts the critical radius at $20.8''$ (or 0.08 Mpc).

Our aim here is to see if we can find a potential profile that would fit the strong lensing constraint, X-ray data and the SZ effect. In order to compare with the observed data we need to simulate both the X-ray and radio observations. In the case of the X-ray data we need to convolve the ROSAT PSF with the X-ray surface brightness deduced from equation (1) (see Figure 2).   Similarly, we multiply the SZ radio surface brightness deduced from equation (2) by the primary beam of the Ryle telescope and Fourier transform the resulting profile to obtain the SZ flux density expected at various baseline lengths of the interferometer (see Figure 3). The

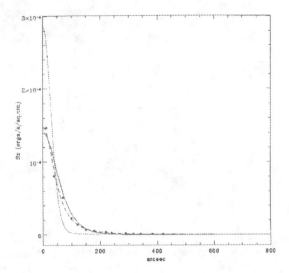

Figure 2. X-ray surface brightness profile. The filled circle is the observed surface brightness from ROSAT PSPC (MB). The solid curve shows the simulated X-ray surface brightness for a cluster potential given in Figure 1 with $T_g = 8 \times 10^7$ K. The dotted curve corresponds to the best fit for the cluster density profile of Model 1 given in MB for $T_g = 8 \times 10^7$ K and the dashed curve corresponds to the best fit for Model 1 in MB with $T_g = 2 \times 10^8$ K.

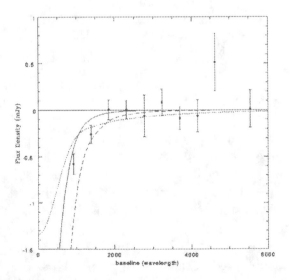

Figure 3. The SZ effect in flux densities versus baseline length. The filled circles are the Ryle data at 15.4 GHz (Jones et al. 1993). The solid curve gives the SZ effect expected for a cluster potential given in Figure 1 (solid curve) with $T_g = 8 \times 10^7$ K. The dotted curve is deduced from the cluster parameters that gave the best fit to the X-ray surface brightness for Model 1 of MB at $T_g = 8 \times 10^7$ K and the dash curve is deduced from the cluster parameters that gave the best fit to the X-ray surface brightness for Model 1 of MB at $T_g = 2 \times 10^8$ K.

position of the giant arc only constrains the innermost part (0.08 Mpc) of the cluster potential. Given the available data, the cluster potential outside 0.08 Mpc is free to vary and we should be able to vary the outer parts of the cluster potential until it fits the X-ray surface brightness profile while keeping the gas temperature fixed to the observed value. Figure 1 shows an example of such a projected density profile. We assumed that the cluster extends to 6 Mpc. The X-ray surface brightness and the SZ effect corresponding to such a potential with $T_g = 8 \times 10^7$ K and central electron density $n_e(0) = 7.3 \times 10^{-3}$ cm^{-3} are shown in Figure 2 and 3 respectively. Both the X-ray surface brightness and the SZ effect thus deduced give reasonable fits to the observed data for $H_0 = 50$ km s^{-1} Mpc^{-1}. We conclude that, given the data available so far, there is no real conflict between the gravitational lensing constraints from the position of the giant arc and that of the observed data, under the assumption of hydrostatic equilibrium. Ideally, the outer parts of the cluster potential can be determined from a weak lensing analysis, but such analysis requires deep optical data taken in sub-arcsec seeing conditions and was not available until very recently (see Kneib et al. and Kaiser et al. in this conference). It should be noted that the potential shown here is just one of the possible solutions that fits all the data and has is yet to be checked against the amount of shear observed by Kneib et al. and Kaiser et al. (in this proceeding).

3. Discussion

In their recent analysis of A2218, MB pointed out that the position of the giant arc in A2218 is in conflict with the X-ray surface brightness distribution measured by the PSPC and the X-ray temperature measured by the *Ginga* satellite (see dotted curve in Figure 2). MB assumed that the clusters are spherically symmetric and in hydrostatic equilibrium. They examined two families of parametrized density profiles. The density profiles were constrained by the position of the arc through equation (3). They found that the central mass implied by lensing is a factor of 2–2.5 too large for the gas at the observed temperature to be in hydrostatic equilibrium solely due to thermal pressure support given the family of cluster density profiles. They found that if the gas temperature is raised to 2.5 times larger than the observed temperature, then they could find within their family of density profile one that yields a good fit to the PSPC X-ray surface brightness profile (dashed curve in Figure 2). However, such a model does not fit very well the data for SZ effect (see dashed curve in Figure 3).

The conclusion of MB differs from ours because MB have restricted themselves to certain functional types for the cluster total mass distribution. On the other hand, we should also point out that the validity of the potential profile found here in Figure 1 is subject to further analysis. We have not so far examined how the velocity dispersion data can constrain the possible density profiles. However, such constraints of the density profiles by the observed velocity dispersion are limited because of the uncertainties in the galaxy orbits. A more useful constraint comes from the weak lensing analysis which are now available from the works by Kneib et al. and Kaiser et al. (in this proceeding). For example, we can predict the amount of shear that can be produced from the above density distribution and compare that with the results from the weak lensing analysis.

4. Conclusions

We found a cluster potential consistent with the "strong" lensing constraint that fits the available data on the SZ effect, the X-ray surface brightness and temperature with $H_0 = 50$ km s^{-1} Mpc^{-1} under the assumption of hydrostatic equilibrium. We have enough information for this cluster to simultaneously determine the shape of the cluster potential and the value of H_0. With the addition of a temperature profile from ASCA, and a weak lensing analysis of the cluster potential out to ~ 3 Mpc, we can deduce directly the cluster matter distribution, dynamical state and the properties of the ICM. In particular, we can obtain a quantitative measure of the difference in dark and luminous matter density as a function of radius and directly test the consistency of the potential deduced from lensing with that of the X-ray surface brightness and temperature under the assumption of hydrostatic equilibrium. The SZ effect then adds extra information on the value of H_0 and the clumpiness factor.

References

Bautz, M., 1995, in New Horizons of X-ray Astronomy, eds. F. Makino & T. Ohashi, (Universal Academy Press)

Blandford, R. D. & Kochanek, C. S., 1988, in Dark Matter in the Universe, eds. J. Bachall, T. Piran & S. Weinberg (Singapore: World Scientific) 133

David, L.P., Slyz, A., Jones, C., Forman, W., Vrtilek, S.D. & Arnaud, K. A., 1993, Ap. J., 412, 479

Dressler, A., 1976, Ph.D. Thesis, Lick Observatory, University of California, Santa Cruz

Jones, M., et al., 1993, Nature, 365, 320

Kaiser, K. & Squires, G., 1993, Ap. J., 404, 441

Kochanek, C. S., 1992, in Gravitational Lenses, eds. R. Kayser, T. Schramm, L. Nieser (Berlin: Springer) 278

Le Borgne, J. F., Pelló, R. & Sanahuja, B., 1992, A&A Suppl., 95, 87

McHardy, I. M., Stewart, G. C., Edge, A. C., Cooke, B., Yamashita, K. & Hatsukade, I., 1990, MNRAS, 242, 2150

Miralda-Escudé, J. & Babul, A., 1995, ApJ, 449, 18

Pelló, R., Le Borgne, J. F., Sanahuja, B., Mathez, G. & Fort, B., 1992, A&A, 266, 6

Syunyaev, R.A. & Zel'dovich, Ya.B., 1981, Astrop & Space Phys Rev, 1, Soviet Scientific Reviews, Section E, 1

THE VELOCITY DISPERSION AND DISPERSION
PROFILE OF ABELL 963

R.J. LAVERY

Iowa State University

Dept. of Physics and Astronomy, Ames IA 50011

AND

J.P. HENRY

Institute for Astronomy, University of Hawaii

2680 Woodlawn Drive, Honolulu, HI 96822

1. Introduction

Abell 963 ($z = 0.206$) is still the best candidate for a true arc-counterarc lens configuration (Lavery & Henry 1988). A simple model explains the positions, lengths and patchy light distribution of the two arcs. Photometry indicates the arcs also have the same B-R color (Lavery & Henry 1988; Ellis et al. 1991). However, spectroscopic observations have not confirmed that these arcs originate from the same background galaxy. Ellis et al. (1991) detected a single emission line in the spectrum of the smaller northern arc, identifying it as [O II] $\lambda 3727$ at a redshift of 0.77. Neither Lavery (1989) or Ellis et al. (1991) detected this emission line in the larger southern arc.

Our overall goal of this program is to compare the various methods of determining the mass distribution profile of A963 in order to determine if the two large arcs are truly from a single background galaxy. Here, we present our preliminary results on determining the velocity dispersion and dispersion profile of this lensing cluster.

2. OBSERVATIONS

These spectroscopic data were obtained over 3 nights in 1993 February on the UH 88-inch telescope of the MKO. Aperture plates were used to obtain spectra of 8 to 13 spectra simultaneously. Spectra and redshifts were

C. S. Kochanek and J. N. Hewitt (eds), Astrophysical Applications of Gravitational Lensing, 163–164.
© 1996 IAU

obtained for a total of 50 objects, as well as several objects with spectra of too low quality for this program. Of these 50, 9 of the galaxies are non-members, 1 is a strong emission line member, and 1 is a star. The remaining 39 objects are red cluster members used in the subsequent analysis.

3. ANALYSIS

We used our highest signal-to-noise spectrum, BOW 2, as our template. Its redshift was determined using six strong absorption features. The wavelength scale of this spectrum was then shifted to the rest-frame.

Recessional velocities for all the red cluster members were determined using the Fourier cross-correlation velocity program in IRAF. These velocities were then averaged to determine the recessional velocity of the cluster. Excluding 2 galaxies, a double system and one galaxy more than 3σ from the mean, we determine $cz = 61450 \pm 260$ and $z = 0.205$ (37 galaxies). Relativistic corrections were applied to each galaxy before calculating the velocity dispersion. We find a mean corrected cluster velocity of 55260 ± 220 km/s and a radial velocity dispersion of $\sigma_r = 1350$ (+210,-150) km/s.

Dividing the present data into three bins with average radii of 60", 130" and 190", we find velocity dispersions of 1380 km/s (12 galaxies), 1320 km/s (14 galaxies) and 1150 km/s (11 galaxies), respectively. While the differences are not significant, there is at least some indication of a decrease with larger radius. We have spectra of additional cluster members from several other observing runs and will incorporate these data once systematic differences between observing sessions are taken into account. These would increase the sample of red cluster members by 25%.

4. DISCUSSION

The double arc system of A963 poses some basic questions to our understanding of cluster lensing. Higher quality spectra of the arcs are needed to confirm their redshifts. More redshifts are certainly needed, but other observational information is available. Recent ROSAT X-ray data have shown the core radius to be ~3 times smaller than previous determinations, with $r_c = 24$" and fairly elliptical contours. This X-ray gas may produce the needed perturbation of the potential from circular symmetry (Kovner 1989).

References

Ellis, R., Allington-Smith, J., & Smail, I., 1991, MNRAS, 249, 184

Kovner, I., 1989, ApJ, 337, 621

Lavery, R.J., 1989, in Gravitational Lenses, ed. J.M. Moran, J.N Hewitt, & K.Y. Lo (New York: Springer-Verlag), 134

Lavery, R.J. & Henry, J.P., 1988, ApJL, 329, L21

THE DISTRIBUTION OF DARK MASS IN GALAXIES

Techniques, Puzzles, and Implications for Lensing

PENNY D. SACKETT

Kapteyn Astronomical Institute
9700 AV Groningen, The Netherlands

Abstract. Gravitational lensing is one of a number of methods used to probe the distribution of dark mass in the Universe. On galactic scales, complementary techniques include the use of stellar kinematics, the kinematics and morphology of the neutral gas layer, kinematics of satellites, and the morphology and temperature profile of X-ray halos. These methods are compared, with emphasis on their relative strengths and weaknesses in constraining the distribution and extent of dark matter in the Milky Way and other galaxies. It is concluded that (1) the extent of dark halos remains ill-constrained, (2) halos need not be isothermal, and (3) the dark mass is probably quite flattened.

1. Introduction

Modeling the gravitational structure of a galaxy, and therefore its lensing properties, requires knowledge of the extent, radial profile, and the geometric form of its mass distribution. The interpretation of microlensing rates and optical depths along different lines of sight through the Milky Way, for example, is strongly dependent on the assumed distribution of total (light and dark) Galactic mass. On larger scales, efficient and reliable image-inversion techniques designed to measure the structure parameters of intervening lensing galaxies require appropriate fitting functions for the lensing mass.

This review focuses on techniques that form a symbiotic relationship with lensing in producing valuable and complementary constraints on galactic potentials, especially in providing partial answers to the following questions about galactic *dark* mass:

- What is the physical extent of dark matter in galaxies?
- Is the distribution of dark mass isothermal?

C. S. Kochanek and J. N. Hewitt (eds), Astrophysical Applications of Gravitational Lensing, 165–174.
© 1996 IAU

– What is the shape of dark "halos"?

2. How Big are Dark Halos?

The size of dark halos controls the galactic "sphere of influence" for lensing, interactions, and accretion. Together with the radial and vertical structure parameters, halo extent determines the total mass of the galaxy. The notion of halos as distinct entities ceases to be useful, of course, on scales larger than half the mean distance to the nearest, comparably-sized neighbor.

If halos are extremely large and isothermal, they cannot be totally baryonic without violating the constraints on $\Omega_B h^2$ from primordial big bang nucleosynthesis (BBN) models. Recent assessments give $0.01 \leq \Omega_B \leq 0.06$ for $0.5 \leq h_{100} \leq 1$ (Walker et al. 1991, Smith, Kawano & Malaney 1993). Since the density of observed baryons is $\Omega_{\mathrm{Lum}} \lesssim 0.007$ (Pagel 1990), the average ratio of dark-to-luminous baryons is $0.4 \leq M_{B,\mathrm{Dark}}/M_{B,\mathrm{Lum}} \leq 8$, and at least some of the Universe's baryons are dark. Rotation curve analysis indicates that $1 \lesssim M_{B,\mathrm{Dark}}/M_{B,\mathrm{Lum}} \lesssim 10$ (cf. Broeils 1992), so that on scales comparable to HI disks ($\sim$30 kpc), halos composed entirely of dark baryons are consistent with BBN. Faint galaxies are more numerous and more dark matter dominated than brighter galaxies, but contribute less to the total luminosity of the Universe. Thus the upper limit placed by BBN on the size of baryonic halos is likely to be considerably larger than 30 kpc (Binney & Tremaine 1987), but its calculation requires a model-dependent integral over the galaxy luminosity function, weighted by $M_{B,\mathrm{Dark}}/M_{B,\mathrm{Lum}}$.

2.1. THE EXTENT AND MASS OF THE MILKY WAY HALO

At large radius, the mass of our galaxy can be estimated from the kinematics of distant, presumably bound, objects — halos stars, satellite galaxies, and group members — and from the kinematics of the Magellanic Clouds + Stream system. The former has been done most recently by Kochanek (1995), who finds that, using a Jaffe model as the global mass distribution for the Galaxy, the total mass inside 50 kpc at 90% confidence is $(5.4 \pm 1.3) \times 10^{11} M_\odot$ if the timing constraints of the Local Group are imposed and Leo I is bound, and somewhat lower at $4.3^{+1.8}_{-1.0} \times 10^{11} M_\odot$ if the timing constraints are not imposed. The corresponding masses within 100 kpc are $7^{+4}_{-3} \times 10^{11} M_\odot$ and $(8 \pm 2) \times 10^{11} M_\odot$, respectively. A recent re-examination of the kinematics and proper motions of the Magellanic Clouds and Stream using two different model potentials for the Milky Way (Lin, Jones & Kremola 1995) yields $(5.5 \pm 1) \times 10^{11} M_\odot$ inside 100 kpc. About one-half this mass must lie *outside* the present Cloud distance (50 kpc) in order to explain the observed infall of the Magellanic Stream.

Since the luminous matter in the Galaxy accounts for $(0.6-1) \times 10^{11} M_\odot$, the full range of dark mass estimates from these two methods is $1.3 \times 10^{11} \lesssim M_{Dark}(< 50 kpc) \lesssim 6.1 \times 10^{11} M_\odot$, with apparent contradictions at the lower end with Local Group timing and at the upper end with Magellanic Stream kinematics. The implied upper limit of $M_{Dark}(< 50\,kpc) \sim 2.7 \times 10^{11} M_\odot$ from the Magellanic Stream model is only just consistent with the lower limit of $\sim 2.3 \times 10^{11} M_\odot$ from the satellite model. For comparison, the spherical isothermal dark halo used by many microlensing teams as a fiducial model contains $4.1 \times 10^{11} M_\odot$ interior to the LMC distance of 50 kpc (Griest 1991). Using this model and its first year of LMC data, the MACHO team concludes with 68% confidence that the total mass in compact dark lenses is $7^{+6}_{-4} \times 10^{10} M_\odot$ (Alcock et al. 1995). (For complete reviews of the mass of the Galaxy and its dependence on assumptions about Leo I, see Fich & Tremaine (1991), Schechter (1993), and Freeman (these proceedings).)

2.2. HALO SIZE OF EXTERNAL GALAXIES

The kinematics of satellites can also be used to study the halos of external galaxies, but since only a small number of satellites are observed per primary, conclusions are based on a statistical analysis of the sample as a whole. Based on satellite velocities and HI rotation curves, Erickson, Gottesman and Hunter (1987) concluded that the primaries in their sample have $M_{Dark}/M_{Lum} < 5$, total $M/L \sim 20$, and potentials that are well-described by a point mass model — all consistent with dark halos that extend no more than 3 disk radii. In a more recent study using a different sample, however, Zaritsky and White (1994) conclude that halos are nearly isothermal, with total $M(< 200kpc) = 1.5 - 2.6 \times 10^{12} M_\odot$ and $110 < M/L < 340$ (for $h_{100} = 3/4$). Their result is primarily due to secondaries at 200-300 kpc, where the orbital times are on the order of a Hubble time, thus necessitating the use of halo formation models to interpret the satellite kinematics. Using their method, Zaritsky and White conclude that the Erickson et al. (1987) sample, which has smaller mean primary-satellite separation, is consistent with both small and large mass halos.

In the future, weak lensing is likely to play a larger role in constraining the extent of dark halos. Recent work by Brainerd, Blandford and Smail (these proceedings) has given the first indication that the tangential distortion of background galaxies due to weak lensing by foreground galaxies is statistically measurable given a large sample (~ 3000) of source-lens pairs. Their measurement of $1.0^{+1.1}_{-0.7} \times 10^{12} h^{-1} M_\odot$ for the total mass within 100 h^{-1} kpc is consistent both with a mass distribution that grows linearly to 100 kpc and one that truncates much sooner with total $M/L \approx 10$.

The ring of HI gas in the M96 group in Leo (Schneider 1985) offers

the rare opportunity to sample galactic potentials at very large radii using
the well-defined orbits of cold gas. The Leo ring has a radius of 100 kpc
and completely encircles the early-type galaxies M105 and NGC 3384. The
radial velocities and spatial distribution of the gas are consistent with a
single, elliptical *Keplerian* orbit with a center-of-mass velocity equal to the
centroid of the galaxy pair, and a focus that can be placed at the barycenter
of the system without compromising the fit. The implied dynamical mass
within 100 kpc is $5.6 \times 10^{11} M_\odot$, (only twice that inferred from the internal
dynamics of the galaxies), giving a total $M/L \approx 25$. The sensitivity of non-
circular orbits to the power law form of the potential suggests that dark
matter does not extend much beyond the ring pericenter radius of 60 kpc.
As a caveat, it is yet clear to what degree M96, a spiral located 60 kpc (in
projection) outside the ring, may perturb the ring kinematics.

3.   Are Dark Halos Isothermal?

The approximate flatness of HI rotation curves is the best observational
evidence that dark matter is present in spirals and has a shallower density
profile than the light; an isothermal halo is as shallow as r^{-2}. Early theoretical studies (Gott 1975) suggested that violent relaxation would cause the
inner regions of galaxies to have steeply falling profiles that would flatten
to $r^{-2.25}$ in the outer parts. More recent CDM models (Navarro, Frenk &
White 1995) indicate that dark halo profiles may be shallower than r^{-2}
in the center and quite steep near the virial radius. Compression by a dis-
sipating gaseous disk may further contract and flatten the dark matter
(Blumenthal et al. 1986), accounting in part for the apparent "conspiracy"
between the dark and luminous mass that produces flat rotation curves.

3.1. MILKY WAY

The radial structure of the mass and light in the Milky Way is less well-
constrained than in external galaxies. Determining the rotation curve of the
Galaxy, in particular, has proven notoriously difficult. On the other hand,
distances and kinematics of old, resolved stars can be used to measure
the vertical restoring force of the local disk — and thus its surface mass
density. In this way, Kuijken and Gilmore (1991) report a mass column of
$71 \pm 6 M_\odot pc^{-2}$ within a 1.1 kpc band from the Galactic plane, with $48 \pm 9 M_\odot pc^{-2}$ due to the disk itself, and the rest contributed by a rounder halo.
Other recent estimates are similar: Gould (1990) weighs in at $54 \pm 8 M_\odot pc^{-2}$,
Bahcall, Flynn and Gould (1992) at $54 \pm 8 M_\odot pc^{-2}$, and Flynn and Fuchs
(1994) at $52 \pm 13 M_\odot pc^{-2}$. The dynamical disk mass thus seems to be in
remarkable agreement with the detectable disk mass of $49 \pm 9 M_\odot pc^{-2}$ — at
least locally, almost none of the disk mass is dark. Since only about one-half

of the local rotation support is provided by the observable disk, this further implies that dark matter in the Galaxy is dynamically important at radii as small as 2.5 disk scale lengths. Stated in the language of §3.2, the Milky Way disk is one-half of its "maximal disk" value. Unfortunately, uncertainties in the outer Galactic rotation curve frustrate attempts to determine the distribution of mass in the outer Galaxy, which is further complicated by a recent suggestion that the generally-accepted local rotation speed, $\Theta_0 = 220$ km s^{-1}, may be overestimated by $\sim$10% (Merrifield 1992). A smaller value would increase the relative dynamical importance of the luminous disk and decrease the slope of the outer rotation curve, to which Θ_0 is tied.

Conclusions drawn from microlensing results about the dark baryonic content of the Milky Way depend on the assumed distribution of dark *and* luminous matter in the Galaxy (cf. Paczyński, these proceedings). Many studies have explored how different assumptions for M/L, rotation curve slope, and the shape, truncation radius and radial profile of the halo affect these conclusions (cf. references in Griest et al. 1995). As an indication of the importance of *luminous* structure, lensing by stellar bars in the Milky Way and the LMC has been held accountable, respectively, for most of the optical depth toward the Galactic center (Zhao, Spergel & Rich 1995) and the LMC (Sahu 1994). On the other hand, if the Galactic disk were "maximal," the MACHO results toward the LMC would be consistent with a dark halo entirely composed of lensing baryons (Alcock et al. 1995).

3.2. RADIAL DISTRIBUTION OF DARK MASS IN EXTERNAL GALAXIES

In contrast to the difficulties in the Milky Way, surface brightness profiles and rotation curves for external galaxies can be measured well, but their disk mass-to-light ratios, M/L, are uncertain. A disk M/L that is constant with radius (but varies from galaxy to galaxy) can explain the kinematics within the optical radius of many spirals (cf. Kalnajs 1983, Kent 1986, Buchhorn 1992), but the high velocities observed at the edges of HI disks can be reproduced only by invoking a rapid radial increase in M/L (cf. Kent 1987, Begeman 1987). Since the age and metallicity gradients inferred from the blueing radial color gradients in spirals do not produce these strong, *positive* gradients in M/L (cf. de Jong 1995), dark matter is implicated.

In order to estimate conservatively the amount of dark matter in a galaxy, the "maximum disk hypothesis" is often adopted (van Albada & Sancisi 1987), which fixes the disk M/L at the value that maximizes the disk mass without violating kinematic constraints. The hypothesis is controversial (cf. Rubin 1987, Casertano & van Albada 1990, Freeman 1993), but when it is used to fit rotation curves, the resulting disk M/L are larger for brighter and earlier type spirals than for fainter and later type spirals

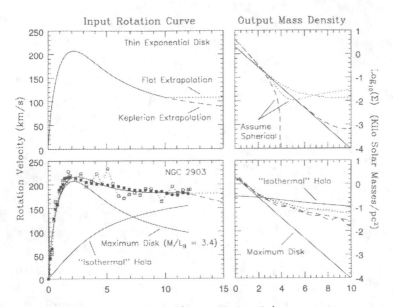

Figure 1. Inversion of rotation curves (left) to derive dynamical surface mass densities, Σ (right). Two extrapolations for $v(r)$ are shown: flat (dotted) and Keplerian (dashed). *Top: Thin, exponential disk.* Both extrapolations overestimate the true Σ (solid) because for an exponential disk $v(r)$ declines faster than Keplerian. Only for inner half of the disk can Σ be determined reliably. A spherically-symmetric mass estimator is unreliable for an exponential disk and can produce negative Σ in the inversion. *Bottom: Sbc NGC 2903.* Real data (solid squares) and artificially noisy data (open squares) are inverted using both extrapolation schemes. M/L increases markedly beyond ~ 2 scale lengths. Adding noise makes little difference. A spherical isothermal halo has a Σ that is $\sim \pi/2$ larger than that of the flat extrapolation, but with similar slope (Sackett, in preparation).

(Broeils 1992, Buchhorn 1992). The correlation appears to be stronger in bluer bands. These trends may be due to the older stellar populations associated with early spirals, a notion supported by comparison with the M/L derived from stellar population synthesis models (Athanassoula, Bosma & Papaioannou 1987) and the observed stellar dispersions in spirals (van der Kruit & Freeman 1986). Alternatively, they may reflect trends in dark matter properties with galaxy type and luminosity that are incorrectly characterized by the application of maximum disk models (van der Kruit 1995).

It should be stressed that *rotation curves do not constrain the dark matter distribution to have a r^{-2} (isothermal) volume density profile.* If (1) rotation curves were perfectly flat, (2) halos were spherical, and (3) the luminous mass were negligible, then indeed dark matter halos could be described by singular isothermal spheres over the radial range of the kinematics. In fact, rotation curves are seldom flat, but instead have slopes that are systematically related to the peak speed or the luminosity concentration of the galaxy (Kent 1987, Athanassoula, Bosma & Papaioannou 1987, Casertano & van Gorkom 1991, Broeils 1992): diffuse, slow rotators have rising rotation curves, while compact, fast rotators have falling curves.

Evidence is mounting that dark halos are not spherical (§4). Finally, since the stellar mass is not strongly constrained, dark halos with asymptotic r^{-3} and r^{-4} density profiles are also consistent with observed rotation curves (Lake & Feinswog 1989). Even when r^{-2} halos are used to fit rotation curves — together with luminous disks of reasonable M/L — a large core radius must be assumed, so that the halo does not achieve its asymptotic (isothermal) speed at the last measured point (Fig. 1). This is especially true of maximum disk fits (cf. Broeils 1992), but often applies to fits that assume smaller disk masses as well (cf. Kent 1987). This suggests that the linewidth of the HI gas used in the Tully-Fisher relation is probably not governed by the asymptotic speed of an isothermal dark halo.

Rotation curve inversion is a step toward a model-independent method for determining the radial distribution of dark mass in galaxies. The technique has been criticized as being sensitive to noise (Binney & Tremaine 1987), but for this application the typical uncertainties of 10-20% are quite tolerable. The method does depend on the assumed geometry of the mass and extrapolation of the rotation curve beyond the last measured point (Fig. 1), but has the advantage of making this dependence explicit rather than camouflaging it by the use of a particular model for the dark mass.

4. What is the Shape of the Dark Mass: Disks or Halos?

Use of term "halo" to describe the distribution of dark matter may be prejudicial: there is no strong theoretical or observational evidence to indicate that dark matter in galaxies is distributed spherically. Dark halos have been favored over dark disks as a means to stabilize galaxies against bar formation (Ostriker & Peebles 1973), but bulges (Kalnajs 1987) and hot disks (Athanassoula & Sellwood 1987) are now believed to be more efficient stabilizers. Traditional rotation curve analysis is insensitive to vertical structure, but accumulating observational evidence from other methods suggests that the dark mass may be considerably flattened toward the stellar plane, while remaining relatively axisymmetric, with an in-plane axis ratio of $(b/a)_\rho > 0.7$ (see review by Rix 1995).

Here, we focus on $(c/a)_\rho$, the vertical flattening of dark matter, since it is likely to have the stronger implications for both microlensing in the Galaxy (cf. Gould, Miralda-Escudé, & Bahcall 1994), and the use of macrolensing as a probe of galaxy structure. The flattening of the dark mass may also provide a clue as to the nature of its constituents. N-body simulations of dissipationless collapse produce strongly triaxial dark halos (Frenk et al. 1988, Dubinski & Carlberg 1991, Warren et al. 1992), but adding a small fraction ($\sim$10%) of dissipative gas results in halos of a more consistent shape — nearly oblate, $(b/a)_\rho \gtrsim 0.8$, but moderately flattened, $(c/a)_\rho \gtrsim 0.6$

(Katz & Gunn 1991, Dubinski 1994). Thus strongly flattened, halos, with $(c/a)_\rho < 0.5$, may imply that dissipation has played an even greater role, perhaps implicating baryonic dark matter.

In order to measure $(c/a)_\rho$ of the dark mass, a probe of the vertical gradient of the potential is required. In the Milky Way, the measured anisotropy of the velocity dispersion of extreme Population II halo stars has been used to estimate the flattening of the mass distribution (Binney, May & Ostriker 1987, van der Marel 1991). Unfortunately, the results depend on the unknown orbital structure of the stellar halo, so that $(c/a)_\rho$ can be confined only to lie between 0.3 and 1 at the solar neighborhood.

In external galaxies, Buote and Canizares (1994, 1995) have used the flattening of extended X-ray isophotes, assuming that the gas is in hydrostatic equilibrium, to place constraints on the flattening of the dark matter in two early-type systems. For the elliptical NGC 720, they find that the dark isodensity contours have axis ratio $0.3 \lesssim (c/a)_\rho \lesssim 0.5$ at 90% confidence; for the lenticular NGC 1332, $0.2 \lesssim (c/a)_\rho \lesssim 0.7$. This suggests that these dark halos are at least as flattened as their corresponding luminous galaxies, which have optical isophotes of axis ratio $q \approx 0.6$.

These values contrast with that of $(c/a)_\rho \geq 0.84$ derived for the S0 NGC 4753 by Steiman-Cameron, Kormendy & Durisen (1992) on the basis of fitting an inclined, precessing disk model to the complicated pattern of the galaxy's dust lanes. Their remarkably good fit is independent of $(c/a)_\rho$; the flattening constraints are based on the assumption that the gas is smoothly distributed and has completed at least 6 orbits at all radii.

Stable rings around galaxies are not observed to have random orientations, but are found preferentially close to the equatorial or polar planes, suggesting that the potential may be oblate. In particular, polar ring galaxies (PRGs) are surrounded by rings of gas and stars in orbits nearly perpendicular to the central stellar plane; these rings can extend to 20 disk scale lengths. Since in an oblate potential closed ring orbits are elongated along the polar axis and have speeds that vary with ring azimuth, the shape and kinematics of a polar ring are excellent extended probes of $(c/a)_\rho$. Early kinematic analyses of three PRGs produced axes ratios for the *potentials* of $0.86 < (c/a)_\Phi < 1.05$ with uncertainties of 0.2 (Schweizer, Whitmore & Rubin 1983, Whitmore, McElroy & Schweizer 1987), corresponding to $0.58 \lesssim (c/a)_\rho \lesssim 1.15$ with very large uncertainties. Subsequent studies using more detailed mass models and higher quality data over a larger radial range have narrowed the range for the dark mass to $0.3 \lesssim (c/a)_\rho \lesssim 0.6$ (Arnaboldi et al. 1993, Sackett et al. 1994, Sackett & Pogge 1995); in each galaxy, $(c/a)_\rho$ is similar to the inferred flattening of the central *stellar* body.

Measurements of $(c/a)_\rho$ for spiral galaxies are rarer, more difficult, and sorely needed. Assuming that gas disks evolve gravitationally toward a discrete bending mode in tilted rigid halos, Hofner and Sparke (1994) find that moderate halo flattening of $0.6 \lesssim (c/a)_\rho \lesssim 0.9$ can reproduce the observed HI warps of five spirals. In principle, $(c/a)_\rho$ can also be constrained by the flaring of the HI layer; in the most detailed study of this type, Olling and van Gorkom (1995) obtain $0.2 < (c/a)_\rho < 0.8$ for the dark halo of the Sc NGC 4244. Since non-gravitational energy sources may be responsible for a substantial fraction of the vertical support of gas (Malhotra 1995, and references therein), this measurement may be an upper limit to $(c/a)_\rho$.

5. Parting Caveats and a Puzzle

Since the mass distribution in cluster galaxies may be modified by the interactions and violent relaxation that shape the evolving cluster potential, we have restricted this review to relatively isolated galaxies that are more likely to be dynamically relaxed. Furthermore, we have largely ignored ellipticals, the inner few kpc of which are thought to be responsible for the strong lensing of distant QSOs and radio sources. Although selection effects operate to favor flattened lenses in multiply-imaged systems (Kassiola & Kovner 1993), image inversion techniques yield lenses that are surprisingly flattened (Kochanek 1995a, and references therein) — the *projected* $(c/a)_\Phi \lesssim 0.8$ corresponds to $(c/a)_\rho \lesssim 0.4$. Can these flat lenses be reconciled with the axis ratio distribution of ellipticals, which peaks at $q = 0.7$ (Ryden 1992), or are disk galaxies implicated?

Acknowledgements: P.D.S. gratefully acknowledges travel support from the Leids Kerkhoven-Bosscha Fonds and the IAU.

References

Alcock et al. 1995, ApJ, submitted (astro-ph/9506113)
Athanassoula, E., Bosma, A. & Papaioannou, S. 1987, A&A, 179, 23
Athanassoula, E. & Sellwood, J. A. 1987, *IAU Sym. 117*, eds. J. Kormendy & G. R. Knapp (Dordrecht: Reidel), 300
Bahcall, J.N., Flynn, C. & Gould, A. 1992, ApJ, 389, 234
Begeman, K. 1987, *Ph. D. thesis,* University of Groningen The Netherlands
Binney, J. & Tremaine, S. 1987, *Galactic Dynamics* (Princeton:Princeton Univ. Press)
Binney, J, May, A. & Ostriker, J.P. 1987, MNRAS, 226, 149
Blumenthal, G. R., Faber, S. M., Flores, R. & Primack, J. 1986, ApJ, 301, 27
Broeils, A. 1992, *Ph. D. thesis,* University of Groningen, The Netherlands
Buchhorn, M. 1992, *Ph. D. thesis,* Australian National University
Buote, D.A. & Canizares, C.R. 1994, ApJ, 427, 86
Buote, D.A. & Canizares, C.R. 1995, to appear in ApJ, astro-ph/9508005
Casertano, S. & van Albada, T.S. 1990, *Baryonic Dark Matter,* (Dordrecht:Kluwer) 159
Casertano, S. & van Gorkom J.H. 1991, AJ, 101, 1231
de Jong, R.S. 1995, *Ph. D. thesis,* University of Groningen, The Netherlands
Dubinski, J., 1994, ApJ, 431, 617
Dubinski, J. & Carlberg, R., 1991, ApJ, 378, 496
Erickson L.K., Gottesman, S.T. & Hunter, J.H., 1987, Nature, 325, 779

Fich, M. & Tremaine, S., 1991, ARA&A, 29, 409

Flynn, C. & Fuchs, B., 1994, MNRAS, 270, 471

Freeman, K.C., 1993, in Physics of Nearby Galaxies: Nature or Nurture?, eds. R.X. Thuan, C. Balkowski & J. Thanh Van, (Editions Frontières) 201

Frenk, C.S., White, S.D.M, Davis, M. & Efstathiou, G., 1988, ApJ, 327, 507

Gould, A., 1990, MNRAS, 244, 25

Gould, A., Miralda-Escudé, J., & Bahcall, J.N., 1994, ApJL, 423, L105

Griest, K., 1991, ApJ, 366, 412

Griest, K., et al., 1995, in Pascos/Hopkins Symposium, in press (astro-ph/9506016)

Gott, J. R., 1975, ApJ, 201, 296

Hofner, P. & Sparke, L.S., 1994, ApJ, 428, 466

Kalnajs, A.J. 1983, in IAU Sym. 100, ed. E. Athanassoula (Dordrecht: Reidel) 87

Kalnajs, A.J., 1987, in IAU Sym. 117, eds. J. Kormendy & G. Knapp (Dordrecht: Reidel) 289

Katz, N. & Gunn, J. E., 1991, ApJ, 377, 365

Kent, S.M., 1986, AJ, 91, 1301

Kent, S.M., 1987, AJ, 93, 816

Kochanek, C.S., 1995a, ApJ, 445, 559

Kochanek, C.S., 1995b, ApJ, in press

Kassiola, A. & Kovner, I., 1993, ApJ, 417, 450

Kuijken, K. & Gilmore, G., 1991, ApJ, 367, 9

Lake, G. & Feinswog, L., 1989, AJ, 98, 166

Lin, D.N.C., Jones, B.F. & Klemola, A.R., 1995, ApJ, 439, 652

Malhotra, S., 1995, ApJ, 448, 138

Merrifield, M.R., 1992, AJ, 103, 1552

Navarro, J.F., Frenk, C.S. & White, S.D.M., 1995, ApJ, submitted (astro-ph/9508025)

Olling, R.P. & van Gorkom, J.H., 1995, in Dark Matter, eds. S. Holt & C. Bennett (New York: AIP) 121

Ostriker, J.P. & Peebles, P.J.E., 1973, ApJ, 186, 467

Pagel, B., 1990, Baryonic Dark Matter, (Dordrecht: Kluwer) 237

Rix, H.-W., 1995, in IAU Sym. 169, in press (astro-ph/9501068)

Rubin, V.C., 1987, in IAU Sym. 117, eds. J. Kormendy & G. Knapp (Dordrecht: Reidel) 51

Ryden, B.S., 1992, ApJ, 396, 445

Sackett, P.D, Rix, H.-W., Jarvis, B.J., & Freeman, K.C., 1994, ApJ, 436, 629

Sackett, P.D. & Pogge, R.W., 1995, in Dark Matter, eds. S. Holt & C. Bennett (New York: AIP) 141

Sahu, K.C., 1994, Nature, 370, 275

Schechter, P., 1993, in Back to the Galaxy, eds. S. Holt & F. Verter (New York: AIP) 571

Schneider, S.E., 1985, ApJ, 288, L33

Schweizer, F., Whitmore, B.C. & Rubin, V.C., 1983, AJ, 88, 909

Smith, M.S., Kawano, L.H. & Malaney, R.A., 1993, ApJS, 85, 219

Steiman-Cameron, T.Y., Kormendy, J. & Durisen, R.H., 1992, AJ, 104, 1339

van Albada, T. & Sancisi, R., 1987 in IAU Sym. 117, eds. J. Kormendy & G. Knapp (Dordrecht: Reidel) 67

van der Kruit, P. C. & Freeman, K. C., 1986, ApJ, 303, 556

van der Kruit, P.C., 1995, in IAU Sym. 164, eds. P.C. van der Kruit & G. Gilmore (Dordrecht: Kluwer) 205

van der Marel, R.P., 1991, MNRAS, 248, 515

Walker, T., Steigman, G., Kang, H. Schramm, D., & Olive, K., 1991, ApJ, 376, 51.

Warren, M. S., Quinn, P. J., Salmon, J. K., & Zurek, W. H., 1992, ApJ, 399, 405

Whitmore, B.C., McElroy, D., & Schweizer, F., 1987, ApJ, 314, 439

Zaritsky, D. & White, S. D. M., 1994, ApJ, 435, 599

Zhao, H.-S., Spergel, D. N. & Rich, R. M., 1995, ApJ, 420, 806

THE MASSIVE DARK CORONA OF OUR GALAXY

K.C. FREEMAN

Mount Stromlo and Siding Spring Observatories
The Australian National University
Canberra, AUSTRALIA

Abstract. From their rotation curves, most spiral galaxies appear to have massive dark coronas. The inferred masses of these dark coronas are typically 5 to 10 times the mass of the underlying stellar component. I will review the evidence that our Galaxy also has a dark corona. Our position in the galactic disk makes it difficult to measure the galactic rotation curve beyond about 20 kpc from the galactic center. However it does allow several other indicators of the total galactic mass out to very large distances. It seems clear that the Galaxy does indeed have a massive dark corona. The data indicate that the enclosed mass within radius R increases like $M(R) \approx R(\mathrm{kpc}) \times 10^{10}\ M_\odot$, out to a radius of more than 100 kpc. The total galactic mass is at least $12 \times 10^{11}\ M_\odot$.

1. Summary

A full version of this paper will appear elsewhere (Freeman 1995). Here I present the main conclusions from the individual indicators of the galactic mass distribution.

- The mass of the known luminous components of the Galaxy is in the range (5 to 12) $\times 10^{10}\ M_\odot$.
- From the rotation curve of the Galaxy, $M(20\ \mathrm{kpc}) \approx 22 \times 10^{10}\ M_\odot$, which is already at least double the estimated mass of the visible components.
- The escape velocity at the solar radius, estimated from high velocity stars in the solar neighborhood, indicates that the galactic mass $> 30 \times 10^{10}\ M_\odot$.

C. S. Kochanek and J. N. Hewitt (eds), Astrophysical Applications of Gravitational Lensing, 175–176.
© 1996 IAU

- From the kinematics of distant stars and satellites, $M(50 \text{ kpc}) \approx 40 \times 10^{10} \ M_{\odot}$.
- The timing arguments from the radial velocities and distances of M31 and Leo I give a consistent asymptotic mass estimate M_{total} of at least $120 \times 10^{10} \ M_{\odot}$.

2. Conclusion

The data are consistent with a mass distribution $M(R) \approx R(\text{kpc}) \times 10^{10}$ $M_{\odot}$ (corresponding to a flat rotation curve with $V_c \approx 220 \text{ km s}^{-1}$), extending out to $R \geq 100$ kpc. The inferred ratio of the mass of the dark corona to the mass of the visible components of the Galaxy is at least 10.

Recently Kochanek (1995) used a Jaffe (1983) model to represent the mass distribution of the Galaxy, and estimated the parameters for this model from the mass indicators taken together. He concludes that the mass within 50 kpc is $(54 \pm 13) \times 10^{10} \ M_{\odot}$, which agrees well with the run of $M(R)$ given here from the mass indicators taken individually.

References

Freeman, K.C., 1995, in "Unsolved Problems of the Milky Way" (IAU Symposium 169), ed L. Blitz (Dordrecht: Kluwer), in press
Jaffe, W., 1983, MNRAS, 202, 995
Kochanek, C.S., 1995, ApJ, in press

GRAVITATIONAL LENSES
AND THE STRUCTURE OF GALAXIES

CHRISTOPHER S. KOCHANEK
Harvard-Smithsonian Center for Astrophysics
60 Garden Street, Cambridge, MA 02138, USA

Abstract. Nearly singular isothermal mass distributions with small core radii are consistent with stellar dynamics, lens statistics, and lens models as a model for E/S0 galaxies. Models like the de Vaucouleurs model with a constant mass-to-light ratio are not. While the isothermal distributions are probably an oversimplification, E/S0 galaxies (at least in projection) must have significant amounts of dark matter on scales of an effective radius.

1. Inferences From Dynamics

The mass distribution in E/S0 galaxies remains unclear despite many years of effort using stellar dynamics and other tracers. It is probable from X-ray studies and the rare polar ring galaxies (see Sackett in this volume) that the outer regions are dominated by dark matter, while the inner regions are consistent with either constant mass-to-light ratio (M/L) models or dark matter models. In addition to any intrinsic interest in the matter distribution of E/S0 galaxies, their structure is of crucial importance in using gravitational lens statistics to determine the cosmological model.

The state of the art, constant mass-to-light ratio, dynamical model for E/S0 galaxies that is fit to observational data is the two-integral axisymmetric model. In a survey of some forty galaxies, van der Marel (1991) derived a mass-to-light ratio of $(M/L)_B = (10 \pm 2)h$ for an L_* galaxy using these dynamical models. The more traditional lensing model is the singular isothermal sphere (SIS), and models of E/S0 galaxies with this mass distribution find that the velocity dispersion of the dark matter is $\sigma_{DM*} \simeq 225 \pm 20$ km s^{-1} (Kochanek 1994, Breimer & Sanders 1993, Franx 1993), which is approximately equal to the central velocity dispersion of

C. S. Kochanek and J. N. Hewitt (eds), Astrophysical Applications of Gravitational Lensing, 177–182.
© 1996 IAU

the stars, σ_c. Earlier models with $\sigma_{DM*} = (3/2)^{1/2}\sigma_c$ (e.g. Turner et al. 1984) are based on oversimplified dynamical models (see Kochanek in this volume). Normal dynamical techniques have difficulty determining which of these two extreme models actually applies to E/S0 galaxies.

2. Inferences From Lens Statistics

Maoz & Rix (1993) and Kochanek (1993, 1995b), made detailed statistical models of the observed lens samples for various mass distributions (also see Rix, Kochanek, and Claeskens et al. in this volume). These studies examined a range of models from de Vaucouleurs models to softened isothermal spheres, emphasizing the limits on the cosmological constant. The cosmological conclusions are independent of the mass distribution used for the lens galaxies.

The normalization of the galaxy masses is determined by the distribution of image separations found in the surveys. Moaz & Rix (1993) first pointed out that de Vaucouleurs models normalized by the mass-to-light ratios estimated from dynamical models of nearby E/S0 galaxies (e.g. van der Marel 1991) produced image separations that were too small to fit the lens data. Kochanek (1995b) demonstrated that a mass-to-light ratio of $(M/L)_B \simeq (25 \pm 5)h$ at 90% confidence is required to fit the separations, compared to $(10 \pm 2)h$ in the dynamical models. For the softened isothermal sphere (Kochanek 1993, 1995b), fitting the separations requires $\sigma_{DM*} = 220 \pm 20$ km s^{-1}, consistent with the dynamical models.[1] The isothermal lens must be nearly singular to avoid the appearance of central images in the lenses (e.g., Wallington & Narayan 1993, Kassiola & Kovner 1993).

At least in theory, large amounts of evolution (Mao & Kochanek 1993, Rix et al. 1994), extreme errors in the selection function, or (for optical lenses) extinction (Tomita 1995, Kochanek 1995b) can invalidate the statistical inferences. In practice, however, the are sufficient constraints in the current data to rule out any dramatic errors. Unfortunately, the statistical models of the lens surveys cannot as yet differentiate between mass distributions except by comparing the normalization required by the lens data to the normalization required by stellar dynamics. Hopefully, models of individual lens systems can both validate the normalization and differentiate between radial mass distributions.

---

[1]Maoz & Rix (1993) added a softened isothermal halo to the de Vaucouleurs models, but the resulting deflection produced by the lens so closely resembles that of a softened isothermal sphere that there is no point in treating them as a separate class of models.

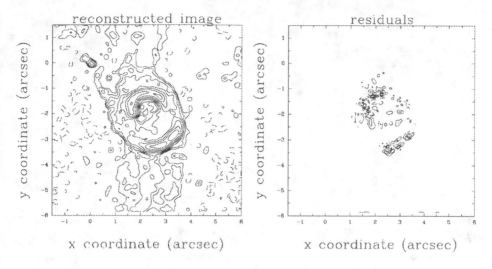

Figure 1. Panel (a) shows the reconstructed image for the best fit expanded ellipsoid model, and Panel (b) shows the residuals. The largest positive and negative residuals are 139 μJy/pixel and -191 μJy/pixel. The dashed contours are drawn at -70 and -35 μJy/pixel and the solid contours are drawn at 35, 70, 140, 280, 560, 1120, 2240, and 4480 μJy/pixel. The estimated noise in the map is 35 μJy/pixel, so the contours lie at ±1, ±2, 4, 8, 16, 32, 64, and 128 times the estimated noise in the map.

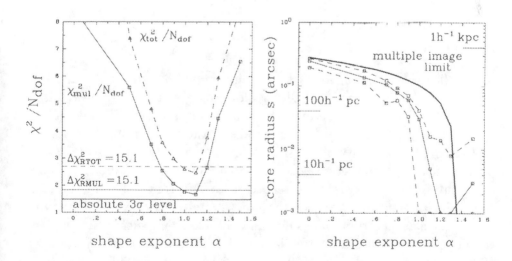

Figure 2. Panel (a) shows the minimum values of χ^2_{mul}/N_{dof} (solid/points) and χ^2_{tot}/N_{dof} (dashed/points) as a function of the exponent α. The core radius of each model has been optimized. The bottom horizontal line shows the formal 3σ deviation of χ^2/N_{dof} from unity. The other two horizontal lines show where $\Delta\chi^2_r = 15.1$ (formally a 99.99% change). Panel (b) shows the optimized value for the core radius (solid/squares) and the limits (dashed/squares) on the core radius for $\Delta\chi^2_r = 15.1$ in the rescaled χ^2_r estimator. The heavy solid line shows the upper limit on the core radius if the multiply imaged region is larger than 1.5 arcseconds.

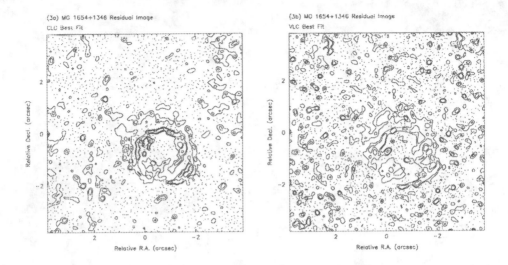

Figure 3. A comparison of the residuals in fitting MG 1654+134 using the Clean map
(a) or the raw visibilities/"dirty map" (b). The contour levels are drawn at ±1, ±2,
4, 8, 16, 32, 64 and 96 percent of the peak in the Clean map. The estimated noise is
59 μJy/Beam. In (a) the minimum, maximum, and rms errors are -411 μJy/Beam, 315
μJy/Beam, and 71 μJy/Beam, while in (b) they are -211 μJy/Beam, 247 μJy/Beam, and
61 μJy/Beam.

3. Inferences from Lens Models

Not all lenses are useful for distinguishing between radial mass distributions. Point lenses without VLBI transformation matrices are particularly
ill-suited to this problem because the fit is either under-constrained (2 image systems) or dominated by the details of the angular structure (4 image systems). The utility of VLBI transformation matrices in avoiding this
problem has yet to be seriously investigated. Some of the radio rings, however, have extended emission spread over a wide range of radii from the
centers of the lenses. These are the most promising systems for studying
the radial mass distribution. Unfortunately, modeling such systems given
the finite instrumental resolution is complicated (see Kochanek & Narayan
1992, Wallington et al. 1994, 1995, and Ellithorpe et al. 1995).

MG 1654+134 was found in the MG survey (Langston et al. 1989, 1990).
One lobe of a double lobed, $z = 1.74$ radio quasar is lensed into a 2 arcsec
diameter ring around a r=18.7 mag, $z = 0.254$ galaxy. The emission in the
ring is very extended, making it an ideal candidate for estimating the radial
mass distribution of the lens. Kochanek (1995a) treated two general classes
of models for the radial profile: the de Vaucouleurs model and softened
power-law density distributions of the form $\Sigma \propto (r^2 + s^2)^{\alpha/2-1}$, where the

isothermal distribution has $\alpha = 1$.

The first question we examine is the normalization of the models or the mass inside the ring. We find that the mass inside radius $r = 0.9$ arcsec from the lens galaxy is $M = (7.75 \pm 0.03)h^{-1}10^{10}M_\odot$ at 90% confidence for $\Omega = 1$. The systematic uncertainty from the cosmological model for the range $0 < \Omega < 1$ is 7%. This corresponds to a blue mass-to-light ratio inside the ring of $(M/L)_B = (20.4 \pm 2.8)(f_e/1.4)h$ where the uncertainty is entirely due to the uncertainties in the enclosed light (also see Burke et al. 1992). Equivalently, the velocity dispersion of an isothermal model must be $\sigma_{DM} = (223 \pm 11)(f_e/1.4)^{-0.28}$ km s^{-1}. The factor f_e corrects for the expected fading of a passively evolving elliptical between the lens redshift and the current epoch. Both of these measurements match the results found in the statistical studies.

Figure 1 shows the best fit model with $\alpha = 1$ for MG1654+134 and its residuals. To the eye, the reconstructed image is indistinguishable from the original images (despite the logarithmically spaced contours), but there is a pattern of residuals surrounding the ring at the level of a few standard deviations above the noise. Figure 2 shows the χ^2 of the fit as a function of α. The allowed models have $\alpha \simeq 1.0 \pm 0.1$, corresponding to an isothermal distribution. Figure 2 also shows the best fit core radius s and its error bars as a function of the exponent α. The core radius must be very compact for isothermal lenses, with $s \lesssim 0.02$ arcsec or approximately $50h^{-1}$ pc, but it must have a large, finite value for the more centrally concentrated models. The best fit de Vaucouleurs model has a χ^2 slightly worse than the models in the permitted range for α. Inside the ring, this model closely matches the deflection profile of the isothermal model, but it drops too rapidly outside the ring. This is the same problem that makes the de Vaucouleurs models incapable of fitting the distribution of lens separations in the statistical studies.

We have repeated the calculations for the nearly isothermal lenses using a variant that fits the raw measured visibilities rather than the processed Clean map (the VLC algorithm, Ellithorpe et al. 1995) and a variant using the maximum entropy method (the LensMEM algorithm, Wallington et al. 1994, 1995) with the same results. While the parameters of the best fit models for each algorithm are mutually consistent, the residuals in the VLC inversions are significantly lower than in the inversions starting from the processed Clean map (see Figure 3). This means that the Clean process can introduce significant and detectable artifacts into maps of lensed images, and accurate models must start from the visibility data. Further experiments with self-calibration showed evidence that the self-calibration step also introduces detectable artifacts (Ellithorpe et al. 1995). A map of a lens produced with a good, if approximate, lens model is generally a better approximation to the true image than a simple Clean map. Even an approximate lens model begins to enforce the constraints required of a real lensed image.

4. Conclusions

As far as we can tell, the only model that is currently consistent with stellar dynamics, lens statistics, and lens models is a mass distribution similar to the nearly singular isothermal sphere. This is, undoubtedly, an oversimplification of the true mass distribution. All is not perfect, however. For example, we have no good model for the ring lens MG1131+0456 (Chen et al. 1995), and there is some evidence that the ellipticities required to fit the lenses are higher than is reasonable for the observed ellipticities of the luminous material. Some of this is due to the oversimplifications of the elliptical structures of the models, but it deserves further attention. A very interesting study that has yet to be done is to obtain data on the velocity dispersion profiles in the ring lenses and directly compare the inferences from stellar dynamics and lens models.

Acknowledgements: This research was supported by NSF grant AST94-01722 and the Alfred P. Sloan Foundation.

References

Breimer, T.G., & Sanders, R.H., 1993, A&A, 274, 96
Burke, B.F., Lehàr, J. & Conner, S.R., 1992, in Gravitational Lenses, eds. R. Kayser, T. Schramm & L. Nieser (Berlin: Springer) 237
Chen, G.H., Kochanek, C.S., & Hewitt, J.N., 1995, ApJ, 447, 62
Ellithorpe, J.D., Kochanek, C.S., & Hewitt, J.N., ApJ, submitted
Franx, M., 1993, in Galactic Bulges, ed. H. Dejonghe et al. (Dordrecht: Kluwer) 243
Kassiola, A., & Kovner, I., 1993, ApJ, 417, 450
Kochanek, C.S., & Narayan, R., 1992, ApJ, 401, 461
Kochanek, C.S., 1993, ApJ, 419, 12
Kochanek, C.S., 1994, ApJ, 436, 56
Kochanek, C.S., 1995a, ApJ, 445, 559
Kochanek, C.S., 1995b, ApJ, submitted
Langston, G.I., et al., 1989, AJ, 97, 1283
Langston, G.I., Conner, S.R., Lehàr, J., et al., 1990, Nature, 344, 43
Mao, S., & Kochanek, C.S., 1993, MNRAS, 268, 569
Maoz, D., & Rix, H.-W., 1993, ApJ, 416, 425
Rix, H.-W., Maoz, D., Turner, E.L., & Fukugita, M., 1994, 435, 49
Tomita, K., 1995, YITP/U94-2 preprint
Turner, E.L., Ostriker, J.P., & Gott, J.R., 1984, ApJ, 284, 1
van der Marel, R.P., 1991, MNRAS, 253, 710
Wallington, S., & Narayan, R., 1993, ApJ, 403, 517
Wallington, S., Narayan, R. & Kochanek, C.S., 1994, ApJ, 426, 60
Wallington, S., Kochanek, C.S., & Narayan, R., 1995, ApJ, submitted

WEAK LENSING BY INDIVIDUAL GALAXIES

TEREASA G. BRAINERD
Department of Astronomy, Boston University
725 Commonwealth Ave., Boston, MA 02215

ROGER D. BLANDFORD
California Institute of Technology
Theoretical Astrophysics 130-33, Pasadena, CA 91125

AND

IAN SMAIL
Observatories of the Carnegie Institution of Washington
813 Santa Barbara St., Pasadena, CA 91101

1. Introduction

In this paper we report on an investigation of statistical weak gravitational lensing of cosmologically distant faint galaxies by foreground galaxies. The signal we seek is a distortion of the images of faint galaxies resulting in a weakly preferred tangential alignment of faint galaxies around brighter galaxies. That is, if the faint galaxies have been gravitationally lensed by the brighter systems, the major axes of their images will tend to lie perpendicular to the radius vectors joining the centroids of the faint and bright galaxies (Fig. 1). Modeling a lens galaxy as a singular isothermal sphere with circular velocity V_c, an ellipticity of $\sim 2\pi V_c^2/c^2\theta$ is induced in the image of a source galaxy at an angular separation θ from the lens. This is of order a few percent for faint–bright galaxy pairs with separations $\theta \sim 30''$ where the lens is a typical bright spiral. Over 1000 pairs must be measured in order to detect such a signal in the presence of the noise associated with the intrinsic galaxy shapes. Given a sufficiently large number of pairs, it may be possible to use the variation of the induced ellipticity with θ to study the angular extent of the halos of the lens galaxies.

Tyson et al. (1984) investigated such galaxy–galaxy lensing using scans of photographic plates (cf. also Webster (1983)) from which they obtained

183

C. S. Kochanek and J. N. Hewitt (eds), Astrophysical Applications of Gravitational Lensing, 183–188.
© 1996 IAU

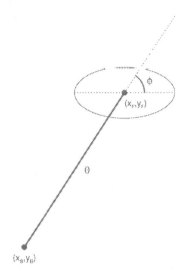

Figure 1.   Orientation of faint galaxies relative to bright galaxies.

images of $\sim 47,000$ faint galaxies ($22.5 < J < 23.5$) and $\sim 12,000$ bright galaxies ($19 < J < 21.5$). For faint–bright galaxy separations greater than $\sim 3''$, no statistically significant deviation from an isotropic distribution of faint image orientations was found.

2. Observational Data

The imaging data used for our analysis is expected to be of sufficient quality, depth, and size to allow a detection of galaxy–galaxy lensing. The data are of a single $9.6' \times 9.6'$ blank field centered on $\alpha(1950) = 17^h21^m07^s$ $\delta(1950) = +49°52'21''$, taken in Gunn r, and were acquired during periods of good seeing ($0.7''$–$0.9''$) using the COSMIC imaging spectrograph (Dressler et al. 1995) on the 5-m Hale telescope. The reduction of the data to a catalogue of detected objects is detailed in Mould et al. (1994).

The final stacked frame consists of a total of 19 individual frames with a cumulative exposure time of 24.0 ksec. The final frame has a 1σ surface brightness limit of $\mu_r = 28.8$ mag arcsec^{-2}, seeing of $0.87''$ FWHM, and total area of 90.1 arcmin^{-2}. Due to the presence of classical distortion in the corners of the frame, all analysis is restricted to those objects which lie within a circle of radius $4.8'$, centered on the chip. There are 4819 galaxies in this area brighter than the $\sim 97\%$ completeness limit of $r = 26.0$.

The probability distribution of the image ellipticities in the sample (which, to linear order, is equivalent to the distribution of the intrinsic source galaxy ellipticities) is adequately fit by $P_\epsilon(\epsilon) = 64\epsilon \exp[-8\epsilon]$ with mean ellipticity $\langle \epsilon \rangle = 0.25 \pm 0.02$.

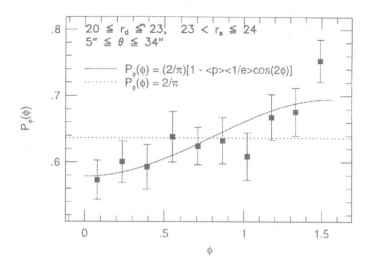

Figure 2.   Probability distribution of faint image orientation relative to bright galaxies.

2.1. POSITION ANGLE PROBABILITY DISTRIBUTION

We have investigated the probability distribution, $P_\phi(\phi)$, of the orientations of the resolved images of faint galaxies ($23 < r_s \leq 24$; 511 objects) relative to brighter galaxies ($20 < r_d \leq 23$; 439 objects) as in Fig. 1. Since we are primarily interested in distinguishing between radial and tangential alignments, positive and negative position angles of the faint galaxies were combined so that ϕ is restricted to the range $[0, \pi/2]$. In Fig. 2, $P_\phi(\phi)$ evaluated using annuli with $5'' \leq \theta \leq 34''$, centered on the bright galaxies, is shown. Error bars were obtained by bootstrap resampling. The inner annulus radius avoids overlapping faint-bright image isophotes and the outer radius should roughly maximize the signal to noise for our data set. With this annulus we obtain 3202 faint–bright pairs and, thus, each faint galaxy is paired with a number of near neighbor bright galaxies. In the case of galaxy–galaxy lensing $\sim 2/3$ of the sources will have encountered 2 or more significant deflectors and, therefore, to optimize the ability to detect the lensing signal it is necessary to average the faint image orientation over all possible deflectors.

Under the assumptions of random intrinsic faint galaxy orientations and no gravitational lensing by the brighter galaxies, $P_\phi(\phi)$ should be consistent with a uniform distribution ($P_\phi(\phi) = 2/\pi$). A χ^2 test performed on the binned distribution $P_\phi(\phi)$ in Fig. 2 rejects a uniform distribution at a confidence level of 97.9%, while a Kolmogorov-Smirnov test performed on the continuous, cumulative distribution of $P_\phi(\phi)$ rejects a uniform distribution at a confidence level of 99.9%. In the case of gravitational lensing $P_\phi(\phi)$

should exhibit a $\cos 2\phi$ variation of the form $P_\phi(\phi) = \frac{2}{\pi}[1 - \langle p \rangle \cos 2\phi \langle \epsilon^{-1} \rangle]$, where $\langle p \rangle$ is the mean image polarization. From the image ellipticity distribution $\langle c^{-1} \rangle = 8.0$. Shown in Fig. 2 is the best-fit $\cos 2\phi$ variation of this form, from which we infer $\langle p \rangle = 0.011 \pm 0.006$ (95% confidence bounds).

A number of null tests were performed to investigate possible systematic effects in the data which would give rise to the observed non-uniform $P_\phi(\phi)$. The tests include (i) ϕ as the orientation of the bright galaxies relative to the faint galaxies, (ii) ϕ as the orientation of the faint galaxies relative to random points, (iii) ϕ as the orientation of the faint galaxies relative to stars, and (iv) a random ϕ was substituted for the true faint image ϕ. In all cases $P_\phi(\phi)$ is consistent with a uniform distribution.

The image polarization of the faint galaxies is robust to splitting of the data into subsamples. Considering (i) positive values of ϕ independently of negative values of ϕ, (ii) objects within $3.4'$ of the center of the chip vs. objects farther than $3.4'$ from the center of the chip, and (iii) objects within each of the north, south, east, and west $1/2$ circles of radius $4.8'$, $P_\phi(\phi)$ is inconsistent with a uniform distribution and $\langle p \rangle$ obtained for the subsamples is consistent with that obtained using the full sample.

From determinations of $P_\phi(\phi)$ for the faint galaxies using independent bins, θ, the variation of $\langle p \rangle$ with lens–source separation was computed. Results are shown in Fig. 3, where the error bars are the formal 1σ error from the least squares fit of the $\cos 2\phi$ variation to $P_\phi(\phi)$.

3. Implications of Image Polarization for Lens Halos

From the variation of $\langle p \rangle$ with θ (Fig. 3), measured properties of local galaxies, and modest extrapolations of the observed redshift distribution of faint galaxies, formal best–fit parameters for the dark halos of the lens galaxies can be derived. Modeling the mass distribution of the halos as

$$\rho(r) = \frac{V_c^2 s^2}{4\pi G r^2 (r^2 + s^2)}, \tag{1}$$

where V_c is the de-projected circular velocity for $r \ll s$ and s is an outer scale radius beyond which $\rho(r) \propto r^{-4}$, we find the image polarization is

$$p(X) = \frac{2\pi V_c^2 D_d D_{ds}}{s D_s c^2} \frac{(2 + X)(1 + X^2)^{1/2} - (2 + X^2)}{X^2 (1 + X^2)^{1/2}}, \tag{2}$$

where D_d, D_s, D_{ds} are angular diameter distances and X is the ratio of the projected lens–source separation and s.

We introduce two characteristic scaling parameters, V^* and s^*, by assuming that the circular velocity scales as the fourth root of the total luminosity in a given band (in agreement with the Tully-Fisher relation) and

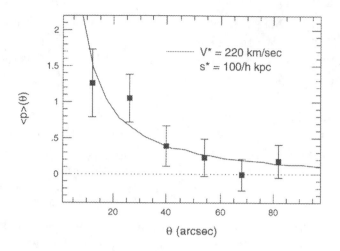

Figure 3. Observed variation of faint image polarization with differential lens–source (points with error bars) and model prediction (solid line).

that the total mass-to-light ratio of a galaxy is a constant independent of its luminosity, obtaining

$$\frac{V_c}{V^*} = \left(\frac{L_\nu}{L_\nu^*}\right)_r^{1/4} \qquad \frac{s}{s^*} = \left(\frac{L_\nu}{L_\nu^*}\right)_r^{1/2} = \left(\frac{M}{M^*}\right)^{1/2}, \qquad (3)$$

where L^* and M^* are the characteristic luminosity and mass, respectively. Allowing for a spectral (or "K") correction we have

$$\frac{L_\nu}{L_\nu^*} = \left(\frac{H_0 D_d}{c}\right)^2 (1+z)^{3+\alpha} 10^{0.4(23.9-r)}, \qquad (4)$$

where $\alpha \equiv -d\ln L_\nu/d\ln\nu \sim 3$, very approximately. We use a parameterized redshift distribution of the form

$$F(z,r) = \frac{\beta z^2 e^{-(z/z_0)^\beta}}{\Gamma(3/\beta)z_0^3} \qquad (5)$$

and for our fiducial model we adopt $\beta = 1.5$. More generally for $20 < r < 24$, $z_0 = k_z[z_m + z_m'(r-22)]$, where $k_z = 0.7$ for $\beta = 1.5$, and $z_m' = 0.1$ fiducially. For $20 \leq r \leq 23$ the parameterized redshift distribution is in good agreement with observation (eg. Lilly 1993; Tresse et al. 1993) and for simplicity we assume that the general form can be extended to $r = 24$.

Using the above relations Monte Carlo simulations of galaxy–galaxy lensing were carried out for sets of parameters (V^*, s^*) for galaxies with magnitudes in the range $20 \leq r \leq 24$. For each source galaxy the net polarization due to all deflectors was determined and the variation of $\langle p \rangle$ with θ for the $23 < r \leq 24$ Monte Carlo galaxies computed and compared to the observed $\langle p \rangle (\theta)$ using a χ^2 test. The values of V^* and s^* were varied until χ^2 reached a minimum, resulting in formal best-fit values for the characteristic halo parameters. An Einstein-DeSitter universe was adopted; however, the results obtained are relatively insensitive to the cosmogony.

From the χ^2 minimization the best-fit characteristic halo parameters are $V^* = 220 \pm 80$ km sec^{-1} (90% confidence bounds) and $s^* \gtrsim 100h^{-1}$ kpc, for which the χ^2 per degree of freedom is of order 0.6. The image polarization is relatively insensitive to the outer scale parameter s^* and, since most of the signal is contributed by lenses that are sufficiently close to the source on the sky that the line of sight passes through the isothermal part of the halo and for large halos the average polarization is approximately independent of s, it is not possible to obtain a unique best-fit value of s^*.

From our limits on the best-fit V^* and s^* we estimate the characteristic masses of the lens halos within a radius r to be $M(100h^{-1}\text{kpc}) \sim 1.0^{+1.1}_{-0.7} \times 10^{12}h^{-1}M_\odot$ and $M(150h^{-1}\text{kpc}) \sim 1.4^{+1.8}_{-1.0} \times 10^{12}h^{-1}M_\odot$, consistent with the dynamical estimates of the masses of field spirals by Zaritsky & White (1994) for which they obtain $M(150h^{-1}\text{kpc}) \sim 1-2 \times 10^{12}h^{-1}M_\odot$. Although our result is somewhat sensitive to the model parameters adopted, the consistency with the dynamical mass estimate is encouraging since the two methods rely on completely different sets of underlying assumptions.

Acknowledgments: We are indebted to Jeremy Mould and Todd Small for acquiring the data used for the analysis and to them, David Hogg, Nick Kaiser, and Tony Tyson for helpful discussions. Support under NSF contract AST 92-23370, the NASA HPCC program at Los Alamos National Laboratory (TGB) and a NATO Advanced Fellowship (IRS) is gratefully acknowledged.

References

Dressler, A., et al., 1995, in preparation
Lilly, S., 1993, ApJ, 411, 501
Mould, J., Blandford, R., Villumsen, J., Brainerd, T., Smail, I., Small, T., & Kells, W., 1994, MNRAS, 271, 31
Tresse, L., Hammer, F., LeFevre, O., & Proust, D., 1993, A&A, 277, 53
Tyson, J.A., Valdes, F., Jarvis, J.F., & Mills, A.P., 1984, ApJL, 281, L59
Webster, R.L., 1983, PhD Thesis, University of Cambridge
Zaritsky, D. & White, S.D.M., 1994, ApJ, 435, 599

A SEARCH FOR DARK MATTER IN THE HALOS OF LENSING GALAXIES USING VLBI

M.A. GARRETT[1], S. NAIR[1], R.W. PORCAS[2] & A.R. PATNAIK[2]
[1] NRAL, Jodrell Bank, UK; [2] MPIfR, Bonn, Germany

1. Introduction

Baryonic Dark Matter (BDM) candidates are segregated into two main mass ranges: (i) sub-solar mass dwarf stars (MACHOS) and (ii) $\sim 10^4 - 10^6 M_\odot$ Very Massive Objects (VMOs). The lower mass range has been the target of the various micro-lensing programs but the first, tentative conclusions (see Stubbs et al. these proceedings) seem to suggest that MACHOs are unlikely to provide the bulk of the dark matter in the galactic halo. Meanwhile the upper mass range ($10^4 - 10^6 M_\odot$) remains largely unexplored. However, Wambsganss & Paczynski 1992 (hereafter WP92), have shown that this mass range is perfectly tuned to a straightforward and *direct* test: gravitational milli-lensing of macro-lensed images (Fig 1).

2. High resolution VLBI Observations of Resolved Lens Systems

We have recently applied the method of WP92 to 1.6GHz Global VLBI observations of 0957+561 A,B (Garrett et al. 1994). We conclude that black holes of mass $m \geq 3 \times 10^6 M_\odot$ do not form a large proportion (> 10 %) of the DM in the halo of the lens galaxy. However, in order to to probe the more interesting mass range of $10^4 - 10^6 M_\odot$ we have embarked on a series of high frequency, sub-mas resolution observations of several radio lens systems. Here we present preliminary sub-mas images of 1830-211 (see Fig 1). The maps are difficult to interpret. While "B" exhibits a 15 mas long jet to the northwest "A" is essentially unresolved. It is not clear how the two images can look quite so different without resorting to contrived relative magnification matrices. Further observations, separated by the image time delay are planned.

C. S. Kochanek and J. N. Hewitt (eds), Astrophysical Applications of Gravitational Lensing, 189–190.
© 1996 IAU

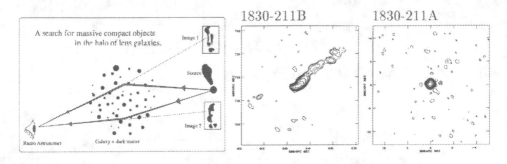

Figure 1. Left: The presence of massive compact objects in the halo of the lens galaxy will produce distortions (*e.g.* kinks, bends, rings, holes, faint multiple images) which will be peculiar to each image (since the line-of-sight is different for each image). Right: 15 GHz VLBA 0.7 mas resolution maps of 1830-211 A,B.

3. Potential Problems

While the method suggested by WP92 is in principle straightforward we anticipate three potential problems:

- significant changes (e.g. $\geq$ beam-size) in the images radio structure on a time scale $\sim$ time delay
- uncooperative relative magnifications — the case in which the lensed images cannot be easily compared
- biasing of the distorted images towards smoothness during the hybrid mapping process

One can overcome the first two of these problems by studying systems which exhibit low variability and for which the time delays are long (*e.g.* 0957+561). We have recently investigated the third problem by simulating realistic visibility data "corrupted" by isolated milli-lensing events; we are satisfied that even subtle distortions can be reliably reconstructed using the standard hybrid mapping techniques.

4. Summary

We conclude that the prospects for the detection of uncorrelated distortions in lensed images remain good. We also note that with the enhanced resolution of the forthcoming Space VLBI programs, detection of compact masses $\geq 10^3 M_\odot$ may be possible.

References

Garrett et al., 1994, MNRAS, 270, 457
Wambsganss, J., & Paczynski, B., 1992, ApJL, 397, L1

OBSERVATIONS AND PREDICTIONS OF THE RATIO OF 3-IMAGE TO 5-IMAGE SYSTEMS IN JVAS

L.J. KING, I.W A. BROWNE AND P.N. WILKINSON
University of Manchester, Nuffield Radio Astronomy Laboratories
Jodrell Bank, Macclesfield, Cheshire SK11 9DL, U.K.

AND

A.R. PATNAIK
Max-Planck-Institut für Radioastronomie
Auf dem Hügel 69, D-53121 Bonn, Germany

Abstract. In JVAS we find more 5–image systems than 3–image systems. On conventional assumptions, we predict the ratio to be ∼5:1.

1. Introduction

The analysis of gravitational lensing events gives unique information about the distribution of matter in the Universe. JVAS (Jodrell/VLA Astrometric Survey; Patnaik et al. (1992)) has examined ∼2500 compact flat spectrum radio sources and is complete to within well–defined limits. We quantify the biases in JVAS and compare the predicted ratio of 3–image to 5–image systems with the observed. In this way we can check if our assumptions about the properties of lensing masses are correct.

2. The biases that might affect the 3:5 image ratio in JVAS

1 **Magnification bias**: The radio luminosity function of quasars shows that there are progressively fewer sources at higher luminosities. When quasars are lensed (magnified) they appear further up the luminosity function and lensed objects are thus over–represented in the high luminosity population (Turner 1980). Average magnifications for 5–image systems are higher than for 3–image systems. Hence when magnifica-

C. S. Kochanek and J. N. Hewitt (eds), Astrophysical Applications of Gravitational Lensing, 191–194.
© 1996 IAU

tion bias is high, i.e. when the quasar luminosity function is steep, one
expects proportionately more 5–image systems than 3–image systems.

2 Bias due to finite source size: Intrinsic 5–image cross–sections are
much smaller than 3–image cross–sections. If the lensed object sizes
are comparable to the 5–image cross-section then this significantly increases the number of 5–image systems expected but leaves the number of 3–image systems substantially unaltered (Kochanek & Lawrence
1990).

3 Bias due to finite core radii: The cross–section for the production of
3–image systems is reduced when the core radius for lensing galaxies
is increased, but the 5–image cross–section remains largely unaffected.

2.1. MAGNIFICATION BIAS

We assume that lensing galaxies have elliptical potentials of the form:

$$\Psi(x,y) = \frac{A}{2\alpha_s}\left(\left[1 + (1-\epsilon)\left(\frac{x}{s}\right)^2 + (1+\epsilon)\left(\frac{y}{s}\right)^2\right]^{\alpha_s} - 1\right) \qquad (1)$$

with parameters (i) ellipticity of the potential (ϵ), (ii) core radius (s), (iii)
softness of the potential (α_s) and (iv) depth of the potential well (A) which
is proportional to σ^2 (eg. Blandford & Kochanek 1987). The core radii of
lensing galaxies are assumed to be small ($\leq$ a few hundred parsecs), as
shown by the observations of Lauer (1985) and supported by the "missing"
central images in lens systems (eg. Wallington & Narayan 1993). For the
source population, we use the Dunlop & Peacock (1990) luminosity function. We use approximations for the probability distributions of magnification (M) associated with 3 and 5 image systems (Kochanek 1992, private
communication):

$$P_3(M)dM \propto M^{-3.5}dM \quad and \quad P_5(M)dM \propto M^{-3}dM. \qquad (2)$$

The intrinsic cross–sections for the productions of three– and five–
images are roughly in the ratio $1:1.5\epsilon^2$ (Kochanek 1991). The flat spectrum
radio luminosity function is shallower than the luminosity function for optically selected sources, so the bias factors and ratio between the bias for
5–image systems as opposed to 3–image systems is smaller.

2.2. SIZE BIAS.

The JVAS lensed systems are all selected from flat spectrum radio sources
which are invariably compact. The systems under discussion have multiple

images of this compact emission. VLBI observations of a subset of these sources show that they are typically much less than 10mas (Taylor et al. 1994) so that the size bias enhancement of the production of 5–image systems relative to 3–image systems is expected to be small; assuming each source has an extent 10 mas × 10 mas only leads to ~ 10% enhancement.

2.3. FINITE CORE RADII

An isothermal lens potential with a softened core (Equation 1) results in a reduction in the total lensing optical depth by a factor of about two when the core radius is increased from zero to several hundred pc. Although this enhances the number of 5–image systems *relative* to 3–image systems, the expected number of 5–image systems is not increased. There are two reasons to think that finite core radii are not affecting the relative numbers of 3–image to 5–image systems by a large factor. The first is that the observed lensing frequencies (see below) indicate an enhancement of the number of 5–image systems rather than a depletion of 3–image systems. The second is that the existing observational evidence, both from lensing and for elliptical galaxies in general, gives no support for large core radii (Wallington & Narayan 1993; Lauer 1985).

2.4. RESULTS OF OUR PREDICTIONS

Without magnification bias, and assuming $\epsilon = 0.1$, the expected ratio is 66:1; with magnification bias, size bias and core radii effects, the expected ratio is decreased to ~ 5:1.

3. JVAS results

In JVAS, five confirmed lens systems, and one strong candidate, have been identified. Of the five secure systems, there are four 5–image systems and only one 3–image system. The strong candidate is a 3–image system. The 1938+666 system probably gets into the sample by virtue of lensed extended radio emission rather than core emission and thus should excluded from the statistics. The conclusion is, however, that 5-image systems are found in radio surveys at least as frequently as 3–image systems – about a factor of five times more frequently than the predictions.

4. A possible resolution?

If the lensing masses were more elliptical than we have assumed, this would result in more 5–image systems relative to 3–image systems (Nair 1993). Some of the known radio selected systems (eg. 1422+231, 1608+654,

2016+12) require highly elliptical or multiple component lenses to reproduce the observed image properties. Thus there may be more higher ellipticity galaxies capable of multiple imaging than has hitherto been assumed, or lensing by multiple (merging?) galaxies may be more common than expected. We may already have clues as to the possible properties of these lenses with the discovery that some lenses are very dusty (Larkin et al. 1993) and that others are likely to be gas rich and dusty disk systems (Patnaik et al. 1993, Myers et al. 1995). The presence of dust biases optical lens surveys against detection of such objects.

5. Conclusions

1 Bias effects amongst lensed systems selected from samples of flat spectrum radio sources are smaller than optically selected sources.
2 Assuming ellipticities of 0.1, our predictions indicate that 3–image systems should outnumber 5–image systems by $\sim$ 5:1.
3 In JVAS, which is the most complete radio radio survey to date, 5–image systems outnumber 3–image systems.
4 The preponderance of 5–image systems is unlikely to be due to 3–image systems being missed. This is because the observed lensing frequency is already as high as expected ($\sim$ 1:500). If the discrepancy were to be removed by a population of missing 3–image systems this would imply a rate of $\geq$ 3:500.
5 A possible resolution may be that some lenses may be much more elliptical than previously thought or may consist of multiple (merging) galaxies.

References

Blandford, R. & Kochanek, C.S., 1987, ApJ, 321, 658
Dunlop, J.S. & Peacock, J., 1990, MNRAS, 247, 19
Kochanek, C.S. & Lawrence, C.R., 1990, AJ, 99, 1700
Kochanek, C.S., 1991, ApJ, 379, 517
Larkin, J.E., et al., 1993, ApJL, 420, L9
Lauer, T.R., 1985, ApJ, 292, 104
Myers, S., et al., 1995, ApJL, 447, L5
Nair, S., 1993, PhD Thesis, University of Bombay
Patnaik, A.R., et al., 1993, MNRAS, 261, 435
Taylor, G.B., et al., 1994, ApJS, 95, 345
Turner, E.L., 1980, ApJL, 242, L135
Wallington, S. & Narayan, R., 1993, ApJ, 403, 517

MG2016+112: A DOUBLE GRAVITATIONAL LENS MODEL

S. NAIR AND M. A. GARRETT

University of Manchester, NRAL, Jodrell Bank, U.K.

MG2016+112, discovered by Lawrence et al. (1984) is one of the best–studied among multiply–imaged systems, but is still only partially under-stood. Ostensibly a three–image system consisting of images A, B and C of a quasar at $z = 3.273$, the observed lensing galaxy D (a giant ellipti-cal at $z = 1.01$) at the centroid of the image system seems inadequate to provide the minimum mass of $\sim 2.5 \times 10^{12}$ M$_\odot$ within 10 kpc of its cen-ter (in projection along the l.o.s.) required to produce the observed 3.″9 image–splitting. C itself appears to consist of two components, radio emis-sion that may be associated with the faint optical image counterpart of A and B (called C_2, see Garrett et al. 1994) and flat–spectrum C_1, which dominates radio observations of the system and apparently consists of at least three linearly stretched subcomponents, C_{11} to C_{13} (see Garrett et al. in these proceedings).

The observationally suggested second lens in region C (e.g. Lawrence et al. 1993) is strongly supported by lens modeling, because of the presence of the faint image C_2. The two elliptical lenses produce a five–image configura-tion with two core–captured images demagnified to levels of undetectability; see Fig.1. Each lens consists of two non–singular oblate spheroidal mass distributions, one compact (the 'galaxy') and the other extended ('dark matter', DM). The DM associated with lens plane D has a scale length of about 25 kpc, and appears to have a high eccentricity (axial ratio about 0.6 in a typical model). Hattori et al. (these proceedings) suggest that there could be a cluster here. The redshift of lens plane C is assumed to be greater than that of D, for definiteness (this is not constrained by the configura-tion). Masses (in M$_\odot$) of the lenses in the present model are: lens plane D: galaxy — 2.7×10^{11}, DM —2.3×10^{12}; lens plane C: galaxy —6.5×10^9 *(high eccentricity)*, DM — 7.0×10^{12} *(scale length of ~ 65 kpc; spherical)*. Lens plane C is at $z_C = 1.2$, and lens plane D is at $z_D = 1.01$. The modeling code used is a version of Narasimha, Subramanian and Chitre (1982, 1984).

C. S. Kochanek and J. N. Hewitt (eds), Astrophysical Applications of Gravitational Lensing, 195–196.
© 1996 IAU

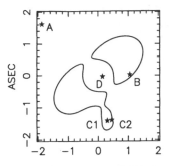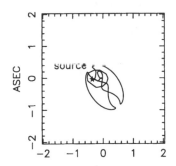

Figure 1. (a) Image and (b) Source Planes for the Two–Lens Model, with the position of the second galaxy, after single–imaging by the foreground lens D, near C_1 in (a). The model is constrained by the image separations (the average error $\sim 10\%$, being the largest with the position of image B), image intensity ratios and vlbi observations of this system.

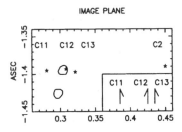

Figure 2. Is C_1 multiply–imaged radio emission? *(Main Fig.:)* The images C_{11} to C_{13} as formed by a 2nd radio source just behind Galaxy C, near a cusp of the 'lips' caustic that develops. C_2 is shown for reference. *(Inset:)* Predicted parity relations between the subcomponent images in the case of C_{11} to C_{13} being formed at the lips caustic. If C_{12} and C_{13} are formed by some relatively extended radio emission (at $z = 3.273$ and related to the core–jet source that gives rise to A, B and C_2) which may be imaged with high magnification between C_{11} and C_2 as it crosses a radial critical curve just east of the source position in Fig.1(b), then the predicted parities for C_{12} and C_{13} are the same as in the previous case. In this picture, C_{11} is the core-captured image near Galaxy C (demagnified version of A, B and C_2). The corresponding images near A and B of this extended radio flux could well be resolved out in VLBI observations. Note high magnification gradient near image C_2; image flux ratios can vary with wavelength.

C_1 could be the second lens as a peculiar (singly–imaged) radio emitting galaxy; else, if it is multiply–imaged background radio emission from a second source or fuzz associated with the source at $z = 3.273$, see Fig.2 for predicted substructure. *Acknowledgements: S.N. thanks the Raman Research Institute, Bangalore, India, for the use of computing facilities.*

References

Garrett, M.A., Muxlow, T.W.B, Patnaik, A.R., & Walsh, D., 1994, MNRAS, 269, 902
Lawrence, C.R., Schneider, D.P., Schmidt, M. et al., 1984, Science, 223, 46
Lawrence, C.R., Neugebauer, G., & Matthews, K., 1993, A.J., 105, 17
Narasimha, D., Subramanian, K., & Chitre, S.M., 1982, MNRAS, 200, 941
Narasimha, D., Subramanian, K., & Chitre, S.M., 1984, MNRAS, 210, 79

A LENS MODEL FOR B0218+357

S. NAIR

University of Manchester, NRAL, Jodrell Bank, U.K.

The smallest radio Einstein Ring, B2018+357, discovered by Patnaik et al. (1993), shows much promise as a tool to constrain the parameters of cosmological models (Refsdal 1964). Dominating this system is a pair of $0.''335$–separation compact radio images, with image A between 2.7 to 3.9 times as strong as image B ($\lambda\lambda$ 18 to 2 cm). Observations have established the lens redshift ($z_l = 0.685$, O'Dea et al. 1992, Browne et al. 1993), a possible source redshift ($z_s = 0.96$, Lawrence, *this conference*), and a tentative value for the time–delay between the highly polarized images A and B (12 ± 3 days, Browne, *this conference*). Recent *mas*–resolution observations have made it possible to understand the imaging of A and B in sufficient detail as to provide constraints on an elliptical lens model for B0218+357; this work presents a model and provides an estimate of Hubble's Constant.

Wilkinson et al. (in preparation) point out that in their VLBI map at 18 cm (Polatidis et al. 1995), the ratio A/B of the apparent sizes of these images far exceeds that of their total flux densities (the map suggests 18.9 as against 2.65, respectively). For a pair of simple, compact but resolved images, the ratios should agree. The central surface brightness of A is also considerably lower than that of image B. The VLBA observations at $\lambda 2$ cm of Patnaik et al. (1995) show a core–knot structure in each image. A is dominated by tangential, and B by radial stretching; I derive a transformation matrix (A to B) from the knot images, which are not likely to be affected by new features developing near the cores. This matrix ($t_{xx} = 0.81, t_{xy} = 0.66, t_{yx} = 0.17, t_{yy} = -.18$) is applied to the VLBI images of A at 6 and 18 cm (Xu et al. 1995, and Polatidis et al. 1995) to obtain image B, and *vice versa* using the inverse matrix. The eigendirections remain roughly the same over $\lambda\lambda 2 - 18$ cm. The relatively minor discrepancies at $\lambda 6$ cm between the observed and transformed images are easily understood as a function of increasing source size with wavelength. The large size of image A at 18 cm is explained if the source is viewed as lying

C. S. Kochanek and J. N. Hewitt (eds), Astrophysical Applications of Gravitational Lensing, 197–198.
© 1996 IAU

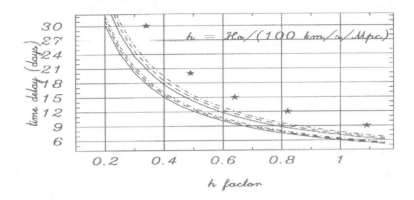

Figure 1. Estimates of Hubble's Constant, H_o, with a q_0=0.5, FRW Universe, having a smoothed–out background matter distribution, for lensed system B0218+357. The plots are for point mass *(lowest three)* and singular isothermal sphere *(middle three)* models, and the elliptical mass model *(stars)* in the present work. The A/B flux ratio is set at 3.6 *(solid)*, 3.8 *(dash)* and 4.0 *(dash-dot)* in the first two cases, and 3.8 for the last case. A time-delay of 12 days puts H_o at 52, 62 and 82 *km/s/Mpc*, respectively, for the three models.

very near the radial caustic in the source plane; as the source size increases, it laps over from the triply–imaged region to the singly–imaged one, and image B therefore consists of *two* merging images at 18 *cm*, which are seen as one compact feature of relatively high central surface brightness, whereas image A contains a significant fraction of singly-imaged source, thus with a lower surface brightness. The picture is consistent with image B lying very near the centre of the radio ring. An elliptical lens mass model for the system (mass = 3.9×10^{10} $M_\odot$; axial ratio 0.65) shows that the source size must grow from about 10 to 20 *mas* between 6 and 18 *cm* in this picture. The model matches the overall image separation to within 1.5% and the VLBA features of Patnaik et al. (1995).

Acknowledgements: Thanks are due to P. N. Wilkinson and A. R. Patnaik for data and discussions, and I.W.A. Browne and R.W. Porcas for discussions. Thanks also to the Raman Research Institute, Bangalore, India, for the use of computing facilities.

References

Browne, I.W.A., Patnaik, A. R., Walsh, D., & Wilkinson, P. N., 1993, MNRAS, 263, L32
O'Dea, C., Baum, S., Stanghellini, C., et al., 1992, AJ, 104, 1320
Patnaik, A. R., Browne, I.W.A., King, L. J., et al., 1993, MNRAS, 261, 435
Patnaik, A. R., Porcas, R.W., & Browne, I.W.A., 1995, MNRAS, 274, L5
Polatidis, A.G., Wilkinson, P. N., & Xu, W. et al., 1995, ApJS, 98, 1
Refsdal, S., 1964, MNRAS, 128, 307
Xu, W., Readhead, A.C.S., Pearson, T. J., et al., 1995, ApJS, submitted

GRAVITATIONAL MICROLENSING,
THE DISTANCE SCALE, AND THE AGES

BOHDAN PACZYŃSKI

Princeton University Observatory

124 Peyton Hall, Princeton, NJ 08544-1001, USA

Abstract. A high optical depth to gravitational microlensing towards the galactic bulge is consistent with current models of the galactic bar. The low optical depth towards the LMC can probably be accounted for by the ordinary stars in our galaxy and in the LMC itself. No conclusive evidence is available yet for the presence or absence of a large number of brown dwarfs or other non-stellar compact objects which might account for the dark matter. There is little doubt that the amount of mass in objects in the range $10^{-8} \leq M/M_\odot \leq 10^6$ will be determined within the next few years with the continuing and expanding searches.

Billions of photometric measurements generated by the microlensing searches have lead to the discovery of $\sim 10^5$ variable stars. In particular, a number of detached eclipsing binaries were discovered in the galactic bulge, in the LMC, and in the globular cluster Omega Centauri. The follow-up observations of these binaries will allow the determination of accurate distances to all these objects, as well as robust age determination of globular clusters.

1. Microlensing

Four groups: DUO (Alard 1995), EROS (Aubourg et al. 1993), MACHO (Alcock et al. 1993) and OGLE (Udalski et al. 1993) are conducting the search for events of gravitational microlensing among millions of stars in the Magellanic Clouds (EROS, MACHO), and in the galactic bulge (DUO, MACHO, OGLE). A total of almost 100 events of microlensing by single objects (Alcock et al. 1993, 1995a,c,d, Aubourg et al. 1993, Udalski et al. 1993, 1994a,b,c, Alard 1995, Stubbs et al. 1995), and at least 3 events

C. S. Kochanek and J. N. Hewitt (eds), Astrophysical Applications of Gravitational Lensing, 199–208.
© 1996 IAU

of microlensing by double object (Udalski 1994d, Alard et al. 1995, Pratt 1995), were reported so far.

1.1. PAST AND PRESENT

This is a very young and rapidly developing field, following a science-fiction like suggestion by Paczyński (1986). The practicality of microlensing searches was first envisioned by K. Freeman, D. P. Bennett, and C. Alcock, and the first dedicated camera was built by C. Stubbs, all of the MACHO collaboration. The detection of the first candidate events was reported almost simultaneously by three groups: EROS (Aubourg et al. 1993), MACHO (Alcock et al. 1993), and OGLE (Udalski et al. 1993).

Originally, data processing lagged behind data acquisition, and microlensing events were discovered on computer tapes many months after the recording had been made. The first single event ever recorded was OGLE #10, which peaked on June 29, 1992 (Udalski et al. 1994b). The first double event was OGLE #7 (Udalski et al. 1994d) which had its complicated light variations in June/July 1993. Double lensing events had been estimated by Mao & Paczyński (1991) to make up to 10% of all events.

In the spring of 1994 the real time data processing became possible for the OGLE collaboration, and the first event detected while in progress was OGLE #11 (Udalski et al. 1994c); it peaked on July 5, 1994. Soon after that the MACHO collaboration, which has yearly data rate ~ 30 times higher than the OGLE's rate, started first partial, and now full real time data processing. The total number of "real time" events is currently 6 for the OGLE (as announced electronically by Udalski), and ~ 30 for the MACHO (Pratt 1995).

Long duration single events should be somewhat time asymmetric because of the acceleration in the Earth's motion around the sun (Gould 1992). This "parallax effect" was first detected by the MACHO group (Alcock et al. 1995d).

The optical depth to microlensing towards the galactic bulge was found to be $\sim 3 \times 10^{-6}$ (Udalski et al. 1994b, Alcock et al. 1995c), while the original theoretical estimates were below 10^{-6} (Paczyński 1991, Griest et al. 1991). Theoretical estimates were increased by Kiraga & Paczyński (1994) to somewhat above 10^{-6} when the importance of self-lensing by the bulge was recognized, and increased even more when the importance of the galactic bar was noticed (Paczyński et al. 1994). Subsequent calculations, using the best available bar model, gave the optical depth of $\sim 2 \times 10^{-6}$ (Zhao et al. 1995), within one standard deviation of the observational results. According to Zhao et al. (1995) the duration distribution of the OGLE events detected towards the bulge is consistent with no brown dwarfs among the lensing objects, though a 50% contribution by brown dwarfs can be ruled out at only 95% confidence level.

In their original paper the EROS group wrote that their two events towards LMC were consistent with all dark matter in the galactic halo being in the form of MACHOs (Aubourg et al. 1993). The MACHO group, on the basis of many more measurements but only three events, estimated that the MACHOs can account for no more than $\sim$ 20% of a standard dark halo (Alcock et al. 1995a). Sahu (1994) pointed out that the dominant stellar contribution to microlensing of stars in the bulge of LMC may be due to stars in the bulge of LMC.

Microlensing events are often advertized as achromatic. However, all searches are conducted in very dense stellar fields, where the detection limit is set by crowding. Therefore, many (perhaps most) images are blends, made of two or more stellar images which appear as one. The apparent images are $\sim$ 1 second of arc across because of the atmospheric seeing, while the cross-section for gravitational lensing is about $(0.001'')^2$. As a result only one star of a blend is likely to be lensed (DiStefano & Esin 1995). As stars may have various colors, the lensing event may not be strictly achromatic. The contribution of blends was first noticed in the analysis of double lensing events (Udalski et al. 1994d, Alard et al. 1995), but it must be common among all events. It may seem surprising that the the observed events are almost achromatic. There may be two reasons for that. First, in the case of the galactic bulge the detection limit is close to the bulge main sequence turn-off point, and stars have only a small range of colors within a few magnitudes of the turn-off point. Second, some genuine events might have been rejected on the grounds that they appeared to be chromatic.

The calibration of microlensing searches may be done at two levels of sophistication. It may be done at the "catalog level", in which the artificial events are introduced into the database of photometric measurements (Udalski et al. 1994b), or at the "pixel level", in which the artificial events are introduced as artificial stars on the CCD frames (Alcock et al. 1995a, Stubbs et al. 1995). The second method is the truly correct way of doing the analysis, but it is also vastly more time consuming than the first. Fortunately, the search sensitivity as calibrated by the two methods differs by no more than $\sim$ 20% (Stubbs et al. 1995) as a result of a near cancellation of the two effects: the number of stars subject to microlensing should be increased because of the blending, but the observed amplitudes are reduced by the blending (Udalski et al. 1994b).

The OGLE and MACHO experiments are fairly efficient in detecting events with time scales in the range 10 days – 100 days, but their efficiency falls rapidly outside these limits. The full range of time scales of microlensing events which might be detectable with the current technology extends from ~ 10 minutes (Paczyński 1986) to ~ 200 years (Paczyński 1995b). The lower limit is set by the requirement that the stellar disk should be smaller than the Einstein ring to make a significant magnification of the apparent brightness possible. This corresponds to the lens of $\sim 10^{-8}\ M_\odot$. The upper limit brings us up to $\sim 10^6\ M_\odot$, the larger masses being detectable by other means (Wambsganss & Paczyński 1992). A sensitive coverage of time scales in the whole range: ~ 10 minutes to ~ 200 years may be expected within a few years, but a really high sensitivity is currently limited to the interval 10 – 100 days. The upper limits on the number of very short time scale events, i.e. the optical depth to very low mass objects, currently available from EROS (Aubourg et al. 1995) and MACHO (Stubbs, private communication) are not very stringent.

There is plenty of evidence that almost all reported events are due to microlensing:

1. The light variations are achromatic or almost achromatic, and the light curves are well described by the theoretical formula.
2. The distribution of peak magnifications is consistent with that expected theoretically.
3. The double lensing events are detected at a frequency consistent with the theoretical expectations.
4. The parallax effect has been detected on the longest event, as expected.
5. The the galactic bar has been "rediscovered" with gravitational microlensing (my personal view).

There is no formal consensus on many other issues, as judged from the published papers:

1. What is the reason for the large optical depth towards the galactic bulge (OGLE vs MACHO).
2. What is the optical depth towards the LMC (EROS vs MACHO).
3. Which objects contribute to the LMC microlensing (Sahu vs MACHO).
4. What is the frequency of double lenses (OGLE & DUO vs MACHO).
5. What can be said about the presence or absence of dark compact objects in the galactic halo (EROS vs MACHO vs author).

There is little doubt that within the next few years all these issues will be resolved.

1.2. FUTURE

There is no lack of bold proposals for future microlensing searches from the ground as well as from space (Gould 1995, and references therein). I shall address here just a few topics.

Currently it is possible to estimate the mass of the lensing object only statistically, as the distance to any lens and its transverse velocity are not known. At the same time our knowledge of the mass function for low mass stars and brown dwarfs is very limited. It should be possible to fix both problems with a new observing program: the search for microlensing events caused by the high proper motion stars (Paczyński 1995a). The most difficult step is the detection, in the densest regions of the Milky Way, of a few hundred or a few thousand stars with proper motion in excess of ~ 0.2 seconds of arc per year in the magnitude range $18 \leq V \leq 21$, or so. The faint high proper motion stars must be nearby, and their distances can be measured directly and accurately with trigonometric parallaxes (Monet et al. 1992). The same observations will determine the proper motion with accuracy high enough to make reliable predictions about the microlensing events of the distant stars in the Milky Way by the faint high proper motion stars or brown dwarfs. The expected time scale for the events is in the range of $1 - 10$ days, i.e. they can easily be followed photometrically. The mass of the lens can be calculated without any ambiguity when the event time scale, as well as the distance and the proper motion of the lens are all measured.

The search for planets (Mao & Paczyński 1991) will require a major upgrade of the current ground based searches, it may even call for space probes capable of monitoring the events from a distance of ~ 1 AU (Gould 1995, and references therein). There is a major problems to solve: how to prove convincingly that a short time scale feature in the light curve is actually caused by a planet, rather than by stellar variability. As the diversity of possible light curves is enormous it is not clear how to conduct statistical tests, like those which can be done for single lensing events.

There is plenty of room for improvement of hardware and software of the ground based searches. On the basis of published OGLE and MACHO results one can find that on average ~ 20 pixels on a CCD are needed to measure brightness of a single (possibly blended) stellar image. It is very likely that higher efficiency might be achieved with some "frame subtraction" technique, most unfortunately referred to as "pixel lensing" (Crotts 1992, Colley 1995, Gould 1995).

Let N be the total number of photometric measurements of all stars monitored in a given experiment, τ the optical depth to microlensing, and n the number of detected microlensing events. Using the published MACHO and OGLE data one finds that $N\tau/n \approx 50 - 100$. A major improvement should be possible if various sources of noise in the data could be better controlled.

The number of stars that can be monitored from the ground is inversely proportional to the square of seeing disk diameter, as the detection limit is set by crowding, not by photon statistics. This implies that the sites with the best possible seeing combined with small, 1-meter class telescopes and the largest possible number of pixels offer the best price to performance ratios. I am not aware of a sound estimate of the number of stars measurable from the ground, but it may be as high as 10^9, and over $\sim 10^3$ microlensing events per year may be within reach with a modest extension of the current technology.

2. Variable stars

The massive photometric searches for gravitational microlensing events lead to the discovery of a huge number of variable stars (Udalski et al. 1994e, 1995, Grison et al. 1995, Alcock et al. 1995b, Cook et al. 1995, Kaluzny et al 1995a,b). The total number of variable stars in the data bases of the four collaborations is $\sim 10^5$, with MACHO having by far the largest number of unpublished objects (Alcock 1995b, Cook et al. 1995), and OGLE having the largest number of published finding charts, light curves, etc. (Udalski et al. 1994e,1995).

I would like to concentrate on just one type of variables: the detached eclipsing binaries, which are also double line spectroscopic binaries, as these are the primary source of the fundamental data about stars: their masses, luminosities, and radii (Andersen 1991, and references therein). Such systems are relatively rare, roughly one out of a few thousand. As the eclipses are narrow it is necessary to accumulate a few hundred photometric measurements to determine the binary period. This means that one needs a total of $\sim 10^6$ photometric measurements to discover one promising candidate for a detached eclipsing binary, which is also likely to be a double line spectroscopic binary. Massive photometric programs are needed, and this is exactly what is provided by the microlensing searches. The light curves and finding charts for many such objects are provided by Udalski et al. (1994a, 1995) for stars towards the galactic bulge, and by Grison et al. (1995) for stars in the LMC.

Theoretical analysis of the light variations during the two eclipses of a detached binary gives the (star size) / (orbit size) ratio for the two stars, R_1/A, and R_2/A, as well as the fraction of total luminosity (in a given photometric band) originating in each component, L_1/L, and L_2/L. The so called "third light", defined as: $L_3 \equiv L - L_1 - L_2 \neq 0$, may also be present and readily measured (Dreshel et al. 1989, Goecking et al. 1994, Gatewood et al. 1995, and references therein). Finally, the inclination of the orbit i can be determined.

If the two stars have similar luminosities then the spectra of both stars, and the amplitudes of radial velocity variations of both stars, K_1 and K_2, can be measured. Given K_1 and K_2, as well as the orbital period and inclination, the size of the binary orbit, A, and the two masses, M_1 and M_2, can be determined. With A as well as R_1/A and R_2/A known, the two stellar radii, R_1 and R_2, become known as well. The practical application of this procedure to real bright stars lead to the determination of all stellar parameters with a $\sim 1\% - 2\%$ accuracy for a few dozen objects with spectral types all the way from O8 to M1 (Andersen 1991).

3. Detached eclipsing binaries as primary distance indicators

The fact that all parameters for the bright binaries can be obtained with $\sim 1 - 2\%$ accuracy implies that it should be possible to measure distances to far away binaries with $\sim 1 - 2\%$ accuracy, provided very high S/N data can be obtained with the large telescopes. There are elements which can make this task difficult to accomplish in practice: unresolved companions and interstellar extinction. There is reason for optimism, as the problem of unresolved companions, i.e. the "third light" problem, is well known among the brightest objects, like Algol (Gatewood et al. 1995), and given high enough S/N data the "third light" contribution can be calculated. The interstellar extinction is a standard problem for any photometric distance determination, and it is always troublesome. It helps to have infrared photometry, as the interstellar extinction in the K-band is several orders of magnitude lower than it is in the visual domain.

In order to measure the distance we have to find out what is the surface brightness of each binary component in our photometric bands. This is the only delicate step in the whole procedure. The surface brightness can be obtained from the observed colors and/or spectra using modern model atmospheres. This relation can be well calibrated empirically with the nearby systems, which have their distances accurately measured either with trigonometric parallaxes (Dommanget & Lampens 1992, Monet et al. 1992, Gatewood 1995, Gatewood et al. 1995), or with a combination of spectroscopic and interferometric (astrometric) orbits (Pan et al. 1990, 1992). The most accurate surface brightness determination is possible in the K-band (Ramseyer 1994).

Given a good estimate of the surface brightness of each star in a selected band, say F_K^*, as well as the direct measurement of the flux in that band at the telescope, F_K, we can calculate the distance as

$$d = R \times \left(\frac{F_K^*}{F_K} \right)^{1/2},$$

where R is the stellar radius, and F_K has been corrected for the interstellar extinction.

This approach was attempted by Bell et al. (1991, 1993) to measure the distance to LMC. Unfortunately, the binaries HV 2226 and HV 5963 had light curves indicating these were semi-detached, i.e. moderately complicated system. Also, the accuracy of radial velocity measurements was rather low. The list of EROS binaries in LMC (Grison et al. 1995) has a number of clearly detached systems, as judged from their light curves. Very accurate measurements of radial velocities of both components of a binary system are now possible with the recently developed TODCOR method (Zucker & Mazeh, 1994). Metcalfe et al. (1995) used TODCOR to measure radial velocity amplitudes for a 13 mag eclipsing binary CM Dra with a precision of $0.15 \ km \ s^{-1}$ with a 1.3 meter telescope.

The detached binaries offer a potential to establish distance to globular clusters, to the galactic center, and to nearby galaxies: LMC, SMC, M31 and M33 (cf. Hilditch 1995), with unprecedented accuracy. Note, that binaries cover a very the wide range of spectral types, and they are numerous, so cross-checks will be possible. If the crowding and interstellar extinction can be handled adequately, the fractional accuracy of distances will be equal to the fractional accuracy of the determination of radial velocity amplitudes, K_1 and K_2, as distances are proportional to stellar diameters, which in turn are proportional to the K values. Eclipsing binaries offer a "single step" distance determination to nearby galaxies, thereby providing an accurate zero point calibration for all pulsating stars, including cepheids – a major step towards very accurate determination of the Hubble constant.

4. Detached eclipsing binaries as primary age indicators

The accurate age determination of globular clusters is one of the most important astronomical issues, as there is a perceived conflict with the determinations of the Hubble constant (Chaboyer et al. 1995). All current age determinations are based on the comparison between the observed and theoretical isochrones in the color-magnitude diagrams (Shi 1995, and references therein). There are at least two problems with this method. First, the age is inversely proportional to the square of distances to a globular clusters, and second, the colors of theoretical models are affected by poorly known and not understood "mixing-length" parameter (Paczyński 1984, Chaboyer 1995, and references therein). Recent discovery of the two detached eclipsing binaries at the main sequence turn-off point in Omega Centauri (Kaluzny et al. 1995b) is a major step to overcome both problems.

The detached binaries in Omega Centauri will be used to measure accurately the distance to this globular cluster with the method described in the previous section. This will considerably reduce the distance uncertainty and the corresponding uncertainty in the classical age determination. Also, for the first time ever it will be possible to measure directly the masses of stars near the main sequence turn-off point, and this in turn will allow the age determination using the mass-luminosity relation, which is not affected by the "mixing-length" parameter (Paczyński 1984). The masses of stars near the turn-off point depend on the age as well as helium content Y, and heavy element content Z. The Z abundance can be measured spectroscopically, but the helium abundance has to be determined from the same mass-luminosity relation as the age. To make this possible it will be necessary to discover detached eclipsing binaries somewhat below the turn-off point, and to measure their masses.

5. Microlensing by Internet

The photometry of OGLE microlensing events, their finding charts, as well as a regularly updated OGLE status report, including more information about the "early warning system", can be found over Internet from the host: "sirius.astrouw.edu.pl" (148.81.8.1), using "anonymous ftp" service (directory "ogle", files "README", "ogle.status", "early.warning"). The file "ogle.status" contains the latest news and references to all OGLE related papers, and PostScript files of some publications. These OGLE results are also available over World Wide Web at: "http://www.astrouw.edu.pl".

Similar information about MACHO results is available over World Wide Web at: "http://darkstar.astro.washington.edu".

6. Acknowledgements

It is a great pleasure to acknowledge the discussions with, and comments by Dr. Dr. J. Kaluzny, A. Kruszewski, D. W. Latham, M. Pratt, and C. Stubbs. This work was supported by the NSF grants AST-9216494 and AST-9313620.

References

Alard, C., 1995, these proceedings
Alard, C., Mao, S., & Guibert, J., 1995, A&A, 300, L17
Alcock, C. et al., 1993, Nature, 365, 621
Alcock, C. et al., 1995a, Phys. Rev. Letters, 74, 2867
Alcock, C. et al., 1995b, AJ, 109, 1653
Alcock, C. et al., 1995c, ApJ, 445, 133
Alcock, C. et al., 1995d, preprint
Andersen, J., 1991, A&AR, 3, 91
Aubourg, E. et al., 1993, Nature, 365, 623
Aubourg, E. et al., 1995, preprint
Bell, S. A. et al., 1991, MNRAS, 250, 119
Bell, S. A. et al., 1993, MNRAS, 265, 1047
Chaboyer, B. et al., 1995, ApJ, 444, L9

Colley, W. N., 1995, AJ, 109, 440

Cook, K. H. et al., 1995, preprint

Crotts, A. P. S., 1992, ApJ, 399, L43

DiStefano, R. & Esin, A. A., 1995, ApJ, 448, L1

Dommanget, J., & Lampens, P., 1992, in PASP Conf. Ser. Vol. 32, eds. H.A. McAlister & W.I. Hartkopf, p. 435

Dreshel, H. et al., 1989, A&A, 221, 49

Gatewood, G., 1995, ApJ, 445, 712

Gatewood, G. et al., 1995, AJ, 109, 434

Goecking, K.-D. et al., 1994, A&A, 289, 827

Gould, A., 1992, ApJ, 392, 442

Gould, A., 1995, these proceedings

Griest, K. et al., 1991, ApJ, 372, L79

Grison, P. et al., 1995, A&AS, 109, 447

Hilditch, R.W.,, 1995, in to appear in "Binaries in Clusters", ASP Conf. Ser., eds. G. Milone and J.-C. Mermilliod.

Kaluzny, J. et al., 1995a, A&A, in press

Kaluzny, J. et al., 1995b, to appear in "Binaries in Clusters", ASP Conf. Ser., eds. G. Milone and J.-C. Mermilliod.

Kiraga, M. & Paczyński, B., 1994, ApJ, 430, 101

Mao, S. & Paczyński, B., 1991, ApJ, 374, L37

Metcalfe, T. S., Mathieu, R. D., Latham, D. W. & Torres, G., 1995, preprint

Monet, D. G. et al., 1992, AJ, 103, 638

Paczyński, B., 1984, ApJ, 284, 670

Paczyński, B., 1986, ApJ, 304, 1

Paczyński, B., 1991, ApJ, 371, L63

Paczyński, B. et al., 1994, ApJ, 435, L113

Paczyński, B., 1995a, Acta Astron., 45, 345

Paczyński, B., 1995b, Acta Astron., 45, 349

Pan, X. et al., 1990, ApJ, 356, 641

Pan, X. et al., 1992, ApJ, 384, 633

Ramseyer, T.F., 1994, ApJ, 425, 243

Pratt, M. R. et al., 1995, these proceedings

Sahu, K., 1994, Nature, 370, 275

Shi, X., 1995, ApJ, 446, 637

Stubbs, C. et al., 1995, these proceedings

Udalski, A. et al., 1993, Acta Astron., 43, 289

Udalski, A. et al., 1994a, ApJ, 426, L69

Udalski, A. et al., 1994b, Acta Astron., 44, 165

Udalski, A. et al., 1994c, Acta Astron., 44, 227

Udalski, A. et al., 1994d, ApJ, 436, L103

Udalski, A. et al., 1994e, Acta Astron., 44, 317

Udalski, A. et al., 1995, Acta Astron., 45, 1

Wambsganss, J. & Paczyński, B., 1992, ApJ, 397, L1

Zhao, H. S., Spergel, D. N., & Rich, R. M., 1995, ApJ, 440, L13

Zucker, S. & Mazeh, T., 1994, ApJ, 420, 806

SEARCHING FOR DARK MATTER WITH
GRAVITATIONAL MICROLENSING:
A REPORT FROM THE MACHO COLLABORATION

C.W. STUBBS[1,2]

[1] *Departments of Astronomy and Physics,*
University of Washington, Seattle, WA 98195
[2] *Center for Particle Astrophysics, UC Berkeley, CA 94720*

WITH

C. ALCOCK, R.A. ALLSMAN, D. ALVES, T.S AXELROD,
A. BECKER, D.P. BENNETT, K.H. COOK, K.C. FREE-
MAN, K. GRIEST, J. GUERN, M. LEHNER, S.L. MAR-
SHALL, B.A. PETERSON, M.R. PRATT, P.J. QUINN, D. REISS,
A.W. RODGERS, W. SUTHERLAND AND D.L. WELCH

(THE MACHO COLLABORATION)

Center for Particle Astrophysics, Berkeley, CA 94720

Lawrence Livermore National Laboratory,
Livermore, CA 94550

Mt. Stromlo and Siding Spring Observatories,
Australian National University, Weston, ACT 2611, Australia

Abstract. Gravitational microlensing is the most straightforward interpretation of the stellar brightenings that have been observed by our team
and other experiments. These data have provided some of the most stringent limits to date on the nature of the Galaxy's dark matter halo. The
number of events seen towards the LMC indicate that our Galaxy is not
surrounded by a "standard" halo of MACHOs in the mass range of 10^{-6} to
0.3 solar masses. The observed optical depth towards the Galactic Center
is an important constraint on the distribution of mass in the plane of the
Galaxy.

C. S. Kochanek and J. N. Hewitt (eds), Astrophysical Applications of Gravitational Lensing, 209–214.
© 1996 IAU

1. Introduction

The MACHO collaboration is searching for gravitational microlensing along three lines of sight, towards the LMC, the SMC and through the disk towards the Galactic center. The main objective of the project is to investigate the amount of halo dark matter that resides in lensing objects. The experiment uses a dedicated 1.3m telescope at Mt. Stromlo with a dual color mosaic CCD camera system that spans 0.5 square degrees. Photometry is performed with a PSF-fitting code that is derived from DoPhot.

I want to make three main points in this talk: 1) Microlensing has in fact been detected, 2) The measured optical depth towards the LMC sets stringent limits on the nature of the Galactic dark halo, and 3) The optical depth towards the Galactic center exceeds early predictions, and understanding this will have significant implications for our understanding of the mass distribution of the Galaxy. The reader is also referred to the talks by my colleagues M. Pratt and D. Bennett for other interesting results from the MACHO project.

2. Microlensing Has Been Detected

When the microlensing surveys were started a few years ago, it was not at all clear that genuine events could be distinguished from background processes that would mimic microlensing. On an event-by-event basis we adopted a number of conditions to consider an excursion to be microlensing. At the time of this writing our collaboration has observed over 75 events that satisfy these criteria. The overwhelming majority are seen in the direction of the Galactic center. Other teams have also detected a significant number of candidate events, as described in other submissions to this volume.

Lack of space precludes my describing all the observed event characteristics in detail. There are two particular events that I think lend substantial credibility to the microlensing interpretation. In September of 1994 we announced (Alcock et al. 1994) the detection of a candidate event well before peak amplification. This allowed the acquisition of spectra during the event (Bennetti et al. 1995). The spectra show no evidence of variation during the course of the event. This achromaticity is a hallmark of gravitational lensing, and is strong support for the microlensing interpretation. Another event that is difficult to account for with any known intrinsic stellar variability is a case with fitted peak amplification of 18, with detected data points at A=15. There is no known process other than microlensing that can give rise to such an increase in stellar flux with the observed temporal character and color-independence. Microlensing is the most straightforward interpretation.

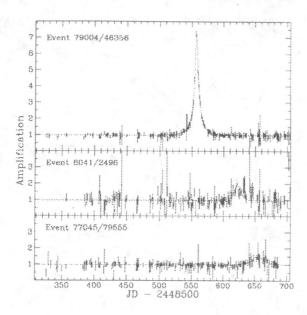

Figure 1. The three microlensing events from the MACHO project's first year LMC data. Relative flux is shown vs. time.

Any set of candidate microlensing events must then exhibit the requisite distributions in amplification and in the CM diagram. Our set of 45 candidate microlensing events from the first year's data towards the galactic bulge satisfy these requirements.

While I am certainly not claiming that *all* of our group's candidate events are due to microlensing, I do think that it has been established that the microlensing phenomenon is responsible for the overwhelming majority of them.

3. Microlensing Towards the LMC, and The Galaxy's Dark Halo

As stressed by Paczynski (1986) the optical depth towards the LMC is a sensitive way to determine the halo content of lensing objects. The LMC is the prime observing target for our experiment, and we have now completed a full analysis of the first year's data. We detected three candidate events in the first year's sample of ~ 8.6 million stars. The three first year LMC events that pass our selection criteria are shown in Figure 1. The experimental detection efficiency was the key ingredient that was needed in order to compare the number of detected events with the number predicted in a given halo model.

We have now completed an in-depth determination of the experiment's

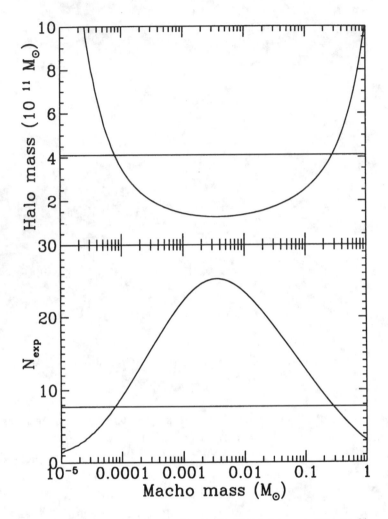

Figure 2.   Constraint curves from the first year LMC data set. The 95 % upper bound on the underlying event rate is 7.7 detected events, shown as the horizontal line in the lower panel. The efficiency-corrected expected number of events is the parabolic curve, assuming a δ function mass model, with mass on the x-axis. A canonical halo of MACHOs of mass between 0.00007 and 0.3 solar masses is ruled out with high confidence. The upper panel shows the experiment's limits on total halo mass.

event detection efficiency, which is described in some detail in Alcock et al. (1995b). The fact that we operate in extremely crowded fields gives rise to two (partially offsetting) effects. First, there are more stars being monitored than one would naively expect from counting the number of "objects" that the photometry code detects. Most "objects" are in fact blends of one or more stars. This has two effects: the likelihood of detecting an event is increased since there are multiple source stars per PSF, but also there will be unamplified flux when a "blended" star is lensed, and this dilutes the observed amplification. We have added artificial stars to a set of frames that spans the observed seeing and sky brightness, and have determined the photometry code's response to adding flux to these objects. This set of transfer functions is used as an element in a Monte Carlo simulation that adds appropriately degraded microlensing events to a random 1% of the data set. This takes into account the actual spacing of the observations as well as blending effects and produces a matrix of detection efficiency data in amplification, event duration, and apparent magnitude of the object. Since the distribution in amplification is known a priori, we integrate out the amplification dependence, and similarly use a completeness-corrected luminosity function to integrate out the magnitude dependence. We are left with a vector of efficiency as a function of event duration that is applicable to the full data set.

The dark matter implications of our first year's LMC data have been described in two publications, Alcock et al. (1995ab). If one adopts the attitude that the detected candidate events set a *upper* limit on the event rate towards the LMC, the 95% upper bound on the mean detected event rate is 7.7 events. Figure 2 compares this with the efficiency-corrected expected rate for a canonical $\rho(r) = \rho_0(a^2 + r^2)^{-1}$ halo, assuming for simplicity a delta function mass distribution in lensing objects.

A "standard" halo of lensing objects between 8×10^{-5} and 0.3 solar masses (assuming a delta function mass distribution) is excluded with high confidence. Taken in conjunction with the null results reported in searches for shorter duration events (from the EROS CCD experiment and an unpublished analysis of the MACHO data for short events), I think it is fair to conclude that the microlensing data preclude a "standard" halo composed exclusively of objects with masses between 10^{-6} and 0.3 solar masses.

Naturally the next question is whether the data can accommodate other halo models. This has been treated in Alcock et al. (1995b), where we analyze a variety of scenarios, ranging from maximal disk models to the standard halo described above. It turns out that the microlensing data provide an almost model-independent constraint on the total amount of *mass* that resides in MACHOs, but if one wants to determine the *halo fraction* that this represents then our ignorance of the halo is the major limitation. It is worth stressing that the amount of dark matter support for the rotation curve at the solar circle is one of the main differences between the various contending models, and this has a direct impact on laboratory direct detection experiments for particle dark matter. I will return to this point below.

4. Microlensing towards the Galactic Center.

The overwhelming majority of the events we detect are along lines of sight towards the Galactic Center. Both our team and the OGLE collaboration have estimated the bulge optical depth to be in the range of 2-4 $\times 10^{-6}$, which exceeds the early predictions that assumed an axisymmetric Galactic mass distribution.

Clearly, one of the main tasks for the microlensing experiments is to map out the optical depth as a function of galactic latitude and longitude. These data must then be used in conjunction with COBE data, the rotation curve observations, high-z tracer stars, faint star counts, and other constraints to refine our knowledge of Galactic structure. Some work along these lines is beginning to appear.

Of particular interest to the dark matter community is the question of whether the maximal disk model is consistent with the data. The amount of dark matter at R=8-10 kpc determines the event rate in *both* microlensing and particle dark matter direct detection experiments. This in turn depends sensitively on the extent to which the local rotation is supported by disk matter. Bulge microlensing will play in important role in establishing this, in my opinion.

The MACHO collaboration's tally of bulge events now exceeds 75, including some exotica like lensing by binary systems (2) and one "parallax" event that shows clear effects of the earth's orbit around the sun.

5. Conclusions

Microlensing has been seen. The LMC data already set stringent and interesting constraints on the nature of the Galaxy's dark halo. The Galactic bulge is giving us an abundance of events that will lead to a deeper understanding of Galactic structure.

References

Alcock, C., et al., 1994, IAUC 6068
Alcock, C., et al., 1995a, PRL, 74, 2867
Alcock, C., et al., 1995b, ApJ, submitted (also astro-ph/9506113)
Benetti, S., Pasquini, L. & West, R., 1995, A&A, 294, L37
Paczynski, B, 1986, ApJL, 304, L1

FIRST RESULTS OF THE DUO PROGRAM

C. ALARD

Centre d'Analyse des Images de l'INSU,
Batiment Perrault, Observatoire de Paris,
61 Avenue de l'Observatoire,
F-75014, Paris, France

Abstract. The DUO (Disk Unseen Objects) program is a project with the main goal of searching for galactic dark matter in a field towards the Galactic Bulge (Alard et al. 1995b). In this paper I present the results obtained from the analysis of half of the data collected during the 1994 season. I start with a brief description of the DUO project, and I will continue with an analysis of the microlensing candidates that I found. In all I found 13 microlensing events in the DUO data, including a double lens event. A model of the double lens light curve predicts a blending effect. The presence of the blended component was confirmed by a gravity center shift during the event. Taking into account the model prediction and the shift, we could predict the exact geometry of the blend. This prediction was recently confirmed by direct imaging of the blend under good seeing conditions. The durations of the single lens events are quite short, but the durations can be seriously affected by blending. I propose a simple experimental test to quantify this duration bias. In the conclusion I emphasize the great importance of a good knowledge of the inner Galactic structure to get a reliable quantitative description of microlensing towards the bulge. Finally I will show that the large number of variables stars found in the data could be a powerful tool to probe this structure.

1. Introduction

The data presented here were collected during the 1994 observing season with the ESO 1m Schmidt telescope. The season lasted from April 5 to October 10. Our sampling contains some gaps due to moon periods and to

C. S. Kochanek and J. N. Hewitt (eds), Astrophysical Applications of Gravitational Lensing, 215–220.
© 1996 IAU

poor weather conditions, especially at the beginning of the season.

The Schmidt telescope provides photographic plates of 28cm×28cm. The scale is 67″/mm, giving a wide field of 5.2x5.2 degrees. Two B plates (IIIaJ) and one R plate (IIIaF) were taken during each photometric night. The mean time interval between two B plates was 1.5 hours, with the R plate taken just in between.

1.1. PLATE SCANNING

All the plates were scanned with the MAMA machine in Paris Observatory. Mama (Machine Automatique a Mesurer pour l'Astronomie, developed and operated by INSU/CNRS) is a high speed microdensimeter providing images with a pixel of 10 μm. All the images were reduced with the standard package of the DUO project (Alard 1995).

1.2. THE LIMITING MAGNITUDES

The bulge stars that can act as sources for microlensing have two different locations in a typical color magnitude diagram: the giant clump, and the region of the bulge turn-off. The DUO field is rather far from the Galactic Center, so the giant clump is not very populated; as a result, most of the bulge stars will come from the turn-off region. In the B band, the limit is just faint enough to reach the turn-off proper, but in R the limit is a few hundredths of a magnitude brighter. This implies that most of the events are found in the B band, and quite close to the limit; in R most of these events are a little bit too faint to provide reliable light curves, especially if the star is badly blended. By chance the double lens event was well separated from its neighbors on the plate and it had a large amplitude, which leads to a rather nice light curve in R. The other events are all blended and very faint on the R plates. Thus, the main problem is that for such very faint fluxes the response of the plate is almost null and highly non-linear. In addition, the response of the plate is seriously affected if a brighter companion lies within a few pixels from the event. It makes the extracted signal very noisy and almost impossible to calibrate.

2. The Microlensing candidates

2.1. THE SELECTION OF THE CANDIDATES

The candidates were selected using a method similar to the one adopted by OGLE (Udalsky et al. 1994). However the definition of the baseline value had to be changed, because OGLE uses the measurements taken in the other season to estimate the baseline, while DUO has only 1 season

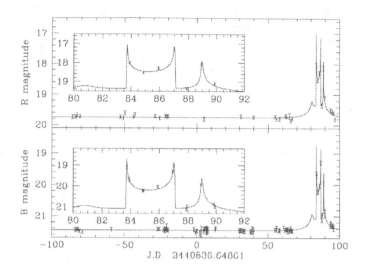

Figure 1. Light curve of DUO2

of observations. I defined the baseline value as the minimum average of 5 consecutive points all over the light curve. All curves with more than 4 consecutive deviating points over 3 sigmas were selected, and fitted to the theoretical light curve. In the last step, a theoretical microlensing light curve was fitted to the data, and the standard rms deviation r1 calculated. The rms r2 to the fit of a constant line was also computed. The final criterion was that the ratio r2/r1 should be greater than 2.

2.2. A DOUBLE LENS

One of the microlensing candidates discovered, DUO 2, shows very unusual variability. The star remained stable for more than 150 days before it brightened by more than two magnitudes in 6 days in both the B and R bands. The light curves are achromatic during the variability (Figure 1).

A binary lens model was fitted to the observations (Alard et al. 1995a). The masses of the lenses are quite small, with the companion possibly in the range of a brown dwarf or even a few times of Jupiter. Our first model did not include any blending, and one point gave a 4 sigma deviation in both the B and R bands. After adding a blending parameter we had a very good fit to the data, although with about 10 data points during the event we can wonder about the significance of this result. But we found other strong evidence for blending: a gravity center shift was observed during the event at a 4 sigma confidence level. This simply means that the gravity center is shifted towards the amplified component during the event. The

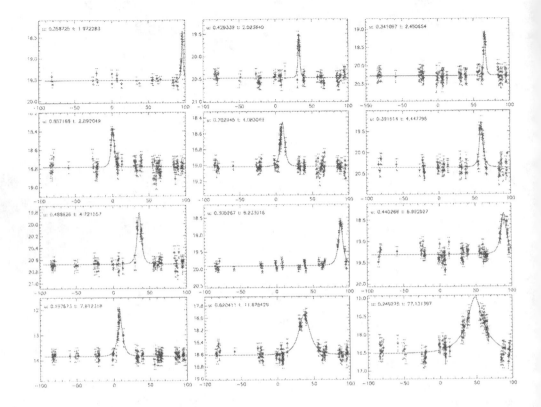

Figure 2. Light curves of the 12 single lens events in the B band (the x axis is graduated in days)

model provides the flux ratio of the components which, combined with the shift in both coordinates, allows us calculate all the blend parameters: the separation, the angle between the components, and their flux ratio.

This blend model was recently tested using two CCD images taken under good seeing conditions (Van Der Hooft 1995; Szymański & Udalski 1995). Both images show two components. Their separation, position angle, and flux ratio are in good agreement with our model.

2.3. THE SINGLE LENS EVENTS

Twelve candidates satisfied all the cuts described in §2.1. In figure 2 we can see their light curves in the B band. The events are in order of increasing durations. Many of them have quite short durations, in the range 1.8 days to 5 days, which could be consistent with lenses like low mass stars or brown dwarfs. However, the duration can be seriously affected by blending, and high resolution images are needed before reaching any definitive conclusions about the lens characteristics of these events.

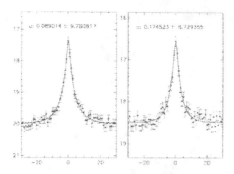

Figure 3. Two simulated light curves with a duration of 10 days: (left) an unblended light curve, and (right) the star blended with a same magnitude companion. The impact parameter u, and the duration t, of the unblended fitted light curve are indicated. Notice the bias in the estimated duration of the right curve.

2.4. BIAS IN THE EVENT DURATIONS

In most cases a theoretical unblended light curve can be fitted to a blended event with a good agreement, as illustrated in figure 3. But we notice that if the fit is consistent with the errors of a typical light curve event like in figure 3, the impact parameter u, and the duration t, are seriously affected. The event durations strongly decrease with the blending effect. This effect would induce an overestimate of the lens mass.

A possible experimental test: Among the two kinds of possible sources for microlensing, one is quite bright: the clump giants. These stars are brighter than most of the other stars in a bulge field, so the effect of blending on the parameter estimate should be negligible. Following this idea, it is quite straight forward to compare the duration of events for these clump stars to the duration of events for the fainter turn-off stars. We should find that the mean duration of the events for clump stars is longer than for the turn off stars. Unfortunately there are only 2 clump events for DUO, but it would be probably rather simple to check this effect in the MACHO experiment.

3. The Inner Galactic Structure

The first step before any serious modeling of microlensing towards the bulge is to determine the structure of the inner Galaxy. At the present time the existence of a bar near the center of our galaxy is well established (Weinberg 1992), but the parameters of the bar model are still poorly known. The inclination of the bar and its size have a very significant effect on the microlensing rate. But some questions about the structure of the disk are also of great importance: does the density of the disk drop at a few kilopar-

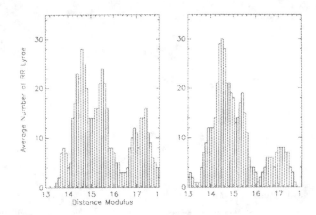

Figure 4. RR lyrae histograms of 2 selected regions (see text for explanations)

sec from the center (Paczyński et al. 1994)? The large number of variable stars found in the data could seriously help answer these questions. I illustrate this idea by showing some histograms of the RR Lyrae variable distribution. Figure 4 presents histograms of two selected regions with low and homogeneous extinction. The bulge is visible as the peak at a distance modulus of 14.5 (8 Kpc), but a strong peak at 15.5 (12.5 Kpc) is also visible, and we can guess that there is another density peak around 6 Kpc. We also notice a clump in the histogram around a distance modulus of 17 (25 Kpc), that is probably associated with the Sagittarius dwarf Galaxy. These results are completely new, and could be the starting point for some new modeling of microlensing towards the bulge.

Acknowledgements

It is a great pleasure to thank B. Paczyński for valuable suggestions and discussions. I would also like to thank G. Gilmore and L. Blitz for helpful discussions.

References

Alard, C., 1995, in preparation
Alard, C. et al., 1995a, A&A, in press
Alard, C. et al., 1995b, in preparation
Paczyński, B. et al., 1994, AJ, 107, 2060
Udalski, A. et al., 1994, Acta Astron, 44, 165
Szymański, M. & Udalski, A., 1995, private communication
Van Der Hooft, F., 1995, private communication
Weinberg, M., 1992, ApJ, 384, 81

REAL-TIME DETECTION OF GRAVITATIONAL MICROLENSING

M.R. PRATT
Department Astronomy,
University of Washington, Seattle, WA 98195

FOR

C. ALCOCK, R.A. ALLSMAN, D. ALVES, T.S. AXELROD,
A. BECKER, D.P. BENNETT, K.H. COOK, K.C. FREEMAN,
K. GRIEST, J. GUERN, M. LEHNER, S.L. MARSHALL, B.A. PE-
TERSON, P.J. QUINN, D. REISS, A.W. RODGERS, C. STUBBS,
W. SUTHERLAND AND D.L. WELCH

(THE MACHO COLLABORATION)

Center for Particle Astrophysics, Berkeley, CA 94720

Lawrence Livermore National Laboratory,
Livermore, CA 94550

Mt. Stromlo and Siding Spring Observatories,
Australian National University, Weston, ACT 2611, Australia

Abstract. Real-time detection of microlensing has moved from proof of concept in 1994 (Udalski et al.1994a, Alcock et al.1994) to a steady stream of events this year. Global dissemination of these events by the MACHO and OGLE collaborations has made possible intensive photometric and spectroscopic follow up from widely dispersed sites confirming the microlensing hypothesis (Benetti 1995). Improved photometry and increased temporal resolution from follow up observations greatly increases the possibility of detecting deviations from the standard point-source, point-lens, inertial motion microlensing model. These deviations are crucial in understanding individual lensing systems by breaking the degeneracy between lens mass, position and velocity. We report here on GMAN (Global Microlensing Alert Network), the coordinated follow up of MACHO alerts.

C. S. Kochanek and J. N. Hewitt (eds), Astrophysical Applications of Gravitational Lensing, 221–226.
© 1996 IAU

1. Introduction

The MACHO project is engaged in an ongoing, time resolved survey of stars in the Magellanic Clouds and galactic bulge to search for microlensing by intervening compact objects[1] (Paczyński 1986, Alcock et al.1995a). The apparent amplification of a lensed star varies over time with the relative motion of star, lens and observer. This is given by

$$A = \frac{u^2 + 2}{u\sqrt{u^2 + 4}} \quad \text{with} \quad u(t) = \sqrt{u_{min}^2 + \left(\frac{2(t - t_{peak})}{\hat{t}}\right)^2} \qquad (1)$$

where $u = b/R_E$ is the impact parameter in units of the Einstein radius

$$R_E = \sqrt{\frac{4GmD_{OL}D_{LS}}{D_{OS}c^2}}. \qquad (2)$$

Here D_{OL}, D_{LS} & D_{OS} are distances between observer, source and lens and m is the mass of the lens. The time scale of an event is given by $\hat{t} = 2R_E/v_\perp$, where $v_\perp$ is the relative perpendicular motion of the unperturbed line of sight and lens. This simple functional form is based on point–source and point–mass approximations and an assumption of uniform linear motion between observer, source and lens. In this regime only $\hat{t}$ provides useful information about the lensing system.

It is possible to extract additional information about the lensing system if one can detect deviations from this simple form. These include:

– microlensing parallax due to the accelerated motion of earth throughout the event
– resolution of the finite size of the source star by the lens
– lensing of binary or coincident sources
– lensing by binary or planetary systems.

At least three of these have been detected (Udalski et al.1994b, Bennett et al.1995, Alard 1995). Both parallax and finite–source models provide additional information about the lens velocity, breaking some of the degeneracy between the physical parameters making up $\hat{t}$. These effects are present at low levels in a significant fraction of events but are usually undetectable with the present photometric precision and time coverage. For example, caustic crossing events and planetary perturbations can produce significant flux changes on time scales of an hour. The survey systems in current operation are designed to maximize the number of detected events and are not well suited for monitoring single events at this rate. As there

---

[1]see also Stubbs et al.and Bennett et al., this volume

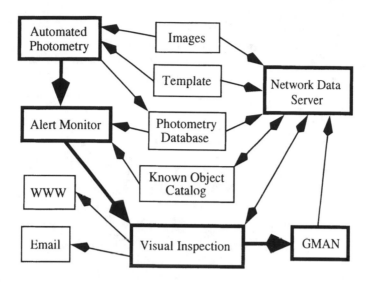

Figure 1. Information flow in the MACHO alert system.

are numerous, under-subscribed, small aperture telescopes with instrumentation adequate for microlensing follow up it is not desirable to divert the survey instruments to monitor individual events in progress. Rather it is beneficial to coordinate observations of both the survey system and follow up telescopes; using all available data to assist in prioritizing observations. In addition, with the development of more sophisticated triggers, the follow up system may become an integral part of the survey in the verification of events.

2. Global Microlensing Alert Network

Global Microlensing Alert Network (GMAN) is the coordinated follow up of MACHO microlensing alerts. The process by which events are identified and scheduled is shown in Figure 1. The two–color images of each night's survey data are processed in real–time at Mount Stromlo Observatory. This is accomplished with SoDophot, a special purpose photometry code that makes use of a template image taken in good conditions to "warm–start" the photometry. If a star has varied significantly from its "template" value it is reported as anomalous and an automated analysis is performed. Stars which have 7σ high-points in both passbands, pass a variety of data quality cuts, and are not listed as variables in the "Known Object Catalog" are reported to collaboration members together with best fit microlensing parameters. Another automated process downloads full light curves of the reported objects to each collaborator's local computer for interactive viewing and analysis. This process produces on the order of ten low level alerts per day. Although aided by the automated analysis, the process of identifying microlensing events is ultimately done by a human after close inspection of the light curve and template image. The event is then posted to the MACHO Alert WWW page[2] and email is sent to subscribers of the MACHO alert email service.[3] The star is also scheduled for observation at GMAN observatories.

Telescopes participating in GMAN are

- MSO 30inch, Australia: priority to microlensing follow up
- CTIO 0.9m, Chile: roughly 1 hour of service observing every night
- UTSO 0.61m, Chile: 50 full nights of service observing in 1995
- Wise Observatory 1.0m, Israel: several hours per week
- Mt. John 24inch, New Zealand: priority to microlensing follow up.

GMAN images are processed at the site where they are acquired. Photometry is accomplished with IRAF scripts calling DaophotII (Stetson 1987). These data are stored on site as photometry files from individual images. Photometry is usually finished within 12 hours of image acquisition and available to team members to aid in subsequent scheduling. Normalization is performed on-the-fly by the "Network Data Server" using a list of reference stars obtained from the MACHO data set.

Figure 2 shows the MACHO data on the 26th alert of the 1995 season together with the GMAN data taken at the CTIO 0.9m telescope. This event was identified approximately 10 days before peak, allowing good coverage throughout the peak from the GMAN site. In most cases, the GMAN measurements exceed the the precision of the survey data and are therefore more likely to expose irregularities in the light curve.

A more spectacular demonstration of the success of GMAN is the first real–time detection of an unusual event in progress. This event, shown in Figure 3, was found to deviate from standard microlensing significantly before the time of the first peak. The object was then placed at the highest priority in the GMAN observing schedule and a spectrum was obtained at the AAT. It is likely that this event is either a binary source, coincident source or binary lens. Fits to multiple source or multiple lens systems obviously require many additional parameters and the multiple lens fit in particular is notorious for it's convergence problems. The additional GMAN data should greatly improve the quality and reliability of these fits. The data in Figure 3 have been normalized to unit baselines using an 11–parameter binary source fit. The fit parameters are still evolving and are not shown.

---

[2]http://darkstar.astro.washington.edu
[3]subscribe by mailing to macho@astro.washington.edu

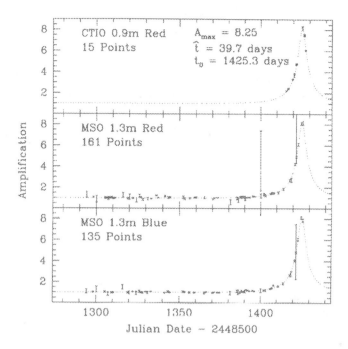

Figure 2. MACHO Object 21161_7671 (alert 95-26). The microlensing fit uses the three shape parameters described in Sect. 1 and an additional baseline parameter for each passband. An additional year of data from the '93 bulge season is used in the fit to better establish baselines but is not visible in this window. The CTIO data are preliminary and subject to change.

3. Conclusion

Both MACHO and OGLE collaborations have demonstrated the ability to distinguish with high reliability microlensing events in progress while discriminating against a very large background of variable stars. The spectroscopic follow up of MACHO alert 94-1 (Alcock et al.1994, Benetti 1995) has provided perhaps the most convincing case for detection of microlensing. More than 30 events have been detected in progress as of this writing. At least one of these events deviates significantly from the standard microlensing model. Development of more robust discrimination and more sensitive triggering should increase the detection rate. In closing, the ability to detect microlensing events in progress is crucial for such future microlensing endeavors as ground based planet searches (Paczyński 1991) and microlensing parallax measurement from a satellite in solar orbit (Gould 1994).

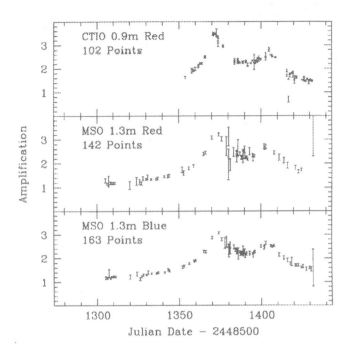

Figure 3.   MACHO Object 21263_1213 (alert 95-12). There are a year of data from the
'93 bulge season that have been used in the fit to provide accurate determinations of the
quiescent luminosity but are not visible in this window. The CTIO data are preliminary
and subject to change.

Acknowledgments

We are very grateful for the skilled support given our project by S. Chan
and the technical staff at the Mt. Stromlo Observatory. We especially thank
J.D. Reynolds for the network software that has made this effort successful.

References

Alard, C., Mao, S., & Guibert, J., 1995, A&A, in press
Alcock, C., et al., 1994, IAU Circulars 6068 & 6095
Alcock, C., et al., 1995, Phys Rev Lett, 74, 2867
Benetti, S., Pasquini, L., & West, R. M., 1995, A&A, 294, L37
Bennett, D.P., et al., 1995, in Dark Matter, Procs. of the 5th Ann. Maryland Conference,
    ed. S. Holt, in press
Gould, A., 1994, ApJL, 421, L75
Paczyński , B., 1986, ApJL, 371, L63
Paczyński , B., 1991, ApJL, 374, L37
Stetson, P.B., 1987, PASP, 99, 191
Udalski, A., et al., 1994a, Acta Astron, 44, 227
Udalski, A., et al., 1994b, ApJL, 436, L103

THE PLANET COLLABORATION

Probing Lensing Anomalies with a world-wide NETwork

M. ALBROW[1], P. BIRCH[2], J. CALDWELL[1], R. MARTIN[2],
J. MENZIES[1], J.-W. PEL[3], K. POLLARD[1], P. D. SACKETT[3],
K. SAHU[4], P. VREESWIJK[3], A. WILLIAMS[2] AND M. ZWAAN[3]
(The PLANET Collaboration)
[1] *South African Astronomical Observatory, Cape Town, South Africa*
[2] *Perth Observatory, Perth, Australia*
[3] *Kapteyn Astronomical Institute, Groningen, The Netherlands*
[4] *European Southern Observatory, Garching, Germany*

Abstract. A newly-formed microlensing monitoring network, the PLANET collaboration, is briefly described.

PLANET (**P**robing **L**ensing **A**nomalies **NET**work) is a worldwide collaboration formed in the spring of 1995 to meet the challenge of microlensing monitoring. Its primary goal is the study of microlensing anomalies — departures from an achromatic point-source, point-lens light curve — through rapidly sampled, multi-band, photometry. Such departures are expected due to blends along the line-of-sight, sources and/or lenses with complex geometries (e.g., binary lenses and sources), resolution of the source star, and complicated relative motion within the lens system (e.g., parallax effects). In particular, microlensing monitoring is a powerful means of searching for extra-solar planets: the caustic patterns arising from the complicated lens geometry of a planetary system could induce sharp peaks in the light curve with durations of a few hours to a few days. Depending on the planetary masses and orbital radii, a significant fraction of detectable caustic crossings are expected (Mao & Paczyński 1991, Gould & Loeb 1992), and in principle, even earth-size planets can be detected in this way.

Our first campaign is now in progress. With dedicated access in June-July 1995 to a suite of southern telescopes, we are monitoring several ongoing bulge events per night in the direction of the Galactic center. The large number of microlensing alerts issued by the MACHO (Pratt et al. 1995) and OGLE (Udalski et al. 1994) teams has kept the PLANET tele-

C. S. Kochanek and J. N. Hewitt (eds), Astrophysical Applications of Gravitational Lensing, 227–228.
© 1996 IAU

Figure 1.   The PLANET 1995 bulge campaign is being conducted with three widely separated southern telescopes: **P**: Perth Observatory 0.6m at Bickley, Australia, **S**: South African Astronomical Observatory 1.0m at Sutherland, South Africa, and **L**: Dutch-ESO 0.9m at La Silla, Chile.

scopes busy whenever the bulge is visible. The telescopes are widely separated in longitude (Fig. 1), giving us the possibility of nearly continuous monitoring and a hedge against bad weather: to date, at least one PLANET telescope has collected useful data in every 24-hour period. In the current campaign, we sample the light curves every 1–2 hours, and are experimenting with higher sampling rates. The monitoring is done in the V and I bands. By performing preliminary crowded-field photometry at each observing site against a set of common secondary standards in each field, we are able to track the progress of the events in real time. Through extensive analysis of this season's large, high-quality data set, we expect the capabilities of PLANET to grow as we refine our strategies for future campaigns.

Acknowledgements: The PLANET collaboration wishes to thank the MACHO and OGLE teams for their dedication in providing prompt, electronic alerts of microlensing events. P.D.S. acknowledges the Leids Kerkhoven-Bosscha Fonds and the International Astronomical Union for travel support to Melbourne.

References

Udalski, A., et al., 1994, Acta Astron, 44, 227
Gould, A. & Loeb, A., 1992, ApJ, 396, 104
Mao, S. & Paczyński, B., 1991, ApJL, 374, L37
Pratt, M.R., 1995, these proceedings.

THE CONTRIBUTION OF BINARIES TO THE OBSERVED GALACTIC MICROLENSING EVENTS

M. DOMINIK AND A. C. HIRSHFELD
Institut für Physik, Universität Dortmund
D-44221 Dortmund, Germany

Abstract. Most of the detected galactic microlensing events are commonly explained as due to microlensing of a pointlike source by a pointlike lens. We have found statistical methods to determine the goodness-of-fit for a specific model of source and lens beyond the assumption that the errors are normal, which does not always hold. In particular, we argue that at least one weak binary lensing event has already been detected (MACHO LMC#1) and thereby confirm our former hypothesis (Dominik & Hirshfeld 1994). We also find that this fit is not unique. We emphasize that knowledge of the fraction of binaries among the observed microlensing events is crucial for estimating the mass distribution of the observed dark objects.

1. Introduction

To draw the correct conclusions from the observed galactic microlensing events it is important to know if the model of source and lens used suffices to describe the data. An insufficient model may lead to errors concerning the time- and mass scales. A standard χ^2-test does not work when the photometry errors do not follow a normal distribution.

Our analysis therefore makes use of the following ideas to decide if a fit is reasonable or not

- Since the light curve should have a constant tail, we use the errors of the data points there to determine a constant rescaling factor.
- We allow distributions with larger tails having the form of Student's t-distribution.
- We look at the correlation between the deviations from the best-fit light curve of the data in two spectral bands.

C. S. Kochanek and J. N. Hewitt (eds), Astrophysical Applications of Gravitational Lensing, 229–230.
© 1996 IAU

2. Results

We find for the MACHO LMC #1 event that

- a goodness-of-fit analysis rejects fits for a pointlike lens with a point source and also with an extended circular source with uniform brightness
- the analysis of the correlation coefficient shows that a binary source is also unlikely to explain the event
- at least 6 binary lens models can be accommodated

We also find that

- OGLE #7 is clearly a strong lensing event.
- We do not see a clear indication for a binary lens in the OGLE #6 event; in particular we have found a binary source fit with a slightly lower χ^2 than that quoted by Mao & DiStefano (1995), which only shows an asymmetry, but no second peak.
- If we apply our rescaling procedure to the OGLE events assuming a normal distribution, we find that the OGLE #4 and OGLE #5 events are not acceptable as being due to microlensing of a point source by a pointlike lens.
- The EROS #1 event is compatible with microlensing of a point source by a pointlike lens.

We think that it should in principle be possible for all of the events to learn about the distribution of the errors from the tails of the light curves in order to enable goodness-of-fit tests between different models of the peak.

The test of the correlation coefficient seems to be another promising tool to reject fits, but its use is limited to the events observed by the MACHO collaboration up to now, since observation in two spectral bands at the same time is necessary.

Acknowledgements

We would like to thank S. Mao for some interesting and fruitful discussions, the MACHO collaboration for making available the data of the MACHO LMC #1 event over the computer network, the OGLE collaboration for making available the data of the OGLE #1...#7 events, the EROS collaboration, esp. A. Milsztajn, for sending us the data of the EROS #1 event, A. Udalski for his help in retrieving a paper and P. Schneider for some advice.

References

Dominik, M., & Hirshfeld, A. C. 1994, A&A, 289, L31
Mao, S., & Di Stefano, R. 1995, ApJ, 440, 22

BINARY MICRO-PARALLAX EFFECTS

S.J. HARDY AND M.A. WALKER
Research Centre for Theoretical Astrophysics
School of Physics A28, University of Sydney
NSW 2006, Australia
Internet: S.Hardy/M.Walker@physics.usyd.edu.au

1. Introduction

Of the large number of microlensing events detected towards the galactic bulge, there has been at least one clear case of a binary lens system, the OGLE 7 event. If this event had been observed simultaneously from three 1 m class ground-based telescopes during its second caustic crossing, an identification of the lens as a bulge or disk object and a determination of the orientation of the lens velocity on the sky could have been made.

Mao & Paczyński (1991) have shown that if the microlensing population is similar to the stellar population, then roughly 10 percent of all microlensing events should be due to binary lenses, and a significant fraction of these will exhibit caustic crossings. Thus, given sufficient warning – a capacity already demonstrated by the OGLE and MACHO teams in detecting microlensing events in real time – it should be possible to acquire the photometric measurements necessary for this experiment on future binary microlensing events. In conjunction with one other measurement, this permits a full determination of the basic parameters of the lens (Hardy & Walker 1995).

2. Expectation, Observation, and Interpretation

The crossing of a fold caustic during a binary lensing event leads to an extremely steep light curve, and this implies a sensitive dependence of the magnification on the source-lens-observer alignment at this time. It is this property that allows the detection of parallax effects from very short observing baselines.

C. S. Kochanek and J. N. Hewitt (eds), Astrophysical Applications of Gravitational Lensing, 231–232.
© 1996 IAU

A convenient means for describing such parallax information is through a temporal offset between essentially identical lightcurves observed at each telescope. For two telescopes separated by an Earth radius, this offset is approximately

$$\delta t \simeq 40 \left(1 - \frac{D_d}{D_s}\right) v_{\perp 200}^{-1} \quad \text{sec} \tag{1}$$

where D_d and D_s are the distance to the lens and source respectively and we have assumed that the relative velocity of the Earth-lens-source system is dominated by the transverse velocity of the lens, $v_\perp \simeq 200 v_{\perp 200}$ km s^{-1}.

Current technology is sufficient to measure such small time differences. Indeed, five minutes of observing a source such as OGLE 7 with a 1m class telescope produces a signal-to-noise of around 150, which translates to an uncertainty of roughly 30 sec in our ability to determine the temporal location of the light curve from a single photometric measurement. Thus, two hours of quasi-simultaneous data from telescopes separated by about an Earth radius would correspond to an error in δt of around 6 sec. For the OGLE 7 event, assuming that the lens was half a solar mass and half way along the line of sight to the source, $\delta t \simeq 70$ sec, implying a detection in excess of 10 standard deviations. Indeed, the experiment should still yield a significant result provided that the lens is not within 2 kpc of the source.

3. Conclusion

Observation of parallax from 3 ground-based observatories serves to constrain two lens parameters: the Einstein ring radius per unit mass, and the velocity of the lens on the sky. In conjunction with other observations this may then lead to a unique characterization of the lens – its mass, distance and velocity. This experiment can be performed with current technology and the experiment may be pursued as soon as the next binary lensing event displaying a fold caustic crossing is identified. Moreover, new programs are now starting which will acquire just this sort of data in search of planets orbiting the microlenses.

References

Hardy, S.J., & Walker, M.A., 1995, MNRAS, in press
Mao, S., & Paczyński, B, 1991, ApJL, 374, L37

MICROLENSING WITH BINARIES AND PLANETS

H.J. WITT[1] AND S. MAO[2]
[1] AIP, An der Sternwarte 16, 14482 Potsdam, Germany
[2] CfA, MS 51, 60 Garden Street, Cambridge, MA 02138, USA

1. Introduction

The ongoing microlensing experiments have now discovered more than 70 candidate events (Alcock et al. 1993, Bennett et al. 1994, Aubourg et al. 1993, Udalski et al. 1994). These experiments have put important constraints on the dark matter content of the Galactic halo (Alcock et al. 1995a) and yielded many interesting results about Galactic structure (Paczyński et al. 1994, Stanek et al. 1994).

A subclass of these lensing events are caused by binary lenses. Here we review the probability, current known events and discuss future prospects.

2. Cross Sections and Light Curves

The cross section of detecting binary systems is closely related to the caustic structures of binary and planetary systems. Depending on the binary and planetary separation and mass ratio distributions, the probability of binary and planetary lensing can be of the order of 5%–10% (Mao & Paczyński 1991) or even higher (Gould & Loeb 1992; Bolatto & Falco 1994).

The light curves of binary events are very diverse (Mao & Paczyński 1991 and references therein). The most dramatic ones are the caustic crossing events with sharp spikes. These can be unambiguously identified. Less dramatic ones will be more difficult to detect. The fitting of a binary light curve involves many parameters (Mao & Di Stefano 1995). Degeneracy may occur when the light curve has large errors and sparse sampling.

3. Candidate Events and Evidence for Blending

The first binary event, OGLE #7, was discovered by Udalski et al. (1994), and confirmed by Alcock et al. (1995b). It lasted for ~100 days and has two

C. S. Kochanek and J. N. Hewitt (eds), Astrophysical Applications of Gravitational Lensing, 233–234.
© 1996 IAU

sharp spikes, the second being resolved (Alcock et al. 1995b). The second binary event (Alard et al. 1995), DUO 2, is much shorter ($\approx$ 8 days) but has three peaks. This candidate is likely to be a very low mass system, with the secondary mass in the range of a brown dwarf or a giant Jupiter.

Remarkably, both events show blending (Udalski et al. 1994, Alard et al. 1995), i.e., an unlensed source is contributing to the total brightness of the light curve. For OGLE #7, the plateau between the spikes has a magnification of 2.4 in the I band, which is below the theoretical minimum of 3, therefore blending must be present (Witt & Mao 1995). For DUO 2, observations under good seeing revealed two components separated by $\approx$ 1″, just as predicted by the model. Blending seems to be rather pervasive. This may have important implications for the estimate of optical depth.

4. The Motion of the Images of a Binary

The motion of the centroid of the images of a binary system yields additional information and would further constrains the binary parameters (§2). For a star crossing a caustic, the light centroid can jump about 0.001–0.01″. This motion can be potentially detected (see e.g. Armstrong et al. 1995).

5. Outlook

The discovery of two binary events provides us with confidence that such events can be identified. Such search can naturally be extended to planetary mass regimes, although the time coverage has to be once or a few times per hour. Such a project is already being pursued (Tytler 1995), with the goal of finding hundreds of Earth-like objects in a decade.

References

Alard, C., Mao, S. & Guibert, J., 1995, A&A, in press
Alcock, C., et al., 1993, Nature, 365, 621
Alcock, C., et al., 1995a, Phys Rev Letters, 74, 2867
Alcock, C., et al., 1995b, in preparation
Armstrong, J.T., et al., 1995, Physics Today, 48 (5), 42
Aubourg, E., et al., 1993, Nature, 365, 623
Bennett, D.P., et al., 1994, in Dark Matter, ed. S. Holt, in press
Bolatto, A. D., & Falco, E. E., 1994, ApJ, 436, 112
Gould, A., & Loeb, A., 1992, ApJ, 396, 104
Mao, S., & Paczyński, B., 1991, ApJ, 374, L37
Mao, S., & Di Stefano, R., 1995, ApJ, 440, 22
Paczyński, B., et al., 1994, AJ, 107, 2060
Stanek, K., et al., 1994, ApJ, 429, L73
Tytler, D., 1995, talk at the Extra-Solar Terrestrial Planet Detection Workshop
Udalski, A., et al., 1994, ApJL, 436, L103
Witt, H.J. & Mao, S., 1995, ApJL, in press

POLARIZATION DURING CAUSTIC CROSSING

ERIC AGOL
Physics Department
University of California, Santa Barbara
Santa Barbara, CA 93106

Abstract. The limbs of hot stars are polarized due to electron scattering, but this polarization cancels out due to rotational symmetry. During microlensing, the star is amplified by different amounts across its surface so that the limb polarization no longer cancels out, but can be observable. Polarization can be much higher for a caustic crossing during microlensing by a binary lens than for a single lens.

1. Introduction

Measuring polarization during binary microlensing can be useful in a number of ways: 1) It will allow another test of stellar atmosphere theory. 2) It can be used to confirm that flux variations are due to microlensing and not due to other variable star phenomena (such as "bumper stars"). 3) It can determine the Einstein radius of the lens. 4) The polarization angle gives the direction of the velocity and the position angle of the binary lens on the sky.

2. Results

Figure 1 shows the amplification and percent polarization during a typical caustic crossing for stars of various radii relative to the Einstein radius. These results assume a pure electron scattering atmosphere and that the caustic is straight and the star is moving perpendicular to the caustic. The same parameters are used as in Schneider and Weiß (1986), their figure 9a. The polarization can be as large as 1% and the polarization angle flips twice as the caustic amplifies different parts of the star. The peak polarization is larger for smaller stars since the part of the star inside the caustic is more highly amplified than for a large star. This is the opposite dependence of polarization for a single lens, in which the larger stars are

235

C. S. Kochanek and J. N. Hewitt (eds), Astrophysical Applications of Gravitational Lensing, 235–236.
© 1996 IAU

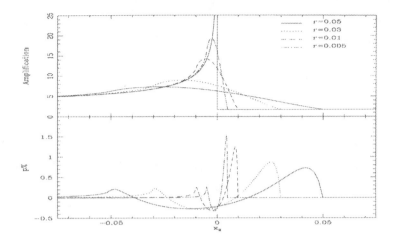

Figure 1. Amplification and percent polarization as a star crosses a caustic. The position of the center of the star is on the x axis for various r, which is the ratio of the radius of the star to the Einstein radius of the lens in the source plane.

polarized more since the change in amplification across the star is larger. The amplitude of the polarization and the separation of the zero crossings both are determined by the size of the star relative to the Einstein radius, so if the stellar radius is known (e.g. spectroscopically), then the Einstein radius of the lens can be determined from the polarization light curve. This can be compared with the size determined from the amplification light curve to see if there are other effects on the polarization, such as absorption.

Polarization is highest for hot stars (O and B) since electron scattering dominates the opacity in their atmospheres at certain wavelengths (especially just above the Lyman and Balmer edges). For these stars, a typical ratio of the stellar radius to Einstein radius is 0.005-0.02. There are no O and B stars in the Galactic bulge, so observations will have to be done towards the LMC or M31.

A caustic crossing will typically last for $\sim$ 10 hours which is long enough to be measurable with a large telescope for hot stars in the LMC. For a B=18 star observed for 100 minutes with a 4 meter telescope (assuming 10% efficiency), the polarization error is 0.05%.

More details on the calculations and further results can be found in Agol (1995).

References

Agol, E. 1995, MNRAS, submitted
Schneider, P., and Weiss, A., 1986, A&A, 164, 237

CHROMATIC AND SPECTROSCOPIC SIGNATURES
OF MICROLENSING EVENTS

DAVID VALLS–GABAUD

URA CNRS 1280, Observatoire de Strasbourg
11, Rue de l'Université, 67000 Strasbourg, France

Abstract. We predict that chromatic and spectroscopic effects are likely to appear for a significant fraction of microlensing events. Differential amplification across a stellar disc produces a time and wavelength dependent signature which constrains the parameters of the lenses and can be used for 3-D mapping of stellar atmospheres.

1. Introduction

The recent detections of microlensing events towards the LMC and the Galactic bulge raise constraints on the distribution of the lenses and sources. The large detection rate in the bulge might be due to lenses in the bulge itself and the low rate of the LMC could be explained by lenses within the LMC. If this is the case, a significant fraction of events is produced by lenses with an angular Einstein radius similar to or smaller than the angular radius of the stars and the structure of the source becomes important.

2. Chromatic and spectroscopic effects

Using the latest (1993) atmosphere models from Kurucz we have computed the limb-darkened profiles for a large variety of stellar models. Figure 1a presents simulated light curves in different photometric bands for a red giant star of solar metallicity. The amplitude of the effect is large enough (4 to 10%) to be be reached with differential photometry.

There is an associated spectroscopic effect due to the systematic variation of the line profile across the stellar disc. It depends on the nature of the line and on the optical depth gradient in the atmosphere. One can

237

C. S. Kochanek and J. N. Hewitt (eds), Astrophysical Applications of Gravitational Lensing, 237–238.
© 1996 IAU

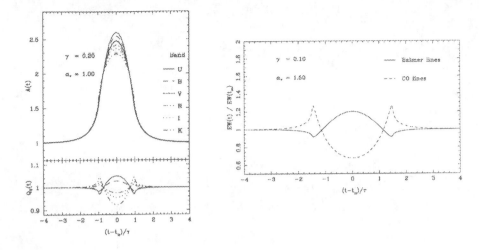

Figure 1.   *(a)* [left] chromatic signatures, $Q_V = A/A_V$, *(b)* [right] spectroscopic effect.

then envision microlensing tomography, a 3-D reconstruction of the stellar atmosphere based on the measure of the time-varying line profiles. Given the present instrumentation, an easier observable is given by the equivalent widths (EWs). We show in Figure 1b the variation of EWs for two different types of lines: the Balmer lines, and the infrared CO lines. Their different behavior reflects the gradient in temperature at the depth of formation.

3. Conclusions

A significant fraction of microlensing events could present a chromatic effect, with a systematic variation with wavelength and an amplitude between 4 and 10% in magnification. The use of photometric bands as widely separated in λ as possible improves the detectability of the effect. The amplitude of the spectroscopic effect depends strongly on the type of line. The variation in EW reaches the 40% level, and should be detectable with current instrumentation. If detected, these effects provide: (1) a unique proof of microlensing, as opposed to a new type of stellar variability; (2) constraints on the Einstein radius and reduced proper motion of the lens (assuming the stellar radius can be inferred); and (3) a new imaging method to map stellar surfaces, not only across the stellar disc, but also with depth.

References

Valls–Gabaud, D., 1994, in the 11th Potsdam Cosmology Meeting, eds. J. Mücket et al.,
     in press
Valls–Gabaud, D., 1995, A&A, in press

A POSSIBLE MANIFESTATION OF MICROLENSING IN PULSAR TIMING

T.I.LARCHENKOVA AND O.V.DOROSHENKO
Astro Space Center of P.N.Lebedev Physical Institute
Profsoyuznaya 84/32, Moscow 117810 RUSSIA

Abstract. Gravitational lensing and the time delay of a pulsar signal in the gravitational field of a mass are General Relativistic effects that may be used as a tool to detect the observational parameters of dark matter in our Galaxy. We propose to use observations of the time delay of pulses from pulsars to detect lensing objects located close to the line of the sight, to study the distribution of dark matter in our Galaxy. We discuss the possibility of finding such an event by measuring the delay of pulses from a pulsar, and apply it to data for PSR B0525+21.

1. Time delay and application to the pulsar B0525+21

Several single pulsars exhibit some unexplained distortions of the observed times-of-arrival (TOA) of their pulses. The differences between the observed TOAs and those calculated using the classical spin-down model of the pulsar rotation are the residuals of the TOA. The residuals are believed to be caused by some instability in the pulsar rotation. They might also be caused by a mass moving close to the line of the sight. The propagation delay of radio signals in the gravitational field of a massive object (the Shapiro effect) causes a sharp growth in amplitude of the TOA near the conjunction of the mass with the line of sight of the observer.

We analyzed the timing data for several pulsars obtained at the JPL by Downs & Reichley (1983) and Downs & Krause-Polstroff (1986) from 1968 to 1983. The data on the arrival times were reduced using the standard fit of the pulsar astrometric and spin parameters based upon the data reduction algorithm developed by Doroshenko & Kopeikin (1990). From an analysis of the observed residuals we concluded that amongst the studied objects

239

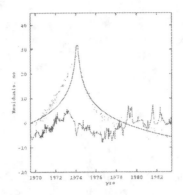

Figure 1.   The TOA residuals for PSR B0525+21. The circles show the observed TOA residuals after fitting for the pulsar spin and astrometric parameters. The solid line is the best fit Shapiro delay for a lensing mass. The dashed line is the post-fit TOA residuals.

PSR B0525+21 may present a case of micro-lensing. After subtracting the best fitting polynomial to the arrival times, we observe significant TOA residuals with a behavior similar to that caused by a mass passing close to the line of sight. The observed residuals of the TOAs of the pulsar are shown as the small circles on the Fig.1.

Our fitted values for the mass and other parameters of the lensing object are found to be: $M = 330 \pm 50\,M_{\odot}$; $V_P/d = (1.0 \pm 0.7) \times 10^{-7}\,\mathrm{s}^{-1}$; $T_0 = 2442040.0$ JD, where M is the lens mass, V_P/d is the ratio of the relative velocity of the pulsar to the impact parameter, and T_0 is the time of conjunction. With these values of the parameters the pre-fit and post-fit residuals are equal to 15ms and 3ms respectively. The results are shown in Figure 1.

Because pulsars are the fastest objects in Galaxy, we can suppose that the pulsar velocity is larger than the lensing object velocity. For B0525+21 $V_P \sim \mu R \approx 200\,\mathrm{km\ s}^{-1}$, so that our value for V_P/d gives the distance of closest approach $d \approx 13$ AU. One can suggest that the observed extra modulation is due to a time delay PSR B0525+21 caused by a mass $M = 330\,M_{\odot}$ passing near the line of the sight. We suggest that the gravitating mass may be a robust association of massive baryonic objects (Moore et al. 1995) or a black hole.

References

Doroshenko, O.V. & Kopeikin, S.M., 1990, SvA, 34(5), 496
Downs, G.S. & Reichley, P.E., 1983, ApJS, 53, 169
Downs, G.S. & Krause-Polstroff, J., 1986, ApJS, 62, 81
Moore, B. & Silk, J., 1995, ApJL, 442, L5

MICROLENSING INDUCED SPECTRAL VARIABILITY IN Q2237+0305

GERAINT F. LEWIS
Institute of Astronomy, Cambridge. UK

MIKE J. IRWIN
Royal Greenwich Observatory, Cambridge. UK

AND

PAUL C. HEWETT
Institute of Astronomy, Cambridge. UK

Abstract. The degree of microlensing induced amplification is dependent upon the size of a source. As quasar spectra consist of the sum of emission from different regions this scale dependent amplification can produce spectral differences between the images of a macrolensed quasar. This paper presents the first direct spectroscopic evidence for this effect, providing a limit on the scale of the continuum and the broad line emission regions at the center of a source quasar (2237+0305). Lack of centroid and profile differences in the emission lines indicate that substructure in the broad emission line region is > 0.05 parsecs.

1. Introduction

Several macrolensed systems exhibit photometric variability consistent with microlensing due to objects of stellar mass located in the lens. The degree of microlensing amplification is dependent upon the size of the source, with smaller sources being more amplified. In general, amplification of sources larger than an Einstein radius projected onto the source plane is negligible. For the quasar Q2237+0305, a quadruple–image lens (Huchra et al. 1985), this radius is 0.05 pc, larger than the predicted size of a continuum–emitting accretion disk, but substantially smaller than the broad line region (Figure 1). This scale difference implies that the continuum will be ampli-

C. S. Kochanek and J. N. Hewitt (eds), Astrophysical Applications of Gravitational Lensing, 241–246.
© 1996 IAU

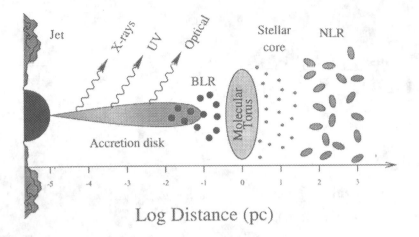

Log Distance (pc)

Figure 1. A schematic representation of the scales of structure at the center of active galactic nuclei (Rees 1984).

fied while the broad line emission remains essentially unchanged during a microlensing event (Sanitt 1971, Kayser et al. 1986).

The broad line emitting region, as a whole, is too large to be microlensed, but substructure on small scales may be significantly amplified. Although the total flux in the line is relatively unchanged, microlensing of substructure can result in changes in the shape of the emission line profiles, and produce measurable shifts in the central wavelength of the line (Nemiroff 1988, Schneider & Wambsganss 1990).

2. Observations and Spectral Modeling

Data were acquired at two epochs, 1991 August 14/15 and 1994 August 17/18, using the William Herschel Telescope at the Roque de los Muchachos Observatory, La Palma. The telescope was equipped with the ISIS double–beam spectrograph, providing a wavelength coverage of $\lambda\lambda 4000 - 8000$, including the emission lines of CIV $\lambda 1549$, CIII] $\lambda 1909$ and MgII $\lambda 2798$ in the quasar. The spectrograph slit was oriented to acquire spectra of pairs of images simultaneously. Direct R–band images were acquired at the Cassegrain auxiliary port, allowing the determination of the magnitudes of the four quasar components (Figure 2).

The spectroscopic CCD frames were debiased, flat fielded and sky subtracted using standard techniques. The resulting frames consisted of pairs of quasar spectra, separated by ~ 1", superposed upon the extended background of the lensing galaxy light. To extract the individual quasar spectra, a model was constructed to describe the spatial distribution of light at each pixel in the dispersion direction. The model consisted of two point spread

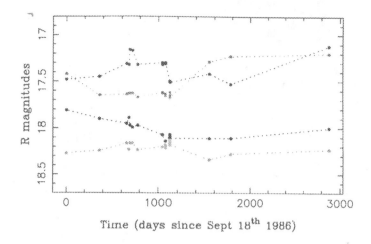

Figure 2. The *R*-band light curve for the images of Q2237+0305. The magnitudes are from the work of Houde and Racine (1994), the HST data of Rix et al. (1992) and our own observations (last two epochs).

functions (the quasars) and an extended profile to represent the galaxy light. Figure 3 presents an example of the model fit to a single spatial cut across the CCD frame.

3. Results

Figures 4 and 5 present the spectra of pairs of images, (A+C, A+D, B+C, B+D), from the 1991 observations. In each case, the spectrum of the fainter image (C or D) is scaled such that its continuum matches the brighter image (A or B). Excess emission line flux of the scaled spectrum over that of the other image is indicated by the solid black regions.

The photometry time series (Figure 2) shows that during the 1991 observations images A and C were faint in comparison to their previous behavior.

Epoch	A	B	C	D
1991	1.10	0.64	1.00	1.73
1994	0.70	0.70	1.00	–

TABLE 1. The equivalent widths of the CIV line. At each epoch the values have been normalized with respect to image C.

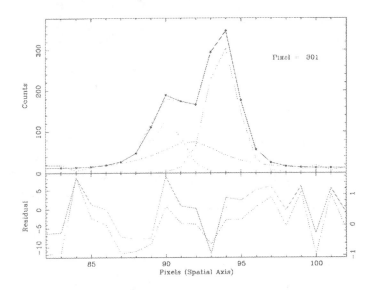

Figure 3. Spatial cross–section from a typical CCD exposure showing the model fit to the data. The model includes three components, two quasars and the galaxy (dashed lines). In the wings, at pixel positions less than 86 and greater than 98, the model consists of a contribution from the galaxy profile only. The χ^2–statistic indicates the model adequately describes the data. The lower panel shows the residuals model – data, in counts (solid line) on the left hand axis, and in counts divided by the noise at each pixel position (dashed line) on the right hand side.

The top panels in Figure 4 reveal that, when scaled so that the continuum levels match, the A and C components possess very similar line strengths in all three emission lines visible.

At the same epoch image D was faint. When scaled and compared to the spectrum of image A (lower panels of Figure 4) image D displays additional flux in all three lines. This can be interpreted as a microlensing deamplification of the continuum in image D.

In 1991 image B was bright, possibly undergoing microlensing, suggesting an enhancement of the continuum should be evident. This is confirmed from the comparison between the scaled spectra of images B and C (top panels of Figure 5).

Table 1 presents the relative equivalent width of the CIV emission line, at both epochs, for all four components. The values are normalized with respect to component C. In 1991 the equivalent width measures confirm the impression gained from Figures 4 and 5, with very similar values for images A and C. Values for 1994 show that image B remained essentially unchanged while the equivalent width in image A dropped substantially.

Cross–correlation of the emission lines in the different components show

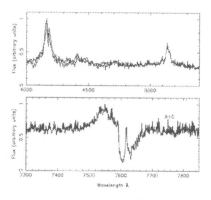

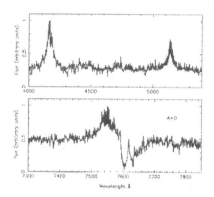

Figure 4. Blue and red ISIS spectra for images A + C (top panel) and A + D (bottom panel) from the 1991 observations. The spectra of C and D are scaled such that their continua match that of image A. The black regions indicate the excess of the scaled spectra over that of image A.

no detectable velocity shifts in the line centroids. Within the limits of the S/N there are also no detectable emission line profile differences. These results indicate that the scale of any structure in the broad line emitting region must be greater than an Einstein radius projected into the source plane ($> 0.05\,pc$). Details of these results will appear in an article currently in preparation.

4. Conclusions

This paper shows evidence for equivalent width differences between the four images of Q2237+0305, providing the first spectroscopic detection of microlensing induced spectral variations in a quasar. The temporal variation of the equivalent widths are consistent with the microlensing hypothesis

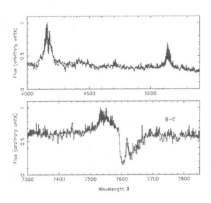

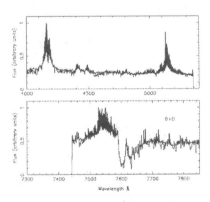

Figure 5.   As for Figure 4, except for images B + C (top panel) and B + D (bottom panel).

and indicate that the continuum in quasars originates in a region small, compared to the scale of an Einstein radius, while the broad line emission emanates from a much larger volume, in accord with the predictions of the standard model for the central regions of active galaxies (Figure 1).

References

Huchra, J., Gorenstein, M., Kent, S., Shapiro, I., Smith, G., Horine, K., & Perley, R., 1985, Astronomical Journal, 90, 691
Houde, M. & Racine, R., 1994, Astronomical Journal, 107, 466
Kayser, R., Refsdal, S. & Stabell, R., 1986, Astronomy and Astrophysics, 166, 36
Nemiroff, R. J., 1988, Astrophysical Journal, 335, 593
Rees, M. J., 1984, ARA&A, 22, 471
Rix, H., Schneider, D. P. & Bahcall, J. N., 1992, Astronomical Journal, 104, 959
Sanitt, N., 1971, Nature, 234, 199
Schneider, P. & Wambsganss, J., 1990, Astronomy and Astrophysics, 237, 42

IS IRAS F10214+4724 GRAVITATIONALLY LENSED?

J. LEHÁR
Harvard-Smithsonian Center for Astrophysics

AND

T. BROADHURST
Johns Hopkins University

Abstract. We show that the $z = 2.3$ IRAS source F10214+4724 is gravitationally lensed by an intervening galaxy, as suggested by the observed near-IR structures. Its many anomalous properties can be explained if the source is an ordinary Seyfert 2 nucleus whose central regions are much more highly magnified than the surrounding host galaxy. Confirming expectations, we find a counterimage to the near-IR arc, and find spectral evidence for the lensing galaxy at $z \sim 1$. We present new optical images which show that the optical source is compact and highly magnified. F10214+4724 may represent a population of lensed AGNs whose central engines are obscured.

1. Introduction

The IRAS source FSC 10214+4724 (hereafter F10214) has attracted a great deal of attention in the past few years. It was discovered in a redshift survey of 1400 IRAS sources (Broadhurst et al. 1995), which were selected from the IRAS Faint Source Catalog. At $z = 2.3$, F10214 has by far the largest redshift in the survey (Rowan-Robinson et al. 1991), with an inferred far-IR luminosity of $L \sim 3 \times 10^{13} h^{-2} L_\odot$, an order of magnitude more luminous than any known source. CO line emission was also detected, also with a very high inferred luminosity (*e.g.* Brown & Vanden Bout 1991). The high inferred far-IR and CO luminosities led many investigators (*e.g.* Rowan-Robinson et al. 1991) to propose that F10214 was undergoing an intense starburst. However, the UV-optical spectral line ratios and the presence of highly polarized spectral emission suggested that the source is an extremely luminous Seyfert 2 nucleus (*e.g.* Lawrence et al. 1993).

C. S. Kochanek and J. N. Hewitt (eds), Astrophysical Applications of Gravitational Lensing, 247–252.
© 1996 IAU

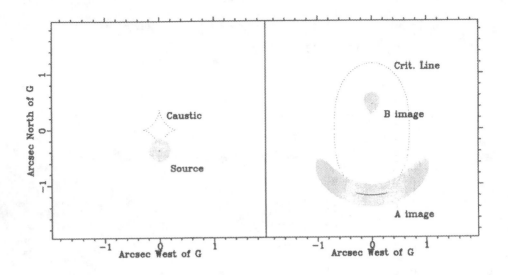

Figure 1.   Gravitational lens model for F10214, showing the source (left) and image (right) planes. The lens is centered on G and elongated N-S, with $\beta = 0.91''$, $\epsilon = 0.18$. The source consists of a small region ($r = 0.2''$) which produces the moderately magnified arc ($m \sim 5$), and a compact core ($r = 0.01''$) which is highly magnified ($m \sim 50$). The source is offset by $b = 0.024''$ from a cusp in the caustic.

Here, we propose that F10214 is an ordinary Seyfert 2 nucleus which has been gravitationally lensed by a foreground galaxy. High resolution near-IR images showed four objects near F10214 (Matthews et al. 1994), of which the brightest (Source 1) forms a $\sim 2''$ arc focused on the fainter Source 2, $1.2''$ away. The middle of Source 1 has a $\sim 0.5''$ core which is dominated by $H\alpha$ line emission. Matthews et al. (1994) suggested that the arc was gravitationally lensed by Source 2, but the marked difference between the continuum and $H\alpha$ structures led them to favor a merger scenario. We show that the structural differences at various wavelengths are the key to a complete understanding of F10214's many unusual properties. A more complete discussion of this is presented in Broadhurst & Lehár (1995).

2. Lens Interpretation

We propose that the light from the distant IRAS source is magnified by the gravitational field of an intervening galaxy. We identify the IRAS source with the arc-shaped object (Source 1), and the lensing galaxy "G" with the central object (Source 2). Sources 3 and 4 cannot be additional lensed images of the IRAS source, and presumably these are just other galaxies close to the line of sight.

We designed a simple lens model which can produce the different struc-

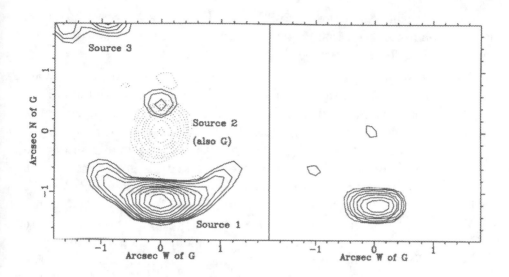

Figure 2. The left panel shows the Near-IR image of F10214 from Matthews et al. (1994) The image was sharpened to 0.3″ resolution, and decomposed into symmetric (dotted contours) and asymmetric (solid contours) parts, by rotating by 180° around the center of G and subtracting. The right panel shows the new optical image, sharpened to 0.2″ resolution. The contours decrease in factors of 1.414 from the map maximum.

tures seen in the near-IR and $H\alpha$ images, and which magnifies the far-IR source enough to make it intrinsically unremarkable. We used a singular isothermal ellipsoidal potential (Blandford & Kochanek 1987), and placed a small source with a compact core just outside a cusp of the astroid caustic (see Figure 1). Two images of this source are formed: an arc-shaped image "A" to the South, and a smaller counterimage "B", just North of the lensing galaxy "G".

A feature of such a lens model is that the magnification m depends strongly on the source angular size r. Basically, small sources will be very highly magnified, while larger sources will be only moderately magnified. Thus in our model, the far-IR source and the $H\alpha$ source are compact objects with $r \sim 0.01″$ and $m \sim 50$, while the near-IR continuum corresponds to a larger source with $r \sim 0.2″$ and $m \sim 5$. The CO and radio emission (Lawrence et al. 1993) would also originate from a smaller source region. Note that unless the far-IR emission is magnified by at least a factor of 20, the source remains intrinsically very luminous, and the fact that it is gravitationally lensed only makes F10214 *doubly unique*.

If F10214 is gravitationally lensed, two simple predictions must be met. First, there should be a counterimage to the arc North of G. We have reanalyzed the near-IR data of Matthews et al. (1994), and find a counterimage $\sim 0.45″$ North of G (see Figure 2). Second, the lensing galaxy must be at a lower redshift than the source. The published spectra of F10214 (*e.g.* Rowan-Robinson et al. 1991) show a distinctive upturn in the continuum flux for $\lambda > 8000$ Å, suggesting a galaxy spectrum at $z \sim 1$.

New optical observations confirm the lensing interpretation, and provide an important constraint to the magnification. We obtained two ten-minute B-filter exposures on the William Herschel Telescope. The images cover the source and two nearby bright stars, and the seeing was $0.6''$. We deconvolved the images using the Lucy-Richardson algorithm, and Figure 2 shows the result of co-adding both images. A single bright arc is visible, coincident with the near-IR "core". Like the radio source, the arc is oriented E-W, with most of its flux toward the Western end. The arc is $\sim 0.5''$ long (FWHM), and its inferred magnification from our model is ~ 50. The arc is unresolved in the N-S direction, setting an upper limit on the source size of $r < 0.5\,h^{-1}$ kpc at $z = 2.3$. The arc's magnitude is $B \approx 21.5$, and no counterimage is detected to a 3σ level of $B \sim 25$. This limits the optical magnification to $m > 20$, which is consistent with the arclength constraint. The lensing galaxy G is also not detected, so it must be very red ($B - K > 7.3$), consistent with an early-type galaxy at $z > 0.5$.

3. Discussion

The near-IR geometry constrains a lens model, which can be used to study the lensing galaxy. Since the arc has a small counterimage and is very symmetric, and since B is closer to G than to A, the mass must be elongated with its major axis within a few degrees of the N-S axis. The positions of A, B, and G constrain the ring radius to $\beta = 0.91''$, and the intrinsic source position to $0.38''$ South of G. The ellipticity ϵ is determined by the requirement that the core of A be highly magnified. The size of the caustic increases with ϵ, reaching the source center when $\epsilon = 0.18$ (or isodensity axial ratio ~ 0.6). The ring radius β provides an estimate of the lensing mass. For a lens at $z_L \sim 1$ and a source at $z_S = 2.3$, the ring size $\beta = 0.91''$ gives an isothermal velocity dispersion of ≈ 320 km s^{-1}, or a mass of $\sim 3 \times 10^{11}\,h^{-1}\,M_\odot$ within β. The near-IR magnitude of G converts to a standard magnitude of $B \sim -20.0$, if we assume an early-type galaxy spectrum. This gives a central mass-to-light ratio of ~ 25, and the predicted velocity dispersion is consistent with the Faber-Jackson relation.

The unusual properties of F10214 are all consistent with the background source being an ordinary AGN. The central engine is obscured by a dusty "torus", and perhaps can only be directly detected with sensitive hard X-ray observations. The obscuring torus itself is expected to be $10 - 100$ pc across. If the source is highly magnified, lensing limits the size to $r < 1.8''\,m^{-1}$. Also, a thermally emitting torus at 80 K must satisfy $r > 0.14''\,m^{-0.5}$. For

$m = 50$, this means that $r \sim 0.03'' \sim 100\,h^{-1}\,\mathrm{pc}$. Comparing F10214 to the nearby Seyfert 2 NGC1028, the obscuring torus itself would radiate predominantly in the IR, with an IR luminosity of $L_{IR} \sim 5 \times 10^{11}\,L_{\odot}$. The apparent far-IR luminosity of F10214 could easily be achieved if the torus were magnified by $m \sim 50$. The optical/IR spectrum is Seyfert 2 like, and is polarized by $\approx 15\%$ (e.g. Lawrence et al. 1993). Such high levels of polarization are found in the "mirror" regions of local Seyfert galaxies, which are generally $50 - 100\,\mathrm{pc}$ across. Sources of this size could produce the $\sim 0.5''$ optical arc if $m \sim 50$, so we suggest that this arc is a highly magnified image of such a mirror. The outer regions of the host galaxy would be less highly magnified. The $2''$ near-IR arc could be produced in the bulge of the host galaxy, with $r \sim 0.2'' \sim 1\,h^{-1}\,\mathrm{kpc}$. The outer galaxy should produce faint ring-like optical structures at $1'' - 2''$ radii.

The view that F10214 is a lensed AGN solves several other problems associated with this source. This source is very "red" in the IR, in that its apparent far-IR to near-IR flux ratio is much higher than for local Seyfert 2 nuclei (Lawrence et al. 1994). This is easily explained if the core is preferentially magnified. If the AGN narrow line region has any ionization structure, the spectral lines will be magnified by different amounts. This may explain the anomalously large high ionization lines, like NV in F10214, compared to other Seyfert 2s. High magnification can explain the unusually bright CO emission. The high temperature deduced from the CO transitions (Solomon et al. 1992) and the observed velocity profile are consistent with conditions in AGN centers. The unexpected polarization angle is also explained. The **E** vector of polarized optical emission is usually perpendicular to the radio elongation in Seyfert 2 sources. In F10214, the polarization shows no special alignment with the radio structure (Lawrence et al. 1993). In our model, the radio source is stretched by a factor of > 20 by lensing, so the observed elongation is unrelated to any intrinsic structure.

The *a priori* probability of finding systems like F10214 in the IR redshift surveys is encouragingly large. The IRAS redshift survey in which F10214 was found covers $0.2\,\mathrm{sr}$ down to $0.2\,\mathrm{Jy}$ at $60\,\mu$, and the redshifts extend over $z < 0.4$. Two evolution models can be fit to the source counts: pure luminosity $\propto (1+z)^5$; and pure density $\propto (1+z)^{2.5}$. We have calculated the total cross-section to lensing, accounting for the high magnification bias from elliptical lenses (Wallington & Narayan 1993). Applying both evolution models, $0.3 - 1.2$ lensed systems are expected, and it is not improbable that lensed systems should have been discovered in this survey. The calculation also gives expectation values for the lens redshift $< z_L > \sim 0.7$, the source redshift $< z_S > = 1.2 - 2.0$, and the source magnification $< m > = 25 - 40$.

4. Concluding Remarks

We have seen that many of its unusual properties can be readily explained if F10214 is a gravitationally lensed Seyfert 2 nucleus. Obscured AGNs could be more useful than QSOs for probing the lens galaxy mass distributions, since their background emission is measurably extended. Many more such systems should be found, and the IRAS source F15307+3252, at $z = 0.93$ (Cutri et al. 1994), is probably another case like F10214. The lensed QSO H1413+117 ($z_s = 2.5$) is a strong CO emitter (Barvainis et al. 1994), and may be a face-on version of F10214. Current searches for lensing have used optical QSOs and active radio sources, which are only a subset of the full AGN population. Selection by far-IR allows a fuller coverage of AGNs but is presently limited by the relatively bright IRAS flux limit.

There is more work to be done on F10214 itself. The counterimage has been confirmed in a recent Keck Telescope image (Graham & Liu 1995), and in new HST observations (Eisenhardt et al. 1995). The HST images also show that the main arc has substructures. Multicolor HST observations could determine whether these are multiple images of a single source or several separate source components. The latter would be very exciting, since those structures could never have been resolved without the aid of gravitational lensing. The most important unresolved issue remaining is the lens redshift. Our initial estimate of $z \sim 1$ is supported by the $z = 0.8$ result of Serjeant et al. (1995), based on similar features. However, Goodrich et al. (1995) identify absorption lines at $z = 1.2$. There have also been reports of $z \sim 0.4$ measurements (Close et al. 1995). Until this issue is settled, the lens mass estimates will be imprecise. For $0.8 < z < 1.2$, the required lensing mass varies by 30%.

References

Barvainis R. et al., 1994, Nature, 371, 586
Blandford R. & Kochanek C., 1987, ApJ, 321, 658
Broadhurst T. & Lehár J., 1995, ApJ, 450, L41
Broadhurst T. et al., 1995, ApJ, in preparation
Brown R. & Vanden Bout P., 1991, AJ, 102, 1956
Close L.M. et al., 1994, to appear ApJL (Oct 10)
Cutri R. et al., 1994, ApJ, 424, L65
Eisenhardt P. et al., 1995, to appear in ApJ
Goodrich R. et al., 1995, in preparation
Graham J. & Liu M., 1995, ApJ, 449, L29
Lawrence A. et al., 1993, MNRAS, 260, 28
Lawrence A. et al., 1994, MNRAS, 266, 41P
Matthews K. et al., 1994, ApJ, 420, L13
Rowan-Robinson M. et al., 1991, Nature, 351, 719
Serjeant S. et al., 1995, to appear in MNRAS
Solomon P. et al., 1992, ApJ, 398, L29
Wallington S. & Narayan R., 1993, ApJ, 403, 517

MICROLENSING IN THE LENSED QUASAR UM 425 ?

F. COURBIN, K. C. SAHU, G. MEYLAN

European Southern Observatory
Karl-Shwarzschild-Straße 2
D-85748 Garching bei München

Abstract. During the ESO key-program on gravitational lenses, the light curves of two images (A and B) of the lensed quasar UM 425 were obtained (from 1987 to 1995). This poster presents a possible interpretation of the light curves in terms of a micro-lensing event in the faint B component of the system, without ruling out a possible intrinsic variation of the source.

1. Long term photometry and interpretation

The photometry of UM 425 was obtained by fitting simultaneously the components A and B with 2-D Moffat profiles (Courbin et al. 1995). In addition to a smooth increase of intensity in both A and B, B shows a peak of intensity suggesting a micro-lensing event, although A seems also to be affected by micro-lensing (Michalitsianos & Oiversen 1995)

Fig. 1 shows the result of the fit of micro-lensing curve on the light curve of UM 425B after subtraction of the smooth variation. Two parameters are free: the impact parameter at maximum amplification U_{min}, and the crossing time τ of the source (time needed for the source to cross the Einstein ring).

Several combinations of U_{min} and of the crossing time τ give a good fit. However, the fitted peak is never higher than 0.3 magnitude, the best fit giving 0.25. The magnification is then 1.26, which indicates that the source does not cross the Einstein ring (where the magnification would be 1.34).

Are these parameters compatible with both the physical properties of micro-deflectors in the lensing galaxy and with its estimated redshift of 0.6 (Meylan & Djorgovski 1995)? Two schemes seem reasonable:

C. S. Kochanek and J. N. Hewitt (eds), Astrophysical Applications of Gravitational Lensing, 253–254.
© 1996 IAU

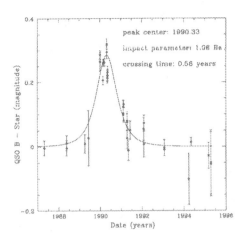

Figure 1. Light curve of UM 425B, "background" subtracted. The results of the fit are indicated as well as the 1σ error bars (photon noise only).

(i) A micro-lensing event by a very low mass object (about $0.01\,M_\odot$) at a redshift close to 0.6 with a transverse velocity of the order of 400-600 $\mathrm{km\,s^{-1}}$. Constraining the velocity to be smaller than $300\,\mathrm{km\,s^{-1}}$ gives lens masses smaller than $0.005\,M_\odot$, corresponding to the masses of big planets, if a redshift of 0.6 is kept for the deflector.

(ii) A micro-lensing event by an object with a mass between $0.01\,M_\odot$ and $0.1\,M_\odot$ at higher redshift, e.g., 1.3-1.4, implying a very high mass for the macro-deflector. If we assume such redshifts, the mass of the lensing galaxy should be as high as $10^{14}\,M_\odot$ in order to produce the separation of 6.5" observed between UM 425 A and B. This is still possible, when considering a cluster of galaxies at high redshift.

Finally, if the peak of intensity observed in UM 425B is due to a micro-lensing event, the size of the magnified source has to be smaller than 10^{-4} pc (diameter of the Einstein ring) in order to reproduce the observed amplification. This small size is compatible with the size of the regions responsible for the continuum emission in quasar spectra. A spectrophotometric monitoring would then allow to decide if the variations observed are due to an intrinsic variation of the quasar or to micro-lensing.

References

Courbin, F., Magain, P., Remy, M., Smette, A., Claeskens, J.F., Hainaut, O., Hutsemékers, D., Meylan, G., & Van Drom, E., 1995, A&A, in press.
Meylan, G., Djorgovski, S.G., 1989, ApJL, 338, L1.
Michalitsianos, A.G., & Oliversen, R.J., 1995, these proceedings.

THE MICROLENSING EVENTS IN Q2237+0305A:

NO CASE AGAINST SMALL MASSES/LARGE SOURCES

STEIN VIDAR HAGFORS HAUGAN

Institute of Theoretical Astrophysics, University of Oslo
Pb. 1029, Blindern
N-0315 OSLO
http://www.uio.no/~steinhh/index.html

1. Introduction

Witt & Mao (1994) claim that the data reported by Racine (1992) contains "a quite well sampled M-shaped double event in image A of 0.3 and 0.4 mag, respectively". They further state that the very low average mass scenario put forward by Refsdal & Stabell (1991) does not predict "well-resolved *asymmetric* events, as have been observed in image A".

The first peak has only six sampling points, with all but one point clustered on one side of the peak. The second peak has 5 sampling points, with *all* points clustered on one side of the peak. The degree of asymmetry is thus very hard to quantify.

2. The Large Source Model

I have studied a large number of large source light curves Haugan (1994) produced with the rayshooting method. A Gaussian source profile $I(\boldsymbol{y}) \propto \exp(-|\boldsymbol{y}|^2/r_s^2)$ was used, where $\boldsymbol{y}$ is the dimensionless position relative to the source center, and r_s is the dimensionless source size.

Large source models do not typically produce asymmetric events, but they certainly do occur. In order to highlight the problems of using isolated events in order to determine the normalized source size, Fig. 1 shows a curve similar to the one appearing in Racine (1992) superposed on simulated light curves with large sources. The light curve parameters κ_*, γ, and r_s, are indicated. Positive γ indicates that the large, elongated caustic structures

C. S. Kochanek and J. N. Hewitt (eds), Astrophysical Applications of Gravitational Lensing, 255–256.
© 1996 IAU

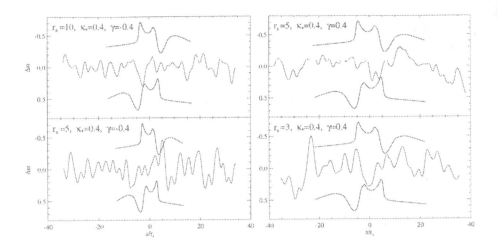

Figure 1. Light curves from simulations with a curve similar to the one in Racine (1992) superimposed. Abscissa values are in units of the source size r_s.

are oriented along the source track. A time-reversed version of the spline curve is also supplied to aid the eye.

Although searching for exact replicas of the observed light curve among simulated light curves is useless, a comparison by eye can easily be done. Although light curves with large sources lack the clear M-shaped events of light curves with small sources ($r_s \ll 1$), the peaks may very well be asymmetric to the extent indicated by the observations.

3. Conclusion

Based on the above arguments, the exclusion of models with a large source or low average masses is *not* justified from the 1988-90 events. Further observations and analysis should therefore not be concentrated solely on interpreting the light curves from the perspective of (very) small sources.

Acknowledgments: The author wishes to thank Rolf Stabell, Sjur Refsdal and Per Barth Lilje for helpful comments on the manuscript.

References

Haugan, S. V. H. 1994, *Master's thesis*, University of Oslo, Institute of Theoretical Astrophysics, Pb 1029 Blindern, N-0315 Oslo, Norway
Racine, R. 1992, ApJ, 395, L65
Refsdal, S., Stabell, R. 1991, A&A, 250, 62
Witt, H. J., Mao, S. 1994, ApJ, 429, 66

FAR-UV SPECTRAL VARIABILITY IN UM425 & PG1115+080

A.G. MICHALITSIANOS AND R.J. OLIVERSEN
NASA Goddard Space Flight Center
Laboratory for Astronomy and Solar Physics

Determining the nature of spectral variability in gravitationally-lensed *BAL* (broad absorption line) quasars is important for time-delay and microlensing studies because the delay-times can be comparable with the time scales of intrinsic QSO ionization changes. Far-UV spectra of the candidate lens UM 425 = Q1120+019A and PG1115+080A (*Triple Quasar*) revealed the presence of strong O VI $\lambda 1033$ emission. In both cases, absorption in the blue-wing of the O VI line profile indicated characteristic *BAL* outflow.

IUE spectra of UM 425 (Image-A, $m_v = 16.2$) revealed complex line profile structure; in particular, the high-ionization emission lines of O VI $\lambda 1033$ and N V $\lambda 1240$ exhibit a broad $\approx 12,000$ km/s *BAL* trough structure (Michalitsianos & Oliversen 1995). The Ly-α $\lambda 1216$ emission centroid corresponds to a $z_{qso} = 1.471 \pm 0.003$, which is slightly greater than the $z_{qso\ Mg\ II} = 1.467$ redshift obtained by Meylan and Djorgovski (1989).

A comparison of our initial *IUE* spectra with data obtained ≈ 10 months later indicates significant changes occurred in both the *BEL* (broad emission line) and *BAL* regions, for example, the O VI emission component increased by a factor ≈ 2. Enhanced *BEL* ionization is also seen in S VI(1) $\lambda\lambda 937,945$, C III(1) $\lambda 977$, N III(1) $\lambda 990$ and S IV(1) $\lambda\lambda 1063,1073$. The Ly-α $\lambda 1216$ flux increased by only a factor ≈ 1.3, where the small increase in Ly-α may be entirely due to a decrease in the N V *BAL* absorption at velocities $v_{BAL} > 4000$ km/s. Increased absorption at velocities $v_{BAL} < 4000$ km/s suggests new material may have been injected into the *BAL* flow, in a region where the acceleration is initiated. We find the greatest change in emission line intensity occurs in high-ionization lines, whereas Ly-α was unchanged. This suggests high-ionization emission lines and lower ionization species, including recombination lines, vary *independently*, similar to results obtained by Dolan et al. (1995) for Q0957+561. The time scale for O VI variations suggests an upper limit for the size of the *BEL* region of ≈ 0.1 pc (correcting for the $1+z$ time dilation), consistent with the size of

C. S. Kochanek and J. N. Hewitt (eds), Astrophysical Applications of Gravitational Lensing, 257–258.
© 1996 IAU

the regions in QSOs in general. However, changes in O VI *BAL* trough absorption occurred on the same time scale, and the same dimensional upper limit thus applies to the *BAL* region. This argues the *BAL* region is comparable in size, perhaps slightly larger, say ≈ 1 pc, than the *BEL* region.

The detection of *BAL* spectral line structure in UM 425A is important because if similar distinct spectral features are found in UM 425B, the low statistical count of *BAL* QSOs and the low probability of finding another *BAL* QSO with the same redshift located $\approx 6''$ from UM 425A would unambiguously show the system is lensed.

IUE spectra of PG1115+080A ($m_v = 15.8$) revealed prominent O VI emission that is superimposed on strong continuum between $\lambda\lambda 900$–1100Å ($z_{qso} = 1.722$ rest frame). A comparison of these data with the only other *IUE* spectra of PG1115+080A taken in this wavelength range indicates O VI emission was not present in 1978; the spectrum showed only a featureless continuum (Green et al. 1980). The detection of O VI resonance line emission suggests a high state of ionization in the *BEL* region, which is emission is accompanied by absorption in the line core, and *BAL* absorption that truncates the blue-wing of the line profile.

After our initial detection, both the O VI emission and *BAL* absorption decreased in flux by $\approx 50\%$ (relative to the local continuum) over ≈ 100 days. Absorption features within the O VI *BAL* trough also changed on time scales of months down to ≈ 1 day. Evidence for rapid time scale O VI absorption variability of ≈ 1 day implies an unreasonably small size of the *BAL* region, considering only the light-travel time for photo-ionization by the continuum source. However, there is growing evidence for rapid variations in the cores of related objects, such as *Active Galactic Nuclei (AGN)*. For example, *ASCA* X-ray spectra of the Seyfert 1 galaxy NGC 3227 by Ptak et al. (1994) indicate rapid changes in the O VI, O VII and O VIII metal absorption edges at 671, 739 and 879 eV, respectively, that occur on time scales of $\approx 10,000$ seconds. It is possible that the ≈ 1 day O VI absorption variations in PG1115+080A are from the same type of process which leads to rapid changes in AGNs. This follows because both the O VI metal absorption-edge and O VI $\lambda 1033$ resonance line are formed in the same gas. Further monitoring of PG1115+080 is required to confirm this result.

References

Dolan, J.F., et al., 1995, ApJ, 442, 87
Green, R. F., et al., 1980, ApJ, 239, 483
Meylan, G. and Djorgovski, S., 1989, ApJL, 338, L1
Michalitsianos, A.G. and Oliversen, R.J., 1995, ApJ, 439, 599
Ptak, A., et al., 1994, ApJL, 436, L31

RESULTS FROM FIVE YEARS OF MONITORING OF THE EINSTEIN CROSS WITH NOT

R. ØSTENSEN

Auroral Observatory, University of Tromsø
N-9037 Tromsø, Norway

S. REFSDAL

Hamburg Sternwarte
Gojenbergsweg 112, D-21029 Hamburg, Germany

R. STABELL

Institute of Theoretical Astrophysics, University of Oslo
Blindern, N-0315 Oslo, Norway

AND

J. TEUBER

Copenhagen University Observatory
Brorfelde, DK-4340 Tølløse, Denmark

After some preliminary observations in 1990, a program was started in 1991 at the Nordic Optical Telescope (NOT) on the island of La Palma for monitoring mainly four well known gravitational lens systems: QSO0142-100, QSO0957+561, QSO1413+117 and QSO2237+0305 (The Einstein Cross).

Here we report results from the monitoring of the Einstein Cross. During the four year period the Monitor Program was active (1990–1993) we gathered 55 high quality observations of this well known lens system. In addition we have collected 32 more observations in 1994 and a further four so far in 1995, a total of 91 new observations of the Einstein Cross.

The observations have all been made with the 2.54 meter Nordic Optical Telescope. For most of the observations the Stockholm CCD camera was used. This camera applies a Tektronix 512^2 chip giving a resolution of 0.2×0.2". From 1994 most observations were made with the new Brorfelde CCD camera whose thinned Tektronix 1024^2 gives an improved resolution of 0.176×0.176", as well as better noise characteristics and quantum efficiency.

C. S. Kochanek and J. N. Hewitt (eds), Astrophysical Applications of Gravitational Lensing, 259–260.

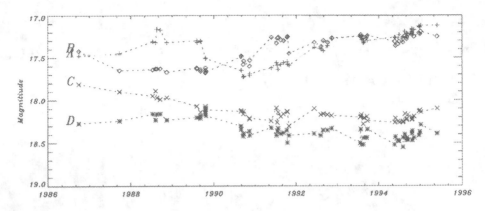

Figure 1.   R-band observations

For the preprocessing of the images the IRAF/ccdred package was used. A processing tool developed especially for precise separation of the close components of the Einstein Cross was used to compute the magnitudes of the four quasar images. The program, which has been named XECClean, applies an interactive CLEAN processing algorithm using four point sources and an analytical galaxy model. A Menu/Button X-windows interface has been implemented using IDL, making XECClean easy to use and able to produce good results in less than five minutes. A complete description of XECClean is given in Østensen's thesis (1994) (available from the author).

The most complete light-curves published until now are in the article by Corrigan et al. (1991), where they presented the results from all the direct image CCD data, of sufficient quality, available at that time. Their best sampled light-curve (R-band) includes 15 data points.

Using the Monitor Program observations to extend these light-curves we get curves like the one shown in Figure 1. Due to the limited space available we show only the best sampled light-curve, namely the R band data. Figure 1 shows the measured magnitudes for the four components. Data points prior to 1990 are all from Corrigan et al. (1991).

The resulting light-curves exhibit several micro-lensing features, and variations on time-scales of several years are found in all four images. For complete tables of magnitudes and error estimates, as well as further interpretations of the features revealed in the light-curves, the reader is referred to Østensen et al. (1995)

References

Corrigan, R.T., Irwin, M.J., Arnaud, J., et al., 1991, AJ, 102, 34
Østensen, R., 1994, Cand. Scient. Thesis, University of Tromsø
Østensen, R., Refsdal, S., Stabell, R., Teuber, J., et al., 1995, A&A, in press

THE CLOVER LEAF QUASAR H1413+117:
NEW PHOTOMETRIC LIGHT CURVES

M. REMY[1], E. GOSSET[1,2] AND D. HUTSEMÉKERS[1,2]
[1] Institut d'Astrophysique, Liège, Belgium
[2] Chercheur Qualifié au FNRS, Belgium

AND

B. REVENAZ[3] AND J. SURDEJ[3,4]
[3] Space Telescope Science Institute, Baltimore, MD, USA
[4] Directeur de Recherches au FNRS, Belgium

Abstract. Direct CCD images of the gravitationally lensed BAL quasar H1413+117 (Bessel B, V, R, I; Gunn r and i), obtained during 1989-1994 in the framework of photometric monitorings at ESO and with the NOT, have been carefully (re-)analyzed. Simultaneous fitting of 4 PSFs, fixing the relative positions of components B, C and D with respect to A, has been achieved successfully. The relative light curves of the 4 lensed QSO images, with respect to a stable nearby comparison star, confirm the general trend already presented by Arnould et al. (1993). The amplitude of the light curves is found to be about 0.5 magnitude, peak-to-peak in the V filter. The four lensed components display brightness variations quasi-simultaneously and in parallel. In addition to these intrinsic light variations due to the source, we find that component D shows extra light variations with respect to the other components. These are very likely caused by micro-lensing.

1. Data, Photometry & Discussion

The optical photometry of the Clover Leaf BAL QSO, already presented in Arnould et al. (1993), has been carefully re-analyzed using multiple PSF fitting of the four lensed components. The relative magnitudes of the field stars with respect to the comparison star already used by Arnould et al. (1993) were checked. The scatter is of the order of 0.03 magnitudes (3σ).

C. S. Kochanek and J. N. Hewitt (eds), Astrophysical Applications of Gravitational Lensing, 261–262.

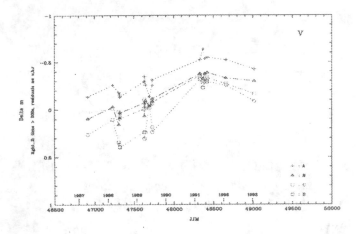

Figure 1. Photometry of H1413+117

The relative magnitudes in the V band of the four components of the Clover Leaf are plotted as a function of JD - 2400000.5 in Figure 1.

The qualitative trends already described by Arnould et al. (1993) appear more clearly in these plots, but some differences exist. The fluxes of the A, B and C components vary in parallel and quasi-simultaneously with a peak in luminosity near 1991-93. The expected time delays between the four lensed components of this quasar are typically shorter than 1 month; intrinsic variations of the source would therefore show up quasi-simultaneously in Figure 1. The D component exhibits the same global trend but with additional light variations. Spectroscopic observations of the D component in March 1989 (Angonin et al. 1990) and June 1993 (with HST + FOS) show that while the continuum is clearly varying in agreement with the photometry, the superimposed emission lines are almost unchanged, providing further good evidence for a microlensing origin to the observed residual variations.

Acknowledgements: We would like to thank the numerous observers at ESO and the NOT. This research was partly supported by ESA under contract PRODEX.

References

Angonin, M.-C., et al., 1990, A&A, 223, L5
Arnould, P., et al., 1993, in Gravitational Lenses in the Universe, eds. J. Surdej et al., (Liège: Université de Liège) 169

THE VARIABILITY OF PG1115+080

PAUL L. SCHECHTER
Massachusetts Institute of Technology

PG1115+080 (Weymann et al. 1980), the first quadruple lens system, remains the most promising for an optical measurement of the Hubble constant through a time delay. Factors contributing to this assessment include the well constrained geometry, the apparent magnitudes of the four components (Christian et al. 1987), the expected time delay (Narasimha & Chitre 1992), and the position of the lensing galaxy (Kristian et al. 1993). There are nonetheless many factors which combine to render this a difficult measurement, among them the small separation of the images and the small variability amplitude and long variability time scale of the "typical" optically selected quasar (Hook et al 1994). The measurement is further complicated by the likelihood of microlensing of the individual components (Witt et al. 1995), which depends upon the degree to which the quasar is resolved and the fraction of the lens mass in microlenses.

We have monitored PG1115 with the 2.4m Hiltner Telescope of the Michigan-Dartmouth-MIT Observatory for the last four seasons, with an eye toward assessing the feasibility of measuring a time delay. Our approach has been to fix the shape of the lensing galaxy and the relative positions of the four images and the galaxy with respect to each other, treating only their fluxes and their overall position as as free parameters (Schechter 1993). Using our best data the galaxy was found to be round to within 10%; we have therefore taken the ellipticity to be identically zero.

We used a set of 14 I band observations obtained over the course of two nights in 1994 April, with seeing ranging from 0.59 to 1.08" (median of 0.74"), to obtain relative positions for the components and the shape of the galaxy. There is substantial covariance between the fluxes for components $A1$ and $A2$ but their sum exhibited an rms scatter, relative to star C (Vanderriest et al. 1986), of 3 millimag. The ratio of component C (the third brightest and most isolated) to $A1 + A2$ had an rms scatter of 8 millimag. The error in the mean would be considerably smaller, though at this level

C. S. Kochanek and J. N. Hewitt (eds), Astrophysical Applications of Gravitational Lensing, 263–264.
© 1996 IAU

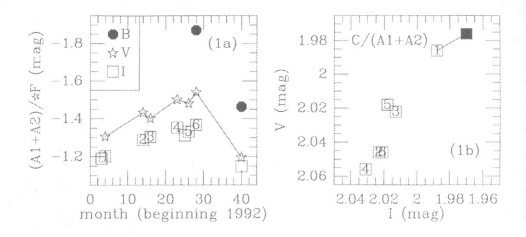

one must worry about systematic errors.

In Figure 1a we show the variability of $A1 + A2$ relative to star F (Vanderriest et al. 1986) over the course of 40 months. In Figure 1b we show the ratio $C/A1 + A2$ as observed at V and I. All of the points are derived from averages of at least 3 frames taken with the Hiltner Telescope, with the exception of 1994 April V point, obtained from a single exposure, and the filled symbol, obtained at CTIO. This latter point was obtained at the same time, 1992 April, as point labeled "1". The agreement of these two, obtained with different seeing and different sampling using different instruments, supports the conclusion that the variation in $C/A1+A2$ is real and not the result of systematic error. This could be due either to differential microlensing or to time delay.

References

Christian, C. A., Crabtree, D., & Waddell, P., 1987, ApJ, 312, 45

Hook, I. M., McMahon, R. G., Boyle, B. J., & Irwin, M. J., 1994 MNRAS, 268, 305

Kristian, J., Groth, E. J., Shaya, E. J., Schneider, D. P., & Holtzman, J. A., et al., 1993, AJ, 106, 1330

Narasimha, D. & Chitre, S. M., 1992, in Gravitational Lenses, eds. R. Kayser, T. Schramm & L. Nieser, (Berlin: Springer-Verlag), 128

Schechter, P. L., 1993, in Gravitational Lenses in the Universe, eds. J. Surdej, D. Fraipont-Caro, E. Gosset, S. Refsdal, & M. Remy, (Liege: Institut d'Astrophysique), 119

Vanderriest, C., Vlerick, G., Lelievre, G., Schneider, J., Sol, H., Horville, D., Renard, L. & Servan, B., 1986, A&A, 158, L5

Weymann, R. J., Latham, D., Angel, J. R. P., Green, R. F., Liebert, J. W., Turnshek, D. ,E., & Tyson, J. A., 1980, Nature, 285, 641

Witt, H. J. , Mao, S., & Schechter, P. L., 1995, ApJ, 443, 18

SUPERLUMINAL GRAVITATIONAL MICROLENSING EVENTS IN PKS 0537-441 AND PROSPECTS FOR FUTURE DETECTION:

G.C. SURPI

Depto. de Física, FCEyN, Universidad de Buenos Aires
Ciudad Univresistaria - Pab. I, 1428 Bs. As., Argentina

G.E. ROMERO

Instituto Argentino de Radioastronomía
CC5, 1894 Villa Elisa, Bs. As., Argentina

AND

H. VUCETICH

FCAyG - Universidad Nacional de La Plata
Paseo del Bosque S/N, 1900 La Plata, Bs. As., Argentina

Abstract. The BL-Lac object PKS 0537-441 has recently displayed strong intraday variability at 1.42 GHz on time scales of $\sim 10^4$ s. Such kind of variability is hardly consistent with an intrinsic-to-source origin, since apparent brightness temperatures higher than 10^{21} K are implied for the emitting region in the source.

An alternative explanation for this ultra-rapid radio variability in PKS 0537-441 is developed based on gravitational microlensing of a superluminal component in the blazar caused by compact objects in the halo of an intervening galaxy. The mass of the microlenses is estimated and the blazar is proposed as a good candidate in a systematic search for extragalactic MACHOs.

1. The Model

The underlying idea in our model, "superluminal microlensing", was originally proposed by Gopal-Krishna and Subramanian (1991). A foreground galaxy with $z = 0.186$ and a close alignment with PKS 0537-441 was reported by Stickel et al. (1988). In the present case, the focusing can be

C. S. Kochanek and J. N. Hewitt (eds), Astrophysical Applications of Gravitational Lensing, 265–266.
© 1996 IAU

approximated by (Romero et al. 1995):

$$k = \frac{\Sigma_c}{\Sigma_{crit}} = \frac{4\pi G D_d D_{ds} \Sigma_c}{c^2 D_s}, \tag{1}$$

where Σ_c is the central surface density in the galaxy and the rest of the symbols have their usual meanings. During a microlensing event the total amplification of the background source is:

$$A = 1 - \frac{1}{(1-k)^2} + \frac{2(1-k) + u^2}{(1-k)^2 u \sqrt{4(1-k) + u^2}}, \tag{2}$$

where u is the distance from the source to the microlens of mass M in the lens plane, in units of the Einstein radius $R_e = [4GM D_d D_{ds}/c^2 D_s]^{1/2}$. If u is expressed as a function of the relative source-lens trajectory, then equation (2) can be used to fit the observed variability reported by Romero et al. (1994), with k restricted to the range $0.2 \leq k \leq 0.6$ imposed by the magnification of the unique macroimage of the blazar. From the results of these fits we estimate the microlenses have masses between $3 \times 10^{-5} M_\odot$ and $3 \times 10^{-3} M_\odot$ for superluminal velocities in the 1-10c range (in the lens plane). Additionally, from the event rate, we estimate a central dark mass density $0.01 \leq \rho_0 \leq 0.04$ under the form of compact objects. This range is of the order of the dark matter densities estimated in our own and other galaxies from rotation curves.

2. Future prospects

We propose the southern blazar PKS 0537-441 as a good candidate in a search for extragalactic MACHOs. Owing to the usual superluminal motions in blazars, the optical depth and the number of events will not significantly depend on the velocity distribution of the lenses, and due to the excellent source-lens-observer alignment in the case of PKS 0537-441 there will be no dependence on peculiarities in the structure and shape of the dark halo. Detailed computations (Surpi et al. 1995) show that if the total range of lens' masses is $10^{-4} M_\odot$ - $1 M_\odot$, then 49 events are expected over one month of continuous monitoring of the blazar at optical frequencies. The time scales goes from 30 minutes to 3 days, in case of a superluminal motion with $3c \leq v_{app} \leq 4c$ in the blazar.

References

Gopal-Krishna, Subramanian, K., 1991, Nature, 349, 766
Romero, G.E., Combi, J.A., Colomb, F.R., 1994, A&A, 288, 731
Romero, G.E., Surpi, G., Vucetich, H., 1995, A&A, in press
Stickel, M., Fried, J.W., Kühr, H., 1988, A&A, 206, L30
Surpi, G., Romero, G.E., Vucetich, H., 1995, in preparation

THE Q0957+561 MICROLENSING

DAVID J. THOMSON
AT&T Bell Labs
Murray Hill, NJ 07974-2079

AND

RUDOLPH E. SCHILD
Harvard-Smithsonian Center for Astrophysics
60 Garden Street, Cambridge MA 02138

Abstract. The TwQSO microlensing shows rich structure with cusp-profiled features on time scales of weeks to months, and with periodic effects on time scales of a few years.

With the TwQSO time delay now converging to a value around 410 days (404 days, Schild & Thomson these proceedings; 440 days, Haarsma et al. these proceedings; 423 days, Pelt et al. 1995), the microlensing on all time scales can be recognized. Some of these results have been discussed in Schild & Thomson (1995, ST95 hereafter) to which we shall frequently refer for illustrations.

The microlensing history is illustrated as Fig 2 in ST95, together with a cubic polynomial fit. In the 15-year history of monitoring of two image components, no strong microlensing feature expected for an unresolved quasar source has been observed. Instead, we find a complex pattern of microlensing fluctuations with a characteristic time scale of 1/4 year.

One of these microlensing cusps, observed in 1988, is shown as Fig 3 of ST95. Plotted is the brightness record of the A image component from 90 nights of observations. The event shows a Lorenzian profile with an amplitude of only 0.04 mag. Superimposed upon this is a pattern of positive and negative cusps with a smaller, 0.01 magnitude amplitude.

These features are entirely unexpected, and may signal the presence of a population of low-mass microlenses in lens galaxy G1. If we use a normal scaling of event duration with square root of mass, we easily determine

C. S. Kochanek and J. N. Hewitt (eds), Astrophysical Applications of Gravitational Lensing, 267–268.

a mass of 10^{-5} solar masses for the responsible objects. Of course other explanations may be possible.

Also unexplained at present is the structure in the Fourier power spectrum, Figure 4 of ST95. We interpret the low-frequency parts of this power spectrum as having two principal components; a continuum extending from 0 to 4 cycles/yr, and several lines indicative of periodicity. The lines may indicate orbital motions of the microlensing masses, and we have found that the lines tend to be strongest in restricted parts of the data set, extending over several years. Note that due to the logarithmic scaling of this plot, the lines have a factor of 10 times the power in the adjacent continuum.

The second feature of the Fourier power spectrum is the continuum with a cutoff at 4 cycles/yr. This could signal the presence of a population of objects in the lens galaxy having planetary mass.

References

Pelt, J., et al., 1995, A&A, in press
Schild, R. & Thomson, D.J., 1995, Dark Matter, eds. S. Holt & C. Bennett (New York: AIP Press)

SCINTILLATIONS AND MICROLENSING

D.B. MELROSE
Research Centre for Theoretical Astrophysics
School of Physics, University of Sydney

Abstract. It is shown how the theory of scintillation may be applied to treat gravitational lensing. The theory is applied to microlensing by a system of N point masses. It is shown that scintillation theory reproduces a known result for the angular broadening due to multiple microlensing. Some unresolved differences between scintillation and microlensing theories for intensity fluctuations are pointed out.

Key words: microlensing, scintillations

1. Introduction

Gravitational microlensing at large optical depth presents a time consuming numerical problem because of the large number of stars that need to be taken into account (e.g., Paczyński 1986). An analytic treatment is desirable and Katz et al.(1986) developed a multiple-scattering theory to describe the associated angular broadening. A similar theory was used by Deguchi & Watson (1988) to treat intensity fluctuations due to a gaussian source. An alternative analytic approach is developed and explored here: the well-established theory of scintillations is used to treat gravitational lensing. In applying the theory to microlensing by a system of stars, the questions addressed are whether scintillation theory reproduces known results for the angular broadening (Katz et al.1986), and whether it provides any new insight into the intensity fluctuations due to microlensing.

2. Review of Scintillation Theory

In the theory of scintillations, the spectrum of fluctuations in the refractive index of the radiation passing through the turbulent medium is assumed

269

C. S. Kochanek and J. N. Hewitt (eds), Astrophysical Applications of Gravitational Lensing, 269–274.

to be given. The theory determines various statistical properties of the emerging radiation, such as the angular broadening of a narrow beam and the spectrum of intensity fluctuations, in terms of the given spectrum of fluctuations in refractive index. Let $\delta\mu(\mathbf{x})$ denote the fluctuating part of the refractive index, and let $\phi(\mathbf{x})$ denote the phase of the wave. Denoting the statistical average by angular brackets, the following two (related) correlation functions are assumed to be given:

$$B_n(\mathbf{x}) = \langle \delta\mu(\mathbf{x})\delta\mu(\mathbf{x}' + \mathbf{x})\rangle, \quad D(\mathbf{x}) = \langle [\phi(\mathbf{x}) - \phi(\mathbf{x}' + \mathbf{x})]^2 \rangle, \quad (1)$$

where $D(\mathbf{x})$ is the phase structure function. The power spectrum of the refractive index fluctuations, $\Phi_n(\mathbf{K})$, is the Fourier transform of $B_n(\mathbf{x})$.

The mean ray direction is assumed to be along the z-axis, and the fluctuations are projected onto a screen, where they are described in terms of two-dimensional vectors, introduced by writing $\mathbf{x} = (\mathbf{r}, z)$ and $\mathbf{K} = (\mathbf{q}, K_z)$. For isotropic turbulence one introduces

$$A(r) = \frac{1}{2\pi} \int_0^\infty dq q \, \Phi_n(q, K_z = 0) J_0(qr), \quad (2)$$

with $r = (x^2 + y^2)^{1/2}$, $q = (q_x^2 + q_y^2)^{1/2}$. The angular broadening and the intensity fluctuations are then given by the theory in terms of $A(r)$ or

$$D(r) = (8\pi^2 L/\lambda^2)[A(0) - A(r)], \quad (3)$$

where λ is the wavelength of the radiation and L is the distance from the image plane to the screen. For a power law $\Phi_n(\mathbf{q}) \propto q^{-\beta}$, one has $D(\mathbf{r}) \propto r^{\beta-2}$.

3. Scintillations due to Gravitational Lensing

Application of the foregoing theory to gravitational lensing proceeds as follows.

3.1. REFRACTIVE INDEX FLUCTUATIONS

The refractive index variation due to a weak gravitational field is given by $\delta\mu(\mathbf{x}) = -2\Phi(\mathbf{x})/c^2$, where $\Phi(\mathbf{x})$ is the Newtonian gravitational potential. The potential is related to the mass density, $\eta(\mathbf{x})$, by $\nabla^2\Phi(\mathbf{x}) = -4\pi G\eta(\mathbf{x})$, whose Fourier transform gives $\tilde{\Phi}(\mathbf{q}) = 4\pi G\tilde{\eta}(\mathbf{q})/|\mathbf{q}|^2$. Thus $A(r)$ in (2) is given by

$$A(r) = \frac{(8\pi)^2 G^2}{c^4} \int \frac{d^2\mathbf{q}}{(2\pi)^2} \exp(i\mathbf{q} \cdot \mathbf{r}) \frac{C_\eta(\mathbf{q})}{|\mathbf{q}|^4}, \quad (4)$$

where $C_\eta(\mathbf{q})$ is the Fourier transform of the correlation function for the density fluctuations.

3.2. A SYSTEM OF POINT MASSES

For a system of N point masses, with the ith mass, M_i, at position $\mathbf{x} = \mathbf{x}_i$, one finds

$$A(r) = \frac{(8\pi)^2 G^2}{\pi R^2 L c^4} \int \frac{d^2\mathbf{q}}{(2\pi)^2} \frac{1}{q^4} \left\langle \sum_{i,j=1}^{N} M_i M_j \exp[i\mathbf{q} \cdot (\mathbf{r} + \mathbf{r}_i - \mathbf{r}_j)] \right\rangle, \quad (5)$$

where $\pi R^2 L$ is the volume of the system and R is its radius.

3.3. CUTOFF

The integral (5) diverges, and needs to be cut off to obtain a finite result. Here the integral is cut off at $q < q_0$, with $q_0 = 1/R$. With $r_{ij} = |\mathbf{r} + \mathbf{r}_i - \mathbf{r}_j|$ the integral gives (Katz et al.1986)

$$\int_{q_0}^{\infty} \frac{dq}{q^3} J_0(qr_{ij}) = \frac{1}{2q_0^2} \left[1 - \frac{q_0^2 r_{ij}^2}{2} \ln\left(\frac{2e^{1-\gamma}}{q_0 r_{ij}}\right) + \cdots \right]. \quad (6)$$

Then (5) gives

$$A(r) = \frac{8G^2}{Lc^4} \left[\left(\sum_i M_i\right)^2 - \sum_{i,j} \frac{M_i M_j r_{ij}}{R^2} \ln\left(\frac{3.05\,R}{r_{ij}}\right) \right]. \quad (7)$$

The first term in the square brackets arises from the $i = j$ term in the sum in (5); this term does not contribute to the scintillations. The remaining term is associated with the $N(N-1)/2$ pairwise combinations of point masses. Thus the scintillations may be attributed to the net effect of lensing by $\sim N^2$ two-point-mass systems.

3.4. STATISTICAL AVERAGING

The average of (5) or (7) over a collection of N identical stars, each of mass M, may be performed by retaining one term, say the ij term, writing $\Delta\mathbf{r} = \mathbf{r}_i - \mathbf{r}_j$, and then averaging over $\Delta\mathbf{r}$. A conventional phase-averaging procedure in scintillation theory is over a random phase ϕ that satisfies $\langle e^{i\phi} \rangle = \exp(-\frac{1}{2}\langle\phi^2\rangle)$. The average over the ij term in (5), with $\Delta\mathbf{r} = \mathbf{r}_i - \mathbf{r}_j$, is then achieved by identifying the phase as $\mathbf{q} \cdot \Delta\mathbf{r}$. One has $\langle(\mathbf{q} \cdot \Delta\mathbf{r})^2\rangle = \frac{1}{2}q^2\langle(\Delta\mathbf{r})^2\rangle$. Thus (5) is replaced by $2N$ times the statistical average of of the ij term, giving

$$A(r) = \frac{64G^2 M^2 N}{R^2 L c^4} \int_{R^{-1}}^{\infty} \frac{dq}{q^3} J_0(qr) e^{-q^2 R_0^2} \approx \frac{64G^2 M^2 N}{R^2 L c^4} \int_{R^{-1}}^{R_0^{-1}} \frac{dq}{q^3} J_0(qr),$$
$$(8)$$

with $R_0^2 = \frac{1}{4}\langle(\Delta \mathbf{r})^2\rangle \sim R^2/N$ of order the mean square separation between the stars. Then using (6), (8) gives

$$A(r) \simeq -\frac{16G^2M^2}{R_0^2Lc^4}\, r^2 \ln(3.05\, N^{1/2}), \tag{9}$$

where the numerical factor in the argument of the logarithm is chosen to facilitate comparison with Katz et al.(1986).

4. Angular Broadening

In scintillation theory angular broadening is described by the mean square fluctuations in the angular deviation of a ray, $\langle(\delta\theta)^2\rangle = -L\nabla^2 A(r)$, which with (9) for $A(r)$ gives

$$\langle(\delta\theta)^2\rangle(r) = 2\theta_R^2 \ln(3.05\, N^{1/2}), \quad \theta_R = \frac{4GM}{R_0c^2} = \frac{4GM}{Rc^2}\, N^{1/2}, \tag{10}$$

where θ_R corresponds to a ray passing a mass M with an impact parameter R_0. The result (10) was derived by Katz et al.(1986) using a model for multiple scattering by point masses. This confirms that scintillation theory reproduces a known result in microlensing. Note that (9) includes only the effect of impact parameters between R_0 and R. Scintillation theory does not apply for impact parameters $\ll R_0$, which corresponds to large-angle scattering by single stars.

5. Fluctuations in intensity

The intensity fluctuations are described by (e.g., Prokhorov et al.1975)

$$B_I(\mathbf{r}) = \frac{\langle I(\mathbf{r}')I(\mathbf{r}'+\mathbf{r})\rangle - \langle I(\mathbf{r})\rangle^2}{\langle I(\mathbf{r})\rangle^2} = \int \frac{d^2\mathbf{q}}{(2\pi)^2}\, e^{-i\mathbf{q}\cdot\mathbf{r}}\, W(\mathbf{q}), \tag{11}$$

$$W(\mathbf{q}) = \int d^2\mathbf{r}\, e^{i\mathbf{q}\cdot\mathbf{r}}\, \left\{ e^{[-D(r)-D(r_F^2q)+\frac{1}{2}D(|\mathbf{r}+r_F^2\mathbf{q}|)+\frac{1}{2}D(|\mathbf{r}-r_F^2\mathbf{q}|)]} - 1 \right\}, \tag{12}$$

where $W(\mathbf{q})$ is the power spectrum of the intensity fluctuations and $r_F = (L/k)^{1/2}$ is the Fresnel scale. The phase structure function (3) corresponding to (7) or (8) contains terms $\propto r^2$ and $\propto r^2 \ln r$. The former cancels and only the latter contributes to (12). This corresponds to scintillations due to a power-law spectrum with $\beta = 4$ (Goodman & Narayan 1985). This case is not amenable to a simple analytic treatment, but several relevant results can be inferred from the existing literature.

5.1. SMOOTH POWER SPECTRUM

For $\beta \neq 4$ a natural scale, $r = r_{\text{diff}}$, is defined by $D(r_{\text{diff}}) = 1$, and then (12) implies peaks in the the power spectrum $W(\mathbf{q})$ at the refractive, $q_{\text{ref}} \sim r_{\text{diff}}/r_F^2$, and diffractive, $q_{\text{diff}} \sim 1/r_{\text{diff}}$, scales. As $\beta = 4$ is approached the refractive and diffractive peaks recede to $q = 0$ and $q = \infty$, leaving a smooth spectrum of intensity fluctuations with no natural scale. Only the end points, $R^{-1} < q \lesssim R_0^{-1}$, can lead to significant features in the power spectrum.

5.2. THE MARGINAL DIFFRACTAL OF BERRY (1979)

The level of the intensity fluctuations for a point source has been estimated by a careful consideration of how the limit $\beta \to 4$ is approached. Berry (1979), who referred to the case $\beta = 4$ as the "marginal diffractal", showed that the asymptotic form (for $L \to \infty$) of the intensity fluctuations gives (cf. also Jakeman & Jefferson 1984)

$$I_2 = B_I(0) + 1 = 2, \tag{13}$$

where I_2 is the second moment of the intensity, with mean intensity $I_1 = 1$.

5.3. INTERPRETATION OF THE MARGINAL CASE

To understand the significance of (13) to gravitational microlensing, one needs to interpret it in terms of a physical model for microlensing. The following remarks describe an unsuccessful attempt to do this.

The intensity fluctuations may be attributed to caustics, and described in terms of the probability distribution $p(A)$ of a magnification A (e.g., Vietri & Ostriker 1983; Rauch et al.1992). The mean intensity (for a source of unity intensity, $I_0 = 1$) is $\langle I \rangle = \langle A \rangle = 1/(1 - \tau)^2$ for $\tau = \pi n r_E^2 \ll 1$, where n is the number density of stars and $r_E = (4GML/c^2)^{1/2}$ is the Einstein radius. A weakness (in the present context) of scintillation theory is that no distinction is made between $\langle I \rangle$ and I_0, and the theory needs to be modified to take account of $\langle A \rangle \neq 0$. The mean square intensity is dominated by individual microlensing events with large amplifications. The probability distribution $p(A) \propto 1/A^3$ for large A applies to both a point-mass lens and (approximately) to the two-point-mass lens systems (Schneider & Weiß 1986) relevant to (5). The integral $\langle I^2 \rangle = \int dA\, A^2 p(A)$ needs to be cut off at some A_{max}, and then $\langle I^2 \rangle$ depends logarithmically on the cutoff value A_{max}. It is not obvious how this model can reproduce (13), that is, $\langle I^2 \rangle = 2$. Further thought needs to be given to the interpretation of (13) in the context of gravitational microlensing.

5.4. INTENSITY FLUCTUATIONS FOR A GAUSSIAN SOURCE

A model for the intensity fluctuations in a gaussian source was presented by
Deguchi & Watson (1988). Their averaging procedure produces an equation,
their (16), of the form (12), but with only the $D(r)$-term in the exponent
(the terms involving r_F do not appear). Deguchi & Watson set $D(\mathbf{r}) \propto r^2$, as
in (9), but $D(r) \propto r^2$ cancels in (12) (e.g., Goodman & Narayan 1985), and
only the term $D(r) \propto r^2 \ln r$ from (7) contributes to the intensity fluctuations (Berry 1979). Thus there is an inconsistency between the statistical
averaging procedure adopted by Deguchi & Watson (1988), in which the
Fresnel scale does not appear, and that used in scintillation theory, which
depends explicitly on r_F.

6. Conclusions

1) Scintillation theory may be used to treat gravitational lensing by identifying the refractive index fluctuations in terms of the gravitational potential.
2) Application to multiple microlensing by a system of point masses reproduces a known result (Katz et al.1986) for the angular broadening.
3) Scintillation theory suggests a smooth power spectrum of intensity fluctuations (the refractive and diffractive scales disappear).
4) There are unresolved difficulties in the treatment of intensity fluctuations. (a) Scintillation theory implies an asymptotic variance of unity (Berry
1979), but a simple model for caustic-induced magnifications does not reproduce this result. (b) The averaging procedure of Deguchi & Watson
(1988) is not compatible with the result (12) of scintillation theory.

In summary, scintillation theory can be used to treat statistical microlensing. However, there are some specific difficulties and inconsistencies
related to intensity fluctuations. The resolution of these difficulties is likely
to provide deeper insight into statistical microlensing theory.

References

Berry, M.V., 1979, J. Phys. A: Math. Gen., 12, 781
Deguchi, S., & Watson, W.D., 1988, ApJ, 335, 67
Goodman, J., & Narayan, R., 1985, MNRAS, 214, 519
Jakeman, E., & Jefferson, J.H., 1984, Optica Acta, 31, 853
Katz, N., Balbus, S., & Paczyński, B., 1986, ApJ, 306, 2
Paczyński, B., 1986, ApJ, 301, 503
Prokhorov, A.M., et al., 1975, Proc. IEEE, 63, 790
Rauch, K.P., Mao, S., Wambsgauss, J., & Paczyński, B., 1992, ApJ, 386, 30
Schneider, P., & Weiß, A., 1986, A&A, 164, 237
Vietri, M., & Ostriker, J.P., 1983, ApJ, 267, 488

SIMULATION OF MICROLENSING LIGHTCURVES
BY COMBINING CONTOURING AND RAYSHOOTING

STEIN VIDAR HAGFORS HAUGAN
Institute of Theoretical Astrophysics, University of Oslo
Pb. 1029, Blindern
N-0315 OSLO
http://www.uio.no/~steinhh/index.html

1. Introduction

The contouring methods described by Lewis et al. (1993) and Witt (1993) are very efficient for obtaining the magnification of a point source moving along a straight track in the source plane. For finite sources, however, the amplification *must* be computed for numerous parallel tracks and then convolved with the source profile. Rayshooting, on the other hand, is an efficient algorithm for relatively large sources, but the computing time increases with the inverse of the source area for a given noise level.

2. The hybrid method

By using the method described in Lewis et al. (1993), all the images of a straight, infinite line in the source plane can be found. The images are the borders between those parts of the lens plane projected above the straight line, and those parts projected below the straight line. After finding the images of one line below the source and one line above the source, it is clear that those parts of the lens plane that are projected between the two infinite lines in the source plane are the areas between the images of the infinite lines.

Furthermore, those segments corresponding to the upper and lower edges of a box surrounding the source may be identified. The end points of these segments are projected onto the corners of the "source box". Starting from the corner points, the contouring method can be "turned around" 90 degrees, and all the lines joining all the corner points of the "source

C. S. Kochanek and J. N. Hewitt (eds), Astrophysical Applications of Gravitational Lensing, 275–276.

box" are found. After this step, all the images of the source box are placed within known, closed polygons. Rayshooting is then performed within all the closed polygons, and the lightcurve is produced in the usual way.

3. Efficiency

The efficiency of the rayshooting part of the method compared to crude, non-optimized rayshooting can be found by comparing the size of the areas where rayshooting has to be performed. A target area in the source plane with length $2l$, and height $2r_s$ gives an effective lightcurve length $L_c = 2l - 2r_s$, where r_s is the source radius. The theoretical efficiency f can be shown to be given by

$$f \approx \begin{cases} (1 + \dfrac{10\sqrt{\kappa_*}}{r_s} + \dfrac{100\kappa_*}{lr_s}) & \text{For } l \gg r_s \\[2em] (1 + \dfrac{20\sqrt{\kappa_*}}{r_s} + \dfrac{100\kappa_*}{r_s^2}) & \text{For } l = r_s, L_c = 0. \end{cases} \tag{1}$$

4. Discussion

The above arguments give a theoretical efficiency factor on the order of 10^5 for e.g. a snapshot of the source with $r_s = 0.01$, $l = r_s$ and $\kappa_* = 0.4$. However, the most time-consuming task for the hybrid method is going to be the contouring itself. For a snapshot like the example above, the contouring amounts to about 10^5 shots (Lewis et al. 1993). This must be compared with the total number of shots necessary to get a specific signal to noise ratio, generally about 10^3 shots. The highest estimates of f thus have to be lowered by roughly a factor of 100, depending on the specific parameters r_s, κ_*, γ, and l.

Even so, the proposed hybrid method has the potential to be a very efficient workhorse for producing accurate model lightcurves for small but extended sources.

Acknowledgments

The author would like to thank Rolf Stabell and Sjur Refsdal for comments during the preparation of this poster.

References

Lewis, G. F., Miralda-Escude, J., Richardson, D. C., Wambsganss, J., 1993, MNRAS, 261, 647
Witt, H. J. 1993, ApJ, 403, 530

SEPARATING INTRINSIC AND MICROLENSING
VARIABILITY USING PARALLAX MEASUREMENTS

STEIN VIDAR HAGFORS HAUGAN
Institute of Theoretical Astrophysics, University of Oslo
Pb. 1029, Blindern
N-0315 OSLO
http://www.uio.no/~steinhh/index.html

1. Introduction

In gravitational lens systems with 3 or more resolved images of a quasar, the intrinsic variability may be unambiguously separated from the microlensing variability through parallax measurements from 3 observers when there is no relative motion of the lens masses (Refsdal 1993). In systems with fewer than 3 resolved images, however, this separation is not straightforward. For the purpose of illustration, I make the following simplifications for the one-dimensional case: The observations consist of well-sampled time series of the observed flux $F_{\mathbf{A}}(t_i)$ and $F_{\mathbf{B}}(t_i)$ at two points in the observer plane. The separation vector of the two points is parallel to the direction of the transverse motion of the source-lens-observer system, and the distance $D_{\mathbf{AB}}$ between the observers is known. Furthermore, the distance $D_{\mathbf{AB}}$ is small compared to the typical length scale of fluctuations in the magnification $\mu(x)$.

It is possible to calculate the ratio of the instantaneous magnification at the two observers as a function of time, defined by

$$r(t_i) = F_{\mathbf{B}}(t_i)/F_{\mathbf{A}}(t_i) \tag{1}$$

where $F_{\mathbf{A}}(t_i)$ and $F_{\mathbf{B}}(t_i)$ are the observed fluxes at observer $\mathbf{A}$ and $\mathbf{B}$ respectively. I am assuming that observer $\mathbf{B}$ is the leading one.

With these assumptions, the magnification history $\mu_{\mathbf{A}}(t_i)$ for observer $\mathbf{A}$, can be reconstructed (apart from boundary conditions) through the formula

$$\mu_{\mathbf{A}}(t_i) = \mu_{\mathbf{A}}(t_i - \Delta t)r(t_i - \Delta t) \quad \text{with} \quad \Delta t = \frac{D_{\mathbf{AB}}}{v_\perp} \tag{2}$$

C. S. Kochanek and J. N. Hewitt (eds), Astrophysical Applications of Gravitational Lensing, 277–278.

where $v_\perp$ is the unknown velocity perpendicular to the line of sight.

Given a velocity $v_\perp$, the microlensing magnification history $\mu_{\mathbf{A}}$ is uniquely determined, and thereby also the intrinsic flux, given by

$$F_{\mathbf{IA}}(t_i) = F_{\mathbf{A}}(t_i)/\mu_{\mathbf{A}}(t_i) \qquad (3)$$

The velocity is chosen by minimizing some measure of the variability (e.g., χ^2) of $F_{\mathbf{IA}}$, given by $\chi^2 = \sum_{i=1}^{N}(F_{\mathbf{IA}}(t_i) - \langle F_{\mathbf{IA}}\rangle)^2$

2. Preliminary results

In order to test the method, dummy data for the intrinsic flux $F_{\mathbf{I}}(t_i)$ and the magnification $\mu(x_i) = \mu(v_\perp t_i)$ were made by simply filtering white noise, $N(t)$, with gaussian low-pass filters with characteristic scales $\tau_{\mathbf{I}}$ and τ_μ, and then exponentiating, e.g.:

$$\begin{aligned} F_{\mathbf{I}}(t) &= \exp(A_{\mathbf{I}}\Phi[N(t); \tau_{\mathbf{I}}]) \\ \mu(t) &= \exp(A_\mu\Phi[N(t); \tau_\mu]) \end{aligned} \qquad (4)$$

where $\Phi[\ldots; \tau]$ denotes gaussian filtering with time scale τ, and then renormalization to make the variance equal to one. $A_{\mathbf{I}}$ and A_μ are the amplitudes of the intrinsic and microlensing variabilities, respectively. For simplicity, but without loss of generality, the units were chosen so that the "true" source-lens-observer transverse velocity $v_\perp$ and the characteristic scale of the magnification fluctuations $\tau_{\mathbf{I}}$ were equal to 1. The observations were simulated according to

$$\begin{aligned} F_{\mathbf{A}}(t_i) &= F_{\mathbf{I}}(t_i)\,\mu(v_\perp t_i) \\ F_{\mathbf{B}}(t_i) &= F_{\mathbf{I}}(t_i)\,\mu\left(v_\perp t_i + \frac{D_{\mathbf{AB}}}{v_\perp}\right) \end{aligned} \qquad (5)$$

The flux ratio $r(t_i)$, the magnification history $\mu_{\mathbf{A}}(t_i)$ and the intrinsic flux $F_{\mathbf{IA}}(t_i)$ were calculated for a range of values for $v_\perp$. For a wide range of parameters, the χ^2 function is fairly well-behaved, with a quadratic minimum, although the minimum may be somewhat displaced compared to the true value of $v_\perp$. The most difficult cases seem to be those where $\tau_{\mathbf{I}} \approx \tau_\mu$ and $A_{\mathbf{I}} \gtrsim A_\mu$.

It is unclear how useful this method is for the two-dimensional case with two observers. This will be the subject of further study. The extension of the method to 3 observers in two dimensions is fairly straightforward. In cases where relative motion of the lensing point masses are important, only a partial separation will be possible.

References

Refsdal, S. 1993, in *Gravitational Lenses in the Universe*, eds. Surdej et al., Université de Liège, Belgium

PROSPECTS FOR THE DETECTION OF MICROLENSING TIME DELAYS

CHRISTOPHER B. MOORE AND JACQUELINE N. HEWITT

Massachusetts Institute of Technology Cambridge, MA, USA

Abstract. Since the image separations produced by microlensing are inaccessible to observation, we are left to observe either a change in total amplitude or the time delay between the images. Spillar (1993) has suggested that a time delay might be observed in the autocorrelation of a microlensed signal. The time delays are of order a few microseconds and are easily accessible to a sufficiently wide bandwidth system. We calculate expected observational results using the Green Bank Telescope with a VLBA recording system, and find that brightness temperatures exceeding the Compton limit are required for detection.

1. Discussion

We consider microlensing of a background source (e.g. an AGN) by a single Schwarzschild potential in some intervening galaxy (or galaxy halo). In what follows, we take $q_0 = 0.5$, $H_0 = 80\,\mathrm{kms^{-1}Mpc^{-1}}$, $z_s = 1.0$, $z_l = 0.05$, lens mass $= M_\odot$, and a uniform disc source of flux density 100 mJy. For a point source lensed by a single Schwarzschild potential the observed electric field is given in terms of the field that would be observed if there were no lens, $E(t)$. The observed field consists of two parts, one for each image: $\sqrt{\mu_1}E(t) + \sqrt{\mu_2}E(t + \Delta)$ where Δ is the gravitational lens time delay and $\mu_{1,2}$ are the magnification factors for the two images. The autocorrelation of this signal has three terms, the first of which appears at $\rho = 0$ and corresponds to the total power of the two images added together. The second two terms appear at $\rho = \pm\Delta$ and allow us to measure the gravitational time delay and $\sqrt{\mu_1\mu_2}$. When observing with limited bandwidth, the observed autocorrelation is convolved with the autocorrelation of the impulse

C. S. Kochanek and J. N. Hewitt (eds), Astrophysical Applications of Gravitational Lensing, 279–280.

response function of the bandpass filter. For the VLBA 16 MHz bandpass, the first zero occurs at a lag of 3.5×10^{-8} sec.

If the source is not point-like and different regions radiate incoherently, then one can divide the source into a large number of point sources and sum their contributions to the autocorrelation signal. We have written software which generates a predicted autocorrelation function from a distribution of flux density on the sky, a bandpass, and the relevant lens parameters.

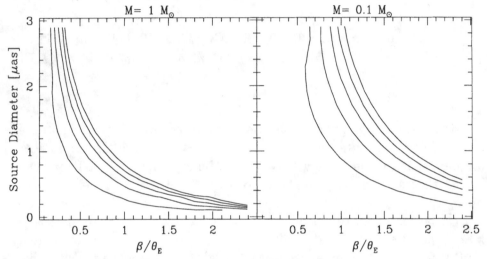

Figures 1 and 2 (above) show the observation time necessary to obtain a 3σ single channel detection for various combinations of source size, lens to source distance, and lens mass. The contours are at intervals of two hours starting at 1 hour. Detection in less than 10 hours observation time requires sources with apparent $T_B \sim 10^{15}$ K which exceeds the Compton limit. Such high brightness temperatures are suggested by studies of intraday variability in AGN which report T_B as high as 10^{19} K inferred from variability (Quirrenbach et al. 1992).

2. Conclusions

We have shown that microlensing time delays should be visible with large telescopes if there are high apparent T_B sources and a sufficient mass density of lensing objects. Remaining to be considered explicitly are the effects of scintillation, the effect of multiple lensing events on detectability, and the special case of a lens with a diffuse object directly behind it.

References

Quirrenbach, A., Witzel, A., Krichbaum, T., Hummel, C., Wegner, R., Schalinski, C., Ott, M., Alberdi, A., & Rioja, M., 1992, A&A, 258, 279
Spillar, E., 1993, ApJ, 403, 20

A CUSP—COUNTING FORMULA FOR CAUSTICS DUE TO MULTIPLANE GRAVITATIONAL LENSING

A. O. PETTERS
Princeton University
Department of Mathematics
Princeton, NJ 08544, USA

1. Main Theorem

Consider a gravitational lens system with k planes. If light rays are traced back from the observer to the light source plane, then the points on the first lens plane where a light ray either terminates, or, passes through and terminates before reaching the light source plane, are "obstruction points." More precisely, tracing rays back to the source plane induces a *k-plane lensing map* $\eta : U \subseteq \mathbf{R}^2 \to \mathbf{R}^2$ of the form $\eta(\mathbf{x}_1) = \mathbf{x}_1 - \sum_{i=1}^{k} \alpha_i(\mathbf{x}_i(\mathbf{x}_1))$. We then define an *obstruction point* of η to be a point $\mathbf{a}$ of U where $\lim_{\mathbf{x}_1 \to \mathbf{a}} |\alpha_i(\mathbf{x}_i(\mathbf{x}_1))| = \infty$ for some "deflection angle" α_i.

Without any additional assumptions on the lensing map η, one readily finds that cusp-counting becomes a hopeless task. In order to have some control on this problem, *we assume that η obeys the following mathematical constraints.* First, η has finitely many obstruction points and is smooth everywhere except at such points. Second, the domain U of η is an open disc punctured by mutually disjoint (small) closed discs that are centered at the obstruction points and disjoint from the critical curves of η. Third, all lenses are isolated, i.e., $\lim_{|\mathbf{x}_i| \to \infty} |\alpha_i(\mathbf{x}_i)| = 0$.

Notation:

(1) If $c(s)$ is a caustic that has no cusps and is parametrized by arc length s, and if $c(0)$ is the initial point of c, then $c(0)$ is an *outside starting point* when there is a line of support at $c(0)$ (i.e., a straight line L through $c(0)$ such that either L coincides with the path traced out by c or the path lies on one side of L). Henceforth, assume each caustic c (possibly with cusps) is piecewise parametrized according to arc length and has a specified outside starting point.

C. S. Kochanek and J. N. Hewitt (eds), *Astrophysical Applications of Gravitational Lensing*, 281–282.
© 1996 IAU

(2) If c is a caustic with an initial point on a fold, and if by a rigid motion the path traced out by c is positioned such that the x-axis is a line of support at $c(0)$, and the path lies on the side of the positive y-direction, then c is called *positive* (resp., *negative*) if the initial velocity vector $\dot{c}(0)$ is in the positive (resp., negative) x-direction. When the initial point $c(0)$ is a cusp and c is positioned through a rigid motion such that the x-axis is a line of support at $c(0)$ and the path lies on the side of the positive y-direction, then c is *positive*. Let $N_{caustics}^{+}$ and $N_{caustics}^{-}$ be, respectively, the total number of positive and negative caustics of η.

(3) If $c(s_1) = c(s_2)$, where $s_1 < s_2$, is a normal self-intersection of c, and if the ordered pair $\{\dot{c}(s_1), \dot{c}(s_2)\}$ is opposite (resp., identical) to the standard orientation of the plane, then the self-intersection is called *positive* (resp., *negative*). Let $N_{self}^{\pm}(c)$ be, respectively, the number of positive and negative self-intersections of a caustic c. Set $N_{self}^{\pm} = \sum_c N_{self}^{\pm}(c)$ and $N_{cusps} = \sum_c N_{cusps}(c)$, where each sum runs over all caustics.

The following theorem applies to generic multiplane gravitational lens systems. It expresses the total number of cusps in terms of the number of light path obstruction points and an algebraic number of caustics and caustic self-intersections.

Theorem 1 *Let η be a locally stable k-plane lensing map. Then the total number of cusps of η is given as follows:*

$$N_{cusps} = 2\left[g_\eta + (N_{caustics}^{+} - N_{caustics}^{-}) + (N_{self}^{+} - N_{self}^{-})\right].$$

A proof and detailed discussion of Theorem 1 will appear in a forthcoming paper on the global geometry of caustics (Petters 1995). In the special case of caustic networks with no self-intersections, Theorem 1 reduces to a formula of Levine, Petters & Wambsganss (1993).

Corollary 2 *Let η be a locally stable k-plane lensing map. Then:*

$$0 \leq N_{cusps} \leq 2[g_\eta + N_{caustics} + N_{self}],$$

where $N_{caustics}$ and N_{self} are, respectively, the total number of caustics and caustic self-intersections (excluding crossings of different caustics).

Acknowledgments: This research was supported in part by NSF Grant DMS-9404522.

References

Petters, A.O., 1995, J Math Phys, in press.
Levine, H., Petters, A.O., & Wambsganss, J., 1993, J Math Phys, 34(10), 4781

NEW CAUSTIC PHENOMENA IN DOUBLE-PLANE LENSING

A. O. PETTERS
Princeton University
Department of Mathematics
Princeton, NJ 08544, USA

AND

F. J. WICKLIN
University of Minnesota
The Geometry Center and School of Mathematics
Minneapolis, MN 55454, USA

Consider two point masses m_1 and m_2 on distinct planes with respective shears γ_1, γ_2 and continuous matter having densities κ_1 and κ_2. It is assumed that the lens equation is as follows:

$$\mathbf{y} = \mathbf{x}_1 - [m_1 \frac{\mathbf{x}_1}{|\mathbf{x}_1|^2} + \kappa_1 \mathbf{x}_1 + \gamma_1(-u_1, v_1)]$$
$$- [m_2 \frac{\mathbf{x}_2(\mathbf{x}_1) - \mathbf{d}}{|\mathbf{x}_2(\mathbf{x}_1) - \mathbf{d}|^2} + \kappa_2 \mathbf{x}_2(\mathbf{x}_1) + \gamma_2(u_2, -v_2)],$$

where

$$\mathbf{x}_2(\mathbf{x}_1) = \mathbf{x}_1 - \beta \left[m_1 \frac{\mathbf{x}_1}{|\mathbf{x}_1|^2} + \kappa_1 \mathbf{x}_1 + \gamma_1(-u_1, v_1) \right]$$

and β is a measure of the distance between the two planes. This lens system is governed by nine parameters: $\mathbf{d} = (\delta_1, \delta_2)$, β, m_1, m_2, κ_1, κ_2, γ_1, and γ_2. Figure 1 displays the caustics for the values $\mathbf{d} = (0.0675, 0)$, $\beta = 0.6$, $m_1 = m_2 = 1$, $\kappa_1 = 1.9995$, $\kappa_2 = 3$, $\gamma_1 = 0.2$, and $\gamma_2 = 0.3$. None of these caustics can be generated by single-plane multiple point-mass lenses (due to the violations of convexity) or a double-plane two point-mass lens (since the latter lens cannot produce lips and swallowtails — Erdl & Schneider (1993)). Also, note the occurrences of two cusps of the second kind (Petters 1995), which are impossible in single-plane, multiple point-mass gravitational lensing.

C. S. Kochanek and J. N. Hewitt (eds), Astrophysical Applications of Gravitational Lensing, 283–284.
© 1996 IAU

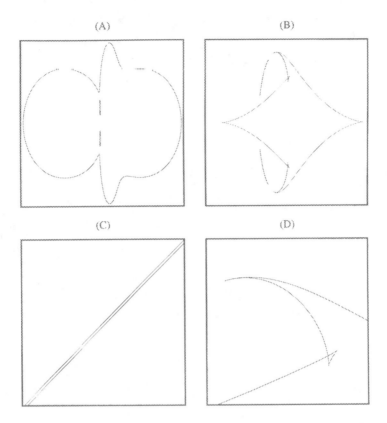

Figure 1. A critical curve and caustic for the double-plane lens with continuous matter and shear. (A) A critical curve with three connected components is transformed (B) into a caustic with lips and swallowtail singularities, and a cusp of the second kind. (C) A close-up of the lips caustic, showing that it is not a double line. (D) A close-up of the cusp of the second kind.

A more extended treatment of caustics due to the double-plane two point-mass lens with continuous matter and shear will appear in the paper Petters & Wicklin (1995).

Acknowledgments: Figure 1 was discovered and produced with the aid of Pisces (Wicklin 1995) and DsTool (Back et al. 1992), respectively. This research was supported in part by NSF Grant DMS-9404522.

References

Back, A., Guckenheimer, J. Myers, M. Wicklin, F., & Worfolk, P., 1992. Notices of the AMS, 39, 303
Erdl, H., & Schneider, P., 1993, A&A, 268, 453
Petters, A.O., 1995, J Math Phys, 36(8), 4276
Petters, A.O., & Wicklin, F.J., 1995, MNRAS, in press
Wicklin, F.J., 1995, Technical Report, The Geometry Center, Univ. of Minnesota

MICROLENSING OF LARGE SOURCES INCLUDING SHEAR TERM EFFECTS

S. REFSDAL
Hamburger Sternwarte
Gojenbergsweg 112, D-21029 Hamburg, Germany

AND

R. STABELL
Institute of Theoretical Astrophysics, University of Oslo
Pb. 1029, Blindern, N-0315 Oslo, Norway

Abstract. We find that the standard deviation for the observed magnitude of a large microlensed source is $\delta m \leq 2.17|\kappa|^{1/2}\theta_o/\theta_s$ even in the presence of non-zero shear.

1. Introduction and Summary

The most spectacular effects of microlensing occur when the angular radius of the source (θ_s) is much smaller than the Einstein Ring (θ_o), and this case has also been investigated in most detail up till now. For large sources $(\theta_s \geq 5\theta_o)$ Refsdal and Stabell (1991) derived analytically a useful formula for the standard deviation of the observed magnitude:

$$\delta m = 2.17\sqrt{|\kappa|}\,\frac{\theta_o}{\theta_s}. \tag{1}$$

Here κ is the optical depth for microlensing and the shear γ is assumed to be zero. The variations in m are mainly due to fluctuations in the smoothed out surface mass density caused by the Poisson fluctuations in the number of stars projected in front of the source. For κ-values equal to 0.1 and 0.4 the value of δm given by Eq. (1) was found to be reasonably accurate (to within 20%) for sources with $\theta_s > 5\theta_o$.

We have carried out more extensive calculations for various values of κ, θ_s and also for some values of $\gamma \neq 0$. The main result is that the value

C. S. Kochanek and J. N. Hewitt (eds), Astrophysical Applications of Gravitational Lensing, 285–286.

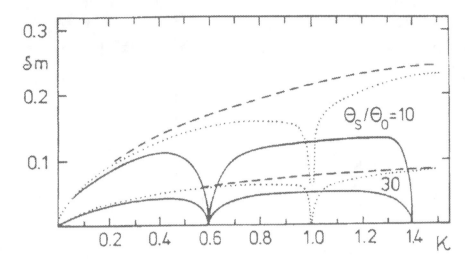

Figure 1.   The standard deviation δm for the case $\gamma = 0.4$ together with the values of δm given by Eq. (1) (dashed) and δm for $\gamma = 0$ (dotted)

of δm given by Eq. (1) is an upper limit to the true standard deviation, regardless of the values of $\theta_s/\theta_o, \kappa$ and γ. This means that Eq. (1) still can be used to estimate an upper limit to the source size as discussed by Refsdal and Stabell (1991). As expected we find that the deviations from Eq. (1) increase with decreasing source size. Furthermore, we find that these deviations typically increase with decreasing value of $(1 - \kappa)^2 - \gamma^2$ (amplification increases, compare to Deguchi & Watson (1987)).

2. Results

For the case $\gamma \neq 0$ we have not succeeded in deriving a simple analytical formula similar to Eq. (1). Some of the results from our numerical calculations are plotted in Fig. 1, ($\theta_s = 10\theta_o$ and $\theta_s = 30\theta_o$, and $\gamma = 0.4$), together with results for $\gamma = 0$ and the values of δm given by Eq. (1).

It is generally found that even for $\gamma \neq 0$, Eq. (1) represents an upper limit for δm. Except for the narrow "forbidden" interval around $\kappa = 1$, we even find that a value of $\gamma \neq 0$ further reduces δm. For a large range of γ-values we see however that the effect of the shear is rather small. An obvious exception is of course when γ approaches $\pm(1 - \kappa)$, since δm then approaches zero.

References

Deguchi, S. & Watson, W.D., 1987, Phys Rev Letters, 59, 2814
Refsdal, S. & Stabell, R., 1991, A&A, 250, 62

GRAVITATIONAL MICROLENSING BY RANDOM MOTION
OF STARS: MOVIE AND ANALYSIS OF LIGHT CURVES

JOACHIM WAMBSGANSS

Astrophysikalisches Institut Potsdam, 14482 Potsdam, Germany

AND

TOMISLAV KUNDIĆ

Princeton University Observatory Princeton, NJ 08544 USA

Abstract. We present a quantitative analysis of the effect of microlensing caused by random motion of individual stars in a galaxy lensing a background quasar. We calculate a large number of magnification patterns for positions of the stars slightly offset from one frame to the next, and thus obtain light curves for fixed quasar and galaxy positions, only due to the change in the relative star positions. These light curves are analyzed to identify microlensing events, which are then classified with respect to height, duration, and slope. These random motion microlensing events are compared with the corresponding ones caused by the bulk motion of the galaxy.

We find that microlensing events produced by random motion of stars are shorter, steeper, and more frequent than bulk motion events, assuming the velocity dispersion of the stars equals the bulk velocity of the galaxy. The reason for this difference is that in the case of random motion, caustics can move with an arbitrarily high velocity, producing very short events, whereas in the comparison case for bulk motion a microlensing event can never be shorter than it takes a fold caustic, which moves with the velocity of the lensing galaxy projected onto the quasar plane, to cross the quasar. An accompanying video illustrates these results. For three different values of the surface mass density κ, it shows time sequences of 1000 magnification patterns for slowly changing lens positions, together with the positions and velocity vectors of the microlensing stars. The full paper including the video can be found in Wambsganss & Kundić (1995). A short version of the video is available as an MPEG movie under anonymous ftp at astro.princeton.edu, in the directory `jkw/microlensing/moving_stars`.

C. S. Kochanek and J. N. Hewitt (eds), Astrophysical Applications of Gravitational Lensing, 287–288.

1. Description of accompanying video

The accompanying video (Wambsganss & Kundić 1995) consists of four sequences: one header sequence, in which the film is described, and three sequences of 1000 magnification patterns calculated for random motion of stars, one sequence each for surface mass densities of $\kappa = 0.2, 0.5, 0.8$. The film shows the changing caustic network for stars moving with velocities drawn from a Maxwellian distribution.

The video screen consists of four panels: the lower left shows the magnification pattern at any given time. The lower right displays positions of stars at that time, so that one can correlate single, double, and multiple lens caustics with the corresponding lenses. Velocity vectors at the lenses – small lines of different length and direction – indicate where the stars are heading. Both of these panels change with time. In addition, there are two panels at the top, indicating microlensing light curves. At the top left we show one light curve for bulk motion: it is just a horizontal cut through the magnification pattern, i.e. so far the "standard" microlensing light curve for fixed positions of the stars and only bulk motion of the lensing galaxy. Because the magnification pattern changes as time goes on, so does this light curve. An example for the new light curves determined for random motion of lenses is shown at the top right panel of the video screen. It is obtained at the central point of the magnification pattern. This light curve is fixed for the whole sequence; it is pre–calculated. All that changes is a black vertical line inside the light curve, indicating "time" or "frame number". In other words: for fixed positions of observer, lensing galaxy and quasar, the observer would see such a light curve with time due to the stars moving inside the galaxy.

After each such sequence for the three values of the surface mass density κ, the changing magnification pattern alone is shown in higher resolution, filling the whole screen (i.e., without the panels with the positions of lensing stars and light curves). This is done in order to allow a more detailed look at the way the caustics move and merge with each other. Finally, the three sequences are shown again, slowed down by a factor of five, so that one can follow the evolution of individual caustics and the motion of the stars producing them. The total duration of the video is about 18 minutes.

References

Wambsganss, J. & Kundić, T., 1995, ApJ, in press

THE EVOLUTION OF QSO SPECTRA:

Evidence for Microlensing?

PAUL J. FRANCIS
School of Physics, University of Melbourne
Parkville, Victoria 3054, Australia
E-mail: pjf@physics.unimelb.edu.au

AND

ANURADHA KORATKAR
Space Telescope Science Institute
3700 San Martin Drive, Baltimore, MD 21218, USA
E-mail: koratkar@stsci.edu

Abstract. We find that the spectra of QSOs evolve: high redshift QSOs ($z > 1.5$) have lower equivalent width emission lines than low redshift QSOs ($z < 0.5$) with the same luminosities and radio properties.

We propose that microlensing by compact objects may account for the apparent evolution. If $\Omega \sim 0.05$ in compact objects, the continuum emission from many high redshift QSOs will be amplified, but not the line emission, leading to the observed decrease in the apparent equivalent widths.

1. Introduction

If much of the missing mass of the universe is in the form of compact objects, it will microlens distant QSOs. If the compact objects have masses comparable to the Sun, their Einstein radii will be comparable in angular size to the continuum sources of the high redshift QSOs. The microlensing can thus strongly amplify the continuum emission from QSOs, but emission coming from more extended regions, such as the emission-line radiation, will not be amplified. Thus abnormally low ratios of emission-line to continuum flux (ie. low equivalent widths) are a signature of microlensing.

Canizares (1982) and Delcanton et al. (1994) showed that for $\Omega \sim 0.05$ in compact objects, many bright high redshift QSOs will be appreciably microlensed, and would show reduced emission-line equivalent widths.

C. S. Kochanek and J. N. Hewitt (eds), Astrophysical Applications of Gravitational Lensing, 289–290.
© 1996 IAU

2. Our Study

We compared the spectra of low redshift ($z < 0.5$) QSOs, drawn from the IUE archives, with those of high redshift ($z > 1.5$) QSOs drawn from the Large Bright QSO Survey (Morris et al. 1991). The samples were carefully matched in luminosity, radio loudness and rest-frame wavelength coverage; see Francis & Koratkar (1995) for details.

3. Results

We see evolution: the mean equivalent width of the Ly-α and C IV emission line is significantly lower in the high redshift sample than in the low redshift sample, as predicted by the microlensing model. The details of the equivalent width distributions also match the microlensing prediction; the change in the mean equivalent width is caused by an increase in the fraction of QSOs having very low equivalent widths at $z > 1$.

For more detail, see Francis & Koratkar (1995).

4. Conclusions

The spectra of QSOs evolve, and the form of the evolution can be explained by microlensing by a cosmological population of compact objects with $\Omega \sim 0.05$. However, we have no way of ruling out intrinsic evolution in QSO spectra, which could also explain our result.

References

Canizares, C. R., 1982, ApJ 263, 495

Delcanton, J. J., Canizares, C. R., Granados, A., Steidel, C. C., & Stocke, J. T. 1994, ApJ, 550

Francis & Koratkar 1995, MNRAS 274, 504

Morris et al. 1991, AJ 102, 1627

QUASAR VARIABILITY FROM MICROLENSING

M.R.S. HAWKINS
Royal Observatory
Blackford Hill
Edinburgh EH9 3HJ

Quasars are known to vary in brightness over a wide range of time scales. Short term intrinsic variability has been well documented, and a strong case can be made that long term variation is due to microlensing. In this paper the effect of time dilation as a means for distinguishing between intrinsic variation and gravitational lensing is discussed.

The idea that small compact bodies might cause quasars to vary has been examined in some detail over the last few years, and extensive numerical simulations have been carried out to explore the effect (Kayser et al. 1986, Schneider & Weiss 1987). Hawkins (1993) suggested that in fact all quasars are being microlensed, and that this is the main mechanism for large amplitude long term variation. Press & Gunn (1973) in a classic paper pointed out that if every line of sight is gravitationally lensed then the mass density of the lensing bodies must be approximately equal to the cosmological critical density (i.e. $\Omega_{lens} \approx 1$). This clearly has important consequences for the dark matter problem, as well as throwing interesting new light on the behavior of quasars.

The question of whether quasar variability is caused by microlensing is explored in more detail in a more recent paper by Hawkins (1995). The point to be decided is the extent to which intrinsic variability in quasars is confined to the well established, short term, small amplitude fluctuations, and whether it can account for the large amplitude long time scale variations which dominate quasar light curves.

Among the arguments used to support the microlensing hypothesis is the apparent lack of a time dilation effect in quasar light curves. One would expect to see a lengthening of the time scale of quasar variation with redshift by a factor $(1 + z)$ if the variation is intrinsic to the quasar. Many attempts have been made to detect this effect but there have been no convincing measurements of it to date. Recently Baganoff & Malkan (1995)

C. S. Kochanek and J. N. Hewitt (eds), Astrophysical Applications of Gravitational Lensing, 291–292.

advanced an ingenious argument to circumvent this difficulty, which raises some interesting questions about the nature of quasar variability.

The main point of Baganoff & Malkan's argument is that at higher redshift the observer is looking progressively further into the ultra-violet, and hence sees an emission region which may well be smaller than at larger wavelengths. They then put forward a model to link the decreasing size of the continuum region with shorter time scale of variability, and claim that this just cancels out the time dilation effect.

The model requires that the time scale of variation in AGN's gets shorter as the wavelength of observation decreases. This prediction can be tested in various ways. The Seyfert galaxy NGC 5548 has been closely monitored in several major observing campaigns with IUE. Peterson et al. (1991) give light curves in the optical at 4870 Å and the ultra-violet at 1350 Å. The correlation between the light curves is almost exact, with only a difference in amplitude. There is clearly no question of a difference in time scales. A similar effect can be seen in the data of Hawkins (1995) where the time varying autocorrelation function (ACF) is evaluated for samples of low and high redshift quasars. For each sample, ACF's for blue and red passbands are shown. Although in each case a small decrease in time scale is seen from low to high redshift, in each redshift bin the time scales for the red and blue passbands are not significantly different. A similar manifestation of this may be seen in the light curves in Hawkins (1995) where the variation is achromatic.

Lack of a time dilation effect thus continues to be an argument against long term intrinsic variability in quasars. There are other aspects of the light curves such as statistical symmetry and achromatic variation which further constrain any model of intrinsic variability.

References

Baganoff, F.K. & Malkan M.A., 1995, ApJL, 444, L13
Hawkins, M.R.S., 1990, Nature, 366, 242
Hawkins, M.R.S., 1995, MNRAS, in press
Kayser, R., Refsdal, S. & Stabell, R., 1986, A&A, 166, 36
Peterson, B.M., et al., 1991, ApJ, 368, 119
Press, W.H. & Gunn, J.E., 1973, ApJ, 185, 397
Schneider, P. & Weiss, A., 1987, A&A, 171, 49

FOREGROUND GALAXIES AND THE VARIABILITY OF LUMINOUS QUASARS

J. VON LINDE, U. BORGEEST, J. SCHRAMM AND S. REFSDAL
Hamburger Sternwarte
Gojenbergsweg 112, D-21029 Hamburg, Germany
email: jlinde@hs.uni-hamburg.de

AND

E. VAN DROM
Université de Liège, Inst. d'Astrophysique
5, Avenue de Cointe, B-4000 Liège, Belgium

In order to look for an amplification bias (AB) by gravitational lensing caused by medium redshift $(0.2 \lesssim z \lesssim 0.8)$ clusters or groups of galaxies, we compare galaxy counts in deep CCD images of highly luminous, high redshift QSOs with those in nearby control fields at a distance of 1 deg at the same galactic latitude. The total sample contains 37 objects up to now, from which one field had to be excluded because of a seeing difference between the QSO and control fields.

Observations were done at the DSAZ 3.5 and 2.2 m telescopes at Calar Alto, Spain, the ESO NTT, Chile, and the *Nordic Optical Telescope* (NOT) at La Palma. Quasars were selected with respect to optical luminosity (12 QSOs, $z \geq 1$ $M_V \leq -29.0$; $H_0 = 50$ km s^{-1} Mpc^{-1}, $q_0 = 0$), and to both optical luminosity *and* radio flux (25 QSOs,, $z \geq 1.5$, $M_V \leq -27.0$ mag, S(6 cm) ≥ 0.8 Jy). All fields were observed in the Johnson R band. Object search and classification was done by an automatic procedure.

Galaxy counts have been analyzed **1.** in the entire fields, **2.** inside a circle with radius 1 arcmin around the quasars and the quasar position in the control fields, respectively, and **3.** inside a circle with radius 15 arcsec around the quasars, one star in the quasar field, and two stars at the same positions in the control fields (as far as possible), respectively. Since the automatic search is not complete in the direct vicinity of the quasars/stars, this was done by eye. The 4 fields observed at the NOT have been excluded from **1.** and **2.** because of the small field.

C. S. Kochanek and J. N. Hewitt (eds), Astrophysical Applications of Gravitational Lensing, 293–294.
© 1996 IAU

The table shows the excess factors N_{qso}/N_{contr} and the number of fields N_f used for the different scales for the entire sample and the 2 subsamples of extremely radio loud ($S_6 \geq 1\,Jy$) and radio quiet QSOs, respectively. One sigma errors based on Poisson statistics are given.

	$q = N_{qso}/N_{contr}$		
Sample	$4'' \leq r \leq 15''$ [1] (N_f)	$4'' \leq r \leq 60''$ (N_f)	entire fields[2] (N_f)
all	1.61 ± 0.44 (36)	1.15 ± 0.10 (32)	1.035 ± 0.033 (32)
S >1 Jy	1.38 ± 0.43 (23)	1.00 ± 0.11 (19)	1.03 ± 0.04 (19)
opt	3.43 ± 2.25 (7)	1.55 ± 0.30 (7)	1.38 ± 0.14 (7)

[1] Background counts are averaged over the 3 stars

[2] Mind that the fields differ in size

We confirm the excess of galaxies around QSOs on small scales reported by other authors (cf. Van Drom et al.1993) which decreases towards larger radii. The strongest effect is found in the subsample of radio quiet objects, in contradiction to the *multiple wave band bias* (Borgeest et al. 1991). The reason for this (if not due to statistical fluctuations) could be the higher average absolute brightness of this sample or selection effects in the optical quasar surveys.

The results cannot be interpreted as evidence for an AB due to clusters of galaxies. However, several faint clusters can be seen in the exposures. Due to the spatial correlation of galaxy clusters, the distance of the control fields might be too small to detect a significant excess in the QSO fields on larger scales.

A very preliminary analyses of light curves of 22 of the objects from the *Hamburg Quasar Monitoring* (HQM) project (Schramm et al.1994 and references therein) gives no evidence for a correlation of distance and magnitude of the foreground galaxies and the variability of the QSOs. The light curves show at least no strong indication for an AB caused by microlensing of the quasars by stars in the foreground galaxies.

The complete poster including light curves and contour plots can be obtained from the authors.

References

Borgeest U., von Linde J., & Refsdal, S., 1991, A&A, 251, L35

Schramm, K.-J., Borgeest, U., Kühl, D., v. Linde, J., & Linnert, M.D., 1994, A&A Suppl, 106, 349

Van Drom E., Surdej J., Magain P., Hutsemèkers D., Gosset E., Claeskens J.F., Shaver P., & Melnick J., 1993 in Gravitational Lenses in the Universe, eds. J. Surdej et al., (Liège: Université de Liège) 301

UV/OPTICAL CONTINUUM VARIABILITY IN AGNS

WEI-HSIN SUN
Institute of Astronomy, National Central University, Taiwan, ROC

CHARLENE A. HEISLER
Anglo-Australian Observatory, Epping Laboratory, Australia

AND

MATTHEW A. MALKAN
Department of Astronomy, UCLA, USA

1. Introduction

The observed strong UV/Optical excess in many Seyfert 1 galactic nuclei and quasars (QSOs) is well described as thermal radiation from the surface of an optically thick accretion disk surrounding a supermassive black hole (Malkan 1983, Sun and Malkan 1989). This scenario naturally leads to a radially symmetric temperature gradient with the innermost regions the hottest. Thus for a perturbation generated in the inner regions propagating outward, sequential variations from higher to lower frequencies should be expected, aside from the bolometric increase of brightness. However, the extremely intensive International AGN watch (Peterson 1993 and references therein) on NGC 5548 and UV/Optical monitoring campaign on Mkn 335 (Sun et al. 1995) point to opposite conclusions on the time lag between UV and Optical continua, with no lag for NGC 5548 and $\sim$ 50 days for Mkn 335. We thus began two monitoring programs at Lick Observatory and Mount Stromlo and Siding Spring Observatory (MSSSO) on a sample of Seyfert 1 galaxies with CCD direct imaging to search for more conclusive evidence on the sequential variabilities.

2. Observations

The variabilities of the target AGNs at 3890Å (U), 5750Å (V), 8020Å(I), and 9750Å(Z), were measured with two-inch square intermediate band

C. S. Kochanek and J. N. Hewitt (eds), Astrophysical Applications of Gravitational Lensing, 295–296.
© 1996 IAU

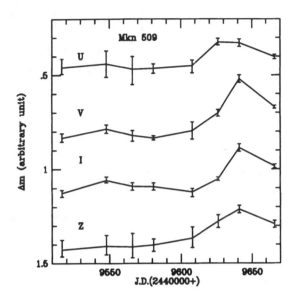

Figure 1. Four-color light curves of Mkn 509

$(250 - 300\text{Å})$ filters via differential photometry with non-variable field stars
at the 1-m telescopes at Lick and MSSSO. The observations are carried out
approximately once every month, starting June 1994 at Lick and October
1994 at MSSSO.

3. Preliminary Results

We find non-simultaneous variabilities in the four selected wavebands on
Mkn 509 with Lick Observations combined with MSSSO data (from 6/94
through 11/94). Three interesting points can be inferred: (i) the variability
amplitudes are in general $0.15 - 0.20$ magnitudes in all four filters over the 6
month period, while the typical error bars are 0.04 magnitude, thus allowing the variability to be accurately measured; (ii) the observed variations
in each filter are not in phase, suggesting the possibility of sequential variation in neighboring wavebands; and (iii) considerable color change from
U through Z in 6 months is observed. While the current dataset show interesting implications in understanding how the optical continuum in AGN
varies, we still need more data points for rigorous time analysis.

References

Malkan, M.A., 1983, ApJ, 275, 477
Peterson, B.M., 1993, PASP, 105 (685), 247
Sun, W.-H. & Malkan, M.A., 1989, ApJ, 346, 68
Sun, W.-H. et al., 1995, in preparation

QUASAR MICROLENSING BY CLUSTER DARK MATTER

A 2dF + MACHO Project

M. A. WALKER AND P. M. IRELAND
Research Centre for Theoretical Astrophysics, A28
School of Physics, University of Sydney, NSW 2006, Australia

1. Introduction

Following Paczyński's (1986) suggestion, many examples of gravitational lensing by Galactic objects have been recorded as a consequence of extensive photometric monitoring programs (e.g. Udalski et al. 1994). Now that such experiments have demonstrated the capability to discover microlensing events against large numbers of target sources, it behooves us to consider whether these techniques can be usefully applied elsewhere.

Moving beyond the Galactic realm, we can identify clusters of galaxies as a distinct environment where we might investigate the nature of the dark matter which is dynamically inferred. In this case the photometric targets could not sensibly be stellar; but quasi-stellar objects (quasars) are sufficiently luminous that one can obtain good photometry even of very distant sources. It is therefore worth exploring the possibility that one could detect compact dark matter in galaxy clusters by looking for microlensing events on background quasars; detailed consideration of this idea can be found in a separate paper (Walker and Ireland 1995). We note that microlensing has already been proposed as the source of observed variability in one sample of quasars (Hawkins 1993).

2. Technique

The observational approach to this experiment is essentially the same as is currently used for the Galactic microlensing experiments, differing mainly in the smaller number of sources which need to be monitored ($\sim 10^2$ vs. $\sim 10^6$). (For an instrument like the MACHO camera observing a rich cluster at redshifts $0.01 < z < 0.05$, the mean optical depth to cluster microlensing is $\sim 10^{-2}$ if all the cluster dark matter is in compact form.) One significant extra requirement is that the target sources need to be identified as

C. S. Kochanek and J. N. Hewitt (eds), Astrophysical Applications of Gravitational Lensing, 297–298.

such. This can be done spectroscopically, using the 400-fiber 2 degree Field instrument on the Anglo-Australian Telescope to isolate the quasars from the faint stars.

For sensible time-scales – between 20 minutes and a few years – the suggested monitoring experiment would be sensitive to dark matter lumps in the mass range $10^{-10} < M/M_\odot < 1$. However, quasars are believed to emit their optical continuum from a region of dimension $\sim 10^{14}$ cm (e.g. Rees 1984), and the magnification would therefore be very small for masses less than $\sim 10^{-4} M_\odot$ where the Einstein ring is smaller than the apparent size of the source. Clearly the experiment is still sensitive to an interesting mass range, including as it does very low mass stars and substellar objects. Weekly monitoring over a period of a few years is sufficient to cover the mass range $10^{-4} < M/M_\odot < 1$.

3. Difficulties

The principal difficulty with the proposed experiment lies in the recognition of microlensing-induced variability. Quasars are variable sources and one needs some means of distinguishing intrinsic variability from microlensing. In the case of Galactic microlensing one usually requires, for example, that an event be achromatic and time symmetric in order for it to be considered a possible lensing event; whereas the anticipated source structure for quasars suggests that asymmetric and chromatic events are quite plausible. In our favor is the observation that typical quasars do not vary profoundly (around 0.2 mag) in the optical band over a decade (Hook, McMahon, Boyle and Irwin 1994). Also in our favor is the fact that, because the optical depth to microlensing is much less than unity, any microlensing should show-up as distinct events on a relatively steady baseline.

These signatures are, however, insufficient as proof of a microlensing origin for any observed variability. Fortunately there is a unique property which cannot be mimicked by intrinsic variability: because the surface density of dark matter varies through the cluster, the microlensing event rate will similarly vary so that the process can, on a statistical basis, be verified as the variability mechanism.

References

Hawkins, M.R.S., 1993, Nature, 366, 242
Hook, I.M., McMahon, R.G., Boyle, B.J. & Irwin, M.J., 1994, MNRAS, 268, 305
Paczyński, B., 1986, ApJ, 304, 1
Rees, M.J., 1984, ARA&A, 22, 471
Udalski, A., Szymański, M., Kałużny, J., Kubiak, M., Mateo, M., & Krzemińskii, W., 1994, ApJL, 426, L69
Walker, M.A. & Ireland, P.M., 1995, MNRAS, in press

OBSERVATIONS OF LENS SYSTEMS WITH KECK I

C. R. LAWRENCE
Jet Propulsion Laboratory 169-506
4800 Oak Grove Drive
Pasadena, CA 91109

Abstract. The extreme difficulty of many essential optical and infrared observations of lens systems has impeded progress and contributed to the popular but erroneous view that lensing is a curiosity rather than an important astrophysical tool. Keck I, with its unprecedented sensitivity for spectroscopy and infrared imaging, will have a major impact on lensing observations.

1. Introduction

The Caltech lensing consortium—R. Blandford, J. Cohen, G. Djorgovski, D. Hogg, J. Larkin, C. Lawrence, K. Matthews, G. Neugebauer, and I. Smail—was organized with both the possibilities and inevitable problems of the early days of a new telescope and instruments in mind. Comprising members the Near Infrared Camera (NIRC) and Low Resolution Imaging Spectrometer (LRIS) teams as well as the Caltech lensing "regulars", the consortium developed prioritized but rather inclusive lists of targets for both imaging and spectroscopy that could be adapted on short notice to available observing time and capabilities. Our goal was to demonstrate the value of Keck by making difficult but important lens observations. Some early highlights are given below.

2. Observations

NIRC (Matthews & Soifer 1994) has a 256×256 InSb array with $0\farcs15$ pixels, giving a $38\farcs4 \times 38\farcs4$ field. Standard filters are available in the 1–$5\,\mu$m range, as well as low-resolution spectroscopic capability.

C. S. Kochanek and J. N. Hewitt (eds), Astrophysical Applications of Gravitational Lensing, 299–304.

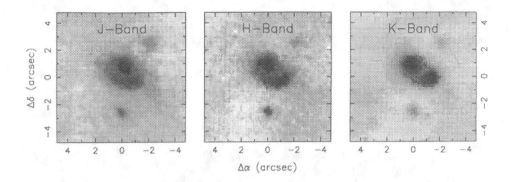

Figure 1. NIRC images of MG 1131+0456 taken on 4 March 1994 (UT). Integration times were 360, 900, and 720 seconds for the J (1.1–1.4 μm), H (1.5–1.8 μm), and K (2.0–2.4 μm) images, respectively. The two quasar images dominate the K image, although the centroid of the one to the east is affected by the nearby galaxy core. The quasar images can still be seen faintly in the H image, but are not detected in the J image. The lensing geometry requires a massive galaxy, but the J image suggests an interacting or merging galaxy rather than a normal, smooth elliptical.

LRIS (Oke et al. 1995; Cohen et al. 1993) has a 2048 × 2048 thinned Tektronix CCD with a peak quantum efficiency of 80% at 6500 Å. Single slits 3′ long, available in several widths, illuminate a 2048 × 800 region on the chip. Various gratings give $800 \lesssim R \lesssim 4300$. The maximum throughput is a high 42%.

MG 1131+0456 was observed at 1.2 and 2.2 μm during the first NIRC engineering run in 1993 March to test our suspicion that the two peaks seen by Annis (1992) at 2.2 μm were the heavily reddened counterparts of the radio core. The images immediately gave strong support to this suspicion, and led us to conclude that $A_V > 4$ mag along two paths through the lens separated by almost 10 kpc (Larkin et al. 1994). Figure 1 shows images taken in 1994 March in better seeing.

MG 0414+0534 (Hewitt et al. 1992) has a unique optical spectrum that is well-fit from 4300–9400 Å by $F_\nu \propto \nu^{-8.8}$. There is a strong absorption feature near 8650 Å. Infrared spectra reveal emission lines, including a strong H α line, that give $z = 2.639$ (Lawrence et al. 1995a). The remarkable overall shape of the spectrum is reproduced very well by a ν^0 continuum, Fe II pseudo-continuum, and standard quasar lines all reddened by $A_V = 5.5$ mag of dust at $z = 0.5$, the theoretically "most likely" lens redshift (Kochanek 1992).

The absorption feature corresponds to ~ 2380 Å at $z = 2.639$. Although Fe II lines near this wavelength are often seen, their equivalent width is usually much smaller. It seemed likely that the absorption feature came

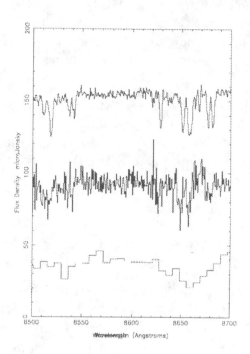

Figure 2. Spectra of absorption in MG 0414+0534. From bottom to top: 4-Shooter spectrum at 27 Å resolution (Hewitt et al. 1992); LRIS spectrum at 2 Å resolution from 1993 October with high readout noise (preprint, never to be published!); and LRIS spectrum from 1994 November (Lawrence et al. 1995b). The top spectrum is fitted beautifully by four Fe II triplets (λ =2343.495, 2373.737, and 2382.039) at z = 2.63172, 2.63487, 2.64268, and 2.64474. The iron is unusually but not unprecedentedly strong.

instead from the lens, and that an unambiguous identification of the feature would give the lens redshift. The strongest absorption lines are from the Na D doublet at z = 0.47 and the Mg II doublet at z = 2.10, which could be easily distinguished by their separations of 9 and 22, Å, respectively, in a high resolution spectrum.

Accordingly, MG 0414+0534 was observed at 2.0 Å resolution during an early LRIS engineering run in 1993 October. Despite readout-noise problems, a clear doublet was seen at 8650 Å with exactly the separation of Na D at z = 0.47. We were sufficiently confident of the identification to submit a paper entitled *The Redshift of the Lens in MG 0414+0534* and to distribute preprints, but sufficiently concerned to reobserve in 1994 November after the readout problems were corrected. Figure 3 shows the original 4-Shooter spectrum along with the two Keck spectra. The answer is unambiguous. The absorption is due entirely to Fe II, and the redshift of the lens remains unknown (Lawrence et al. 1995b).

Lawrence et al. (1995a) and Larkin et al. (1994) discuss three explanations for the extremely red colors of MG 0414+0534 and MG 1131+0456: dust in the lens; dust near the quasar; and intrinsically red quasars. None is free of improbabilities, nor are the three mutually exclusive. The following reasons summarize why on balance we prefer the dust-in-the-lens explanation.

1. The spectrum of MG 0414+0534 is fitted very well by standard quasar "components" reddened by dust at $z \sim 0.5$, significantly less well by dust at $z = 2.64$ with ultraviolet extinction as observed in the Small Magellanic Cloud (i.e., no 2200 Å feature), and hopelessly badly by dust at $z = 2.64$ with standard Galactic ultraviolet extinction.

2. The systematic variation of flux ratios with wavelength but not with time (but see Vanderriest et al. this volume for a possible temporal change) is explained easily by differential extinction along paths through the lens separated by up to 10 kpc, and not easily by microlensing.

3. MG 0414+0534 and MG 1131+0456 are the reddest quasars known to us in complete radio samples. It *might* be just coincidental that both lie behind galaxies. It is much more likely that the quasars are red *because* they lie behind galaxies.

2016+112 was observed with LRIS during the summer of 1994. The primary goal was to measure the redshift of the second lens, which Lawrence et al. (1993) estimated at 2 from colors. We do not yet have a redshift for the galaxy, but we got nice spectra of the three images, including for the first time C', shown in Figure 4. The mean flux density of C' from 5500–7000 Å is only 1.0 μJy!

1422+231 was observed at 2.2 μm with NIRC in 1993 March. 1422+231 itself was saturated because of a bug in the acquisition software, but the image confirmed Hogg & Blandford's (1994) prediction of additional mass to the southeast with the detection of two faint galaxies.

1413+117 was observed at 2.2 μm with NIRC in 1995 June, with the goal of detecting the lens (Figure 5). Neither inspection nor preliminary PSF subtraction reveals the lens, but the 0."3 FWHM of the images (!!) shows Keck I's outstanding optical performance.

0218+357 was observed with LRIS in the summer of 1994 to determine the redshift of the blazar (Figure 5). Hints of an emission line near 5500 Å had been reported (Browne 1993). An emission line is confirmed near 5500 Å, with an associated absorption doublet. The most likely identification is Mg II$\lambda\lambda$2795.528,2802.705 at a redshift of 0.96. Spectra extending further to the red (to cover Hβ and the [O III] doublet) or to the blue (to cover [O II]λ3727) should confirm this result.

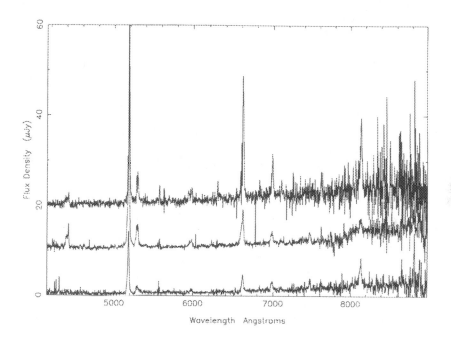

Figure 3. Spectra of 2016+112 A (top), B (middle), and C′. A and B are offset by 20 and 10 μJy, respectively, for clarity. The slit covered galaxy D and C′. Total integration time was 15,000 s, but telescope tracking glitches meant that C′ was not always in the slit and A and B sometimes were. Light from galaxy D appears in B redward of ∼ 8000 Å.

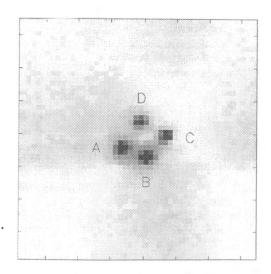

Figure 4. Image of 1413+117 obtained with NIRC in 1995 June. The total integration time was 2220 s. Tick marks are 1″ apart.

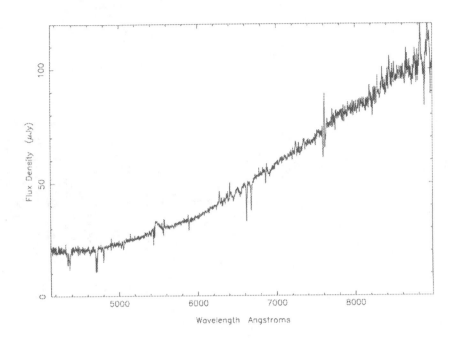

Figure 5. Spectrum of 0218+357, showing an unambiguous emission line near 5560 Å with associated absorption doublet.

References

Annis, J., 1992, ApJL, 391, L17

Browne, I.W.A., Patnaik, A.R., Walsh, D., & Wilkinson, D., 1993, MNRAS, 263, L32

Cohen, J.G., Cromer, J.L., Southard, S.Jr., & Clowe, D., 1993, in Astronomical Data Analysis Software and Systems III, PASP Conference Series 61, eds. D.R. Crabtree, R.J. Hanisch, & J. Barnes, 469

Hewitt, J.N., Turner, E.L., Lawrence, C.R., Schneider, D.P., & Brody, J.P., 1992, AJ, 104, 968

Hogg, D.W., & Blandford, R.D., 1994, MNRAS, 268, 889

Kochanek, C.S., 1992, ApJ, 384, 1

Larkin, J.E. et al., 1994, ApJL, 420, L9

Lawrence, C.R., Neugebauer, G., & Matthews, K., 1993, AJ, 105, 17

Lawrence, C.R., Elston, R., Jannuzi, B.T., & Turner, E. L., 1995a, AJ, in press

Lawrence, C.R., Cohen, J.G., & Oke, J.B., 1995b, AJ, in press

Matthews, K., & Soifer, B.T., 1994, in Infrared Astronomy with Arrays: the Next Generation, ed. I. McLean, (Dordrecht: Kluwer) 239

Oke, J.B., Cohen, J.G., Carr, M., Cromer, J., Dingizian, A., Harris, F., Labrecque, S., Lucinio, R., Schaal, W., Epps, H., & Miller, J., 1995, PASP, 107, 375

MILLIARCSECOND STRUCTURES IN GRAVITATIONALLY LENSED SYSTEMS

ALOK R. PATNAIK AND RICHARD W. PORCAS
Max-Planck-Institut für Radioastronomie,
Auf dem Hügel 69, D-53121 Bonn, Germany.

Abstract. VLBI studies of small-scale structures in gravitationally lensed systems can provide important constraints on the lens mass distribution. We review the current status of VLBI observations of a number of lensed systems.

1. Introduction

Angular resolutions down to sub-mas scales are routinely achieved at radio wavelengths using the technique of Very Long Baseline Interferometry (VLBI). Although the wide fields of lens systems create special problems for VLBI (Porcas 1994; Garrett et al. 1994b), there are many areas where it can make a unique contribution:
1) Searching for new systems in the separation range 1–500 mas (see the paper by Patnaik et al. in these proceedings).
2) Confirming lens candidates, especially those without image redshift measurements, by revealing matching source structure in the images.
3) Measuring the time delay from structural variations in the images.
4) Investigation of the lens mass distribution by determining the relative magnification matrix between images and changes across extended images.
5) Looking for 'granularity' in the mass distribution (e.g. milli-lensing due to $10^6 M_\odot$ black holes) by detailed comparison of different images (see the paper by Garrett et al. in these proceedings).
6) Studying the lensing process itself by examining the surface brightness of resolved features and wavelength-dependent flux ratios between images.
7) Studying the imaged source under a magnifying glass, and with multiple versions of the same source (e.g. this can lead to a unique registration of maps made at different frequencies and epochs, Porcas and Patnaik 1995).

C. S. Kochanek and J. N. Hewitt (eds), Astrophysical Applications of Gravitational Lensing, 305–310.
© 1996 IAU

2. Review of individual lens systems

0957+561 The earliest observations (Porcas et al. 1979, 1981; Haschick et al. 1981; Gorenstein et al. 1984, 1988) established the existence of core-jet structures in both A and B images and resulted in a relative magnification matrix used in subsequent modeling of the complex galaxy/cluster mass distribution. Recently, Garrett et al. (1994a) and Campbell et al. (1994) have used the extended jet structures to determine the magnification gradient over the images. Campbell et al. (1994, 1995) have also attempted to measure structural changes in both image cores; although no separation changes could be detected, a rough (but independent!) estimate of ca. 1 yr. for the time delay was determined from core flux density changes. The weak radio component G, which may be emission from the galaxy G1 or a possible third image of the quasar, was first detected with VLBI by Gorenstein et al. (1983), and has also been observed on the EVN at 1.7GHz, where it is unresolved (Garrett 1990). Higher resolution 2.3GHz observations (Rogers et al. 1989) indicate a possible core-jet structure for G, with size ca. 7mas.

2016+112 Heflin et al. (1991) have studied this system, which comprises 2 image components A, B and a third component C. At 1.7GHz with 3mas resolution A and B must each be modeled with at least 2 sub-components, although apparently not easily related by a magnification matrix. Garrett et al. (these proceedings) have also mapped this system at 1.7GHz, using the EVN. Component C breaks up into 4 sub-components: a colinear triple (C11, C12, C13) and a separated component, C2, which may be a third image, the radio counterpart to the source of Lyman α emission near C. Global 1.7GHz observations show that C12, C13 and C2 are compact on the 5mas scale (Garrett et al. 1994b).

MG0414+0534 This 'quad' comprises two bright, close components (A1, A2) and weaker components B and C (Hewitt et al. 1992). Our 1.7GHz EVN full-field map of this system is shown in Fig. 1 with a resolution of 25mas. Images A1, A2 and B are clearly resolved into subcomponents, with a bright 'core' and fainter extended regions. Component C is extended in the N-S direction. In our higher resolution 5GHz EVN maps, the cores become more pronounced, and image C is also resolved into a core and secondary component. Unpublished global observations at 5GHz (Hewitt, private communication) show similar structures in the images.

PKS1830-211 This system contains two compact image components (NE, SW) separated by 1", embedded in an elliptical ring structure. No optical identification of lens or background object exists for this low Galactic latitude object. Southern Hemisphere VLBI observations (Jauncey et al. 1991) showed compact structure in both image components. Our 1.7GHz EVN map is shown in Fig. 2(left) with a resolution of 30mas. The NE compo-

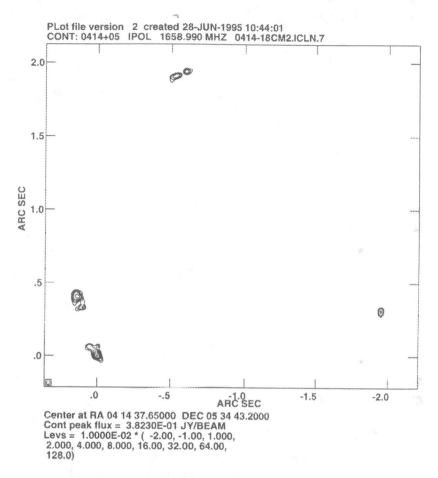

PLot file version 2 created 28-JUN-1995 10:44:01
CONT: 0414+05 IPOL 1658.990 MHZ 0414-18CM2.ICLN.7

Center at RA 04 14 37.65000 DEC 05 34 43.2000
Cont peak flux = 3.8230E-01 JY/BEAM
Levs = 1.0000E-02 * (-2.00, -1.00, 1.000,
2.000, 4.000, 8.000, 16.00, 32.00, 64.00,
128.0)

Figure 1.   1.7GHz EVN map of MG0414+0534.

nent has the brightest peak, and a jet-like extension pointing to the SE. The SW component also comprises a peak and a curving, jet-like extension pointing NW, as expected if it is the counterpart of that in the NE. Both components are resolved out on the long VLBI baselines between Europe and Hartebeesthoek, South Africa at this frequency. Jones (1994) presents 5GHz maps of the two compact components observed at two epochs. These also exhibit extensions along the jet directions, and apparent changes in their lengths between epochs; if this corresponds to real motion, it must be highly superluminal. These maps, and our own EVN 5GHz map, also show a prominent jet-like extension of the NE component in the NW direction, i.e. opposite to the 1.7GHz jet; no obvious counterpart to this feature is seen in the SW component, which has a higher peak brightness than the NE component at 5GHz. Garrett et al. (these proceedings) report VLBA observations at 15GHz which show clearly this NW-pointing jet in the NE

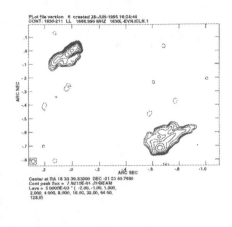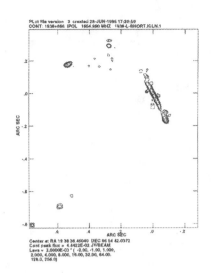

Figure 2. Left: 1.7GHz EVN map of PKS1830-211. Right: Global 1.7GHz map of B1938+666 using the short baselines, resolution 15mas.

component, but only a compact core in the SW. Jones et al. (these proceedings) speculate that interstellar scattering may cause the apparent size of the SW component to increase with wavelength. This may also explain our lack of detection on long baselines at 1.7GHz.

B0218+357 This system consists of an Einstein ring and two image components, both of which show compact, core-jet structures (Patnaik et al. 1993, 1995). A detailed description of this, the smallest known system, is given in Porcas and Patnaik (these proceedings).

B1938+666 The MERLIN map (Patnaik 1994) shows a complex morphology, with a number of components spread around an 'arc', and a further 3 isolated point-like features. We have used our 1.7GHz global observations to make a 'low' resolution map (15mas resolution) using just the short baselines (Fig. 2, right) since the source was largely not detected on long baselines. All three point-like components are detected, as are point-like components at the extremes of the arc. However, the strongest feature is a long, thin, slightly curved portion of the arc. A cut along the arc at PA 24° reveals at least two separate peaks along its length. At 5GHz we detect only the brightest feature in the arc.

B1422+231 This system is a 'quad' of overall size 1.3arcsec (Patnaik et al. 1992) with two equally bright radio images A and B (1:1), and weaker images C (1/2) and D (1/30). Our global 1.7GHz observations detect all 4 components, and A, B and C show faint extensions. Maps of all the images from our global 5GHz observations (1mas resolution) are shown in Figs. 3 and 4. Both A and B show elongated, core-jet like structures,

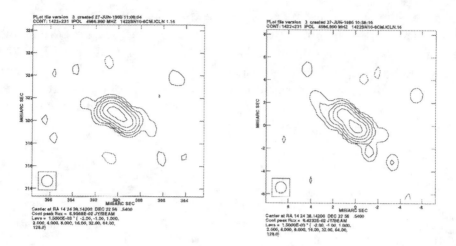

Figure 3.   Global 5GHz map of B1422+231 A(left) and B (right), resolution 1mas.

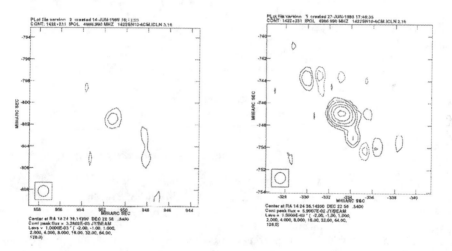

Figure 4.   Global 5GHz map of B1422+231 D (left) and C (right), resolution 1mas.

indicating the expected tangential stretching. Component C also shows extended components. The brightness peaks of the 5GHz images coincide with the positions of 1.7GHz images.

3. Conclusions

Although the seven lens systems investigated with VLBI present a wealth of detail which can help constrain lens masses and examine the lensing process, it is clear that a number of puzzles remain. The interpretation of the mas radio structures of 2016+112 and PKS1830-211 is far from clear, and we still do not know to what extent magnification gradients across spectrally inhomogeneous images are affecting the observed radio image flux density ratios. The present radio searches will hopefully produce many new lens systems which can be studied at mas resolution using VLBI.

Acknowledgements We wish to thank Michael Garrett and Ian Browne, who are our collaborators in some of the unpublished work presented here, and Jackie Hewitt, Dayton Jones and Josef Lehàr for communicating results in advance of publication. We also thank Sunita Nair for many interesting discussions.

References

Campbell, R.M., et al., 1994, ApJ, 426, 486

Campbell, R.M., Lehàr,J., Corey, B.E., Shapiro, I.I., & Falco, E.E., 1995, AJ, submitted

Garrett, M.A., 1990, Doctoral Thesis, U. Manchester, 90

Garrett, M.A., et al., 1994a, MNRAS, 270, 457

Garrett, M.A., Patnaik, A.R., & Porcas, R.W., 1994b, in EVN/JIVE Symposium, Torun, ed. Kus et al., TRAO, 73

Gorenstein, M.V. et al., 1983, Science, 219, 54

Gorenstein, M.V. et al., 1984, ApJ, 287, 538

Gorenstein, M.V. et al., 1988, ApJ, 334, 42

Haschick, A.D., et al., 1981, ApJL, 243, L57

Heflin, M.B., Gorenstein, M.V., Lawrence, C.R. & Burke, B.F., 1991, ApJ, 378, 519

Hewitt, J.N. et al., 1992, AJ, 104, 968

Jauncey, D.L. et al., 1991, Nature, 352, 132

Jones, D.L., 1994, in Proc. Compact Extragalactic Sources, ed. Zensus and Kellermann, NRAO, 135

Patnaik, A.R., et al., 1992, MNRAS, 259, 1p

Patnaik, A.R., 1994, in Gravitational Lenses in the Universe, Ed. Surdej et al., (Liège: Université de Liège), 311

Patnaik, A.R., et al., 1993, MNRAS, 261, 435

Patnaik, A.R., Porcas, R.W., & Browne, I.W.A, 1995, MNRAS, 274, 5p

Porcas, R.W. et al., 1979, Nature, 282, 385

Porcas, R.W. et al., 1981, Nature, 289, 758

Porcas, R.W., 1994, in Compact Extragalactic Sources, ed. Zensus and Kellermann, NRAO, 125

Porcas, R.W. & Patnaik, A.R., 1995, in 10th working Meeting on European VLBI for Geodesy and Astrometry, Matera (in press)

Rogers, A.E.E., 1989, in Gravitational Lenses, eds., Moran et al. (Dordrecht: Springer) 77

MULTI-FREQUENCY VLBI OBSERVATIONS OF B0218+357

RICHARD W. PORCAS AND ALOK R. PATNAIK
Max-Planck-Institut für Radioastronomie
Auf dem Hügel 69, D–53121 Bonn, Germany.

Abstract. We present the results from VLBI observations at three frequencies of the gravitational lens system B0218+357. From the source double structure, seen in both the A and B images at 15 GHz, we have derived a relative magnification matrix, and we show that the lens mass distribution must be non-spherical. We investigate how far the matrix parameters derived from 15 GHz observations can be used to relate the A and B images at 1.7 and 5 GHz, where the image sizes are much larger.

1. Introduction

The gravitational lens system B0218+357 (Patnaik et al. 1993) consists of two flat-spectrum radio image components separated by 335 mas, and an Einstein ring of similar diameter whose centre is slightly offset from the B (fainter) image. Since its discovery as a lensed system it has been subject to intensive study at both optical and radio wavebands, especially with the hope that the lens mass - of single galaxy size - may have a simple distribution. Measurement of this, taken together with a time delay between image variations, could lead to a direct determination of the Hubble constant (Refsdal, 1964). The redshift of the lens is well established as 0.6847 from optical and radio absorption measurements, and a preliminary estimate of the background object redshift is 0.96 (e.g. Lawrence in these proceedings). Corbett et al. (these proceedings) has reported a value of ca. 12 days for the time delay, from VLA measurements of variations in the image percentage polarization.

2. VLBI observations at 15 GHz

Determination of the image relative magnification matrix by VLBI studies of the mas structure of the images can help constrain the lens mass dis-

C. S. Kochanek and J. N. Hewitt (eds), Astrophysical Applications of Gravitational Lensing, 311–316.
© 1996 IAU

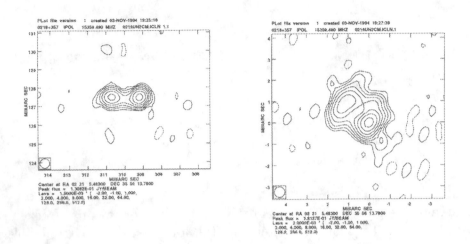

Figure 1.   First epoch 15GHz map of B0218+357 B(left) and A (right), resolution 0.5mas.

tribution, and a number of VLBI observations at 1.7 and 5GHz have been published (Patnaik et al. 1993, 1994, Shaffer 1994, Polatidis et al. 1995, Xu et al. 1995). At these frequencies, however, it seems impossible to disentangle the contributions from intrinsic source structure and lens distortion in the observed image morphologies. We made observations with the VLBA in 1994 October at 15GHz, a frequency a factor 3 higher, and a resolution a factor 2 higher, than any previous VLBI observations. Details of the observations are described in Patnaik et al. (1995), and maps of the A and B images with 0.5mas resolution are shown in Fig. 1. The intrinsic double structure of the source at 15GHz is clearly seen in both images, the component flux ratio being 1.64 in both A and B. Some striking features of these VLBA maps are the difference in the component-separation position angles in the two images, and also the elongation of both components in PA −40° in the A image. This extension of image A is also seen in maps at 1.7 and 5GHz, and corresponds to the expected 'tangential stretching' expected for the exterior image in this system (Narasimha and Patnaik 1994).

These observations of B0218+357 seem to provide a 'textbook' case which exemplifies gravitational lens properties. The equality of the component ratio in the two images shows that the magnification is uniform over the region of the 15GHz structure. Gaussian fits to the more resolved component yield deconvolved sizes whose ratio between the images (A/B = 3.9±0.6) agrees well with the flux ratio (A/B = 3.64), showing that the surface brightness of this component is the same in A and B. This system also demonstrates the interesting property that, although the action of the lens produces a change of the component-separation position angle between the images, the position angle of linear polarization (corrected for

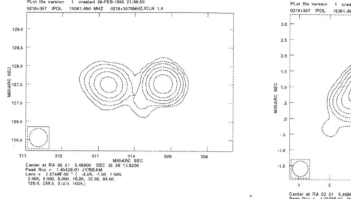

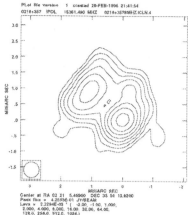

Figure 2. Second epoch 15GHz map of B0218+357 B(left) and A (right), resolution 0.5mas.

Faraday rotation) is the same in both images. Finally, we have been able to determine a relative magnification matrix, describing the mapping of the B image into A, using the image flux ratio and the transformation of the component separation vector (Patnaik et al. 1995). The properties of this matrix show that the lens mass distribution is non-spherical.

Further observations of B0218+357 were made with the VLBA at 15GHz in 1995 January, as part of the survey of Patnaik et al. (these proceedings). The maps of A and B are presented in Fig 2; they have a resolution of 0.5mas. They show morphologies very similar to those from the 1994 October observations, and, in particular, there is no significant change in the component separations in either image. However, the faint extension of the 'core' (western) component at the lowest contour levels in PA ca. $-10°$ seen in 1994 October has become a much more prominent, well-defined feature in 1995 January, presumably indicating the ejection of a new 'jet' component. Traces of this new component can also now be seen in the B image, as an extension of the lowest contours in PA ca. $135°$. (on the opposite side of the main component separation line, due to opposite parity of the images). We have used this exciting new development to furnish a new constraint to refine our previous values for the elements of the image relative magnification matrix.

3. VLBI observations at 1.7 and 5.0GHz

We have also made global VLBI (MK3) observations of B0218+357 at 1.7 and 5.0GHz, in 1992 June and March. Details of the observations will be published elsewhere. The 5.0GHz maps of images A and B (resolution 1

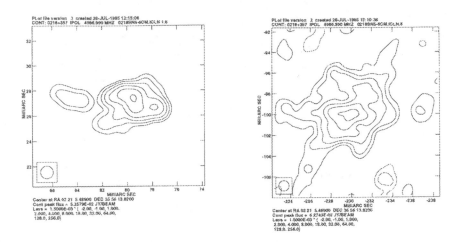

Figure 3. 5GHz maps of B0218+357 B(left) and A(right) with 1mas resolution.

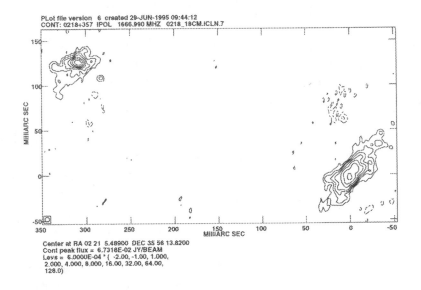

Figure 4. 1.7GHz map of B0218+357 with 5mas resolution

mas) are presented in Fig.3 and the 1.7GHz map of both images (resolution 5mas) is presented in Fig.4. At these frequencies the angular resolution is not sufficient to easily identify the 1.4mas double structure seen at 15GHz, but both observations show a 'radial' extension of image B, and the tangential extension of image A in PA −40°, so prominent at 15 GHz. The image sizes also increase with increasing wavelength.

Although the registration of VLBI maps made at different frequencies usually presents difficulties, knowledge of the relative magnification matrix, and measurement of the precise relative separation between corresponding features in two images, can lead to a unique registration (Porcas and Patnaik 1995). Using this method we find that the peaks of the 5GHz image maps lie close to a point between the two components seen at 15GHz. At 1.7GHz, the separation of the centroid of the brightness distributions of the images (which are corresponding points) locates them very close to the eastern, more resolved component at 15GHz.

The method used above assumes that the same relative magnification matrix can be used for these maps at all 3 frequencies; this assumption may break down when the image sizes become large compared with the scale size of the lens potential. As a crude test, we measured the size of the 5GHz images (at the 3rd contour down from the peak) in the two directions corresponding to the 15GHz matrix eigenvector directions. The ratio of the A and B image sizes in these directions do indeed agree very well with the ratios from the matrix. This confirms that the achromatic property of gravitational lensing, and small source size, result in the same matrix being applicable to our 5GHz maps. At 1.7GHz the image sizes are very much larger. We have measured the ratio between the areas of image A and B at two different contour levels. At the 2.3mJy/beam level, the ratio (2.65) agrees well with the flux density ratio (2.61) between A and B at 1.7GHz, showing the same surface brightness in the images at the largest angular scales. This ratio is known to be frequency dependent, however, and presumably demonstrates the breakdown of the assumption of small image size at 1.7GHz. For surface brightness > 5mJy/beam, the ratio of areas (3.56) differs significantly from the flux ratio (2.77), however, and the peak brightness in image B is higher than in image A. Although this can arise, in principle, if there is compact structure in both A and B which is unresolved in different directions, it is unclear whether this explanation applies here. Nair (these proceedings) has suggested an alternative explanation for this effect.

Acknowledgements We thank Drs. Ian Browne and Sunita Nair for lively conversations regarding the image structures of this intriguing lens system.

References

Narasimha, D. & Patnaik, A.R., 1994, in Gravitational lenses in the Universe, eds. Surdej et al., (Liège: Université de Liège), 295

Patnaik, A.R., Browne, I.W.A., King, L.J., Muxlow, T.W.B., Walsh, D. & Wilkinson, P.N., 1993, MNRAS, 261, 435

Patnaik, A.R., Porcas, R.W., Browne, I.W.A., Muxlow, T.W.B., Narasimha, D., 1994, in Compact Extragalactic Sources, eds. Zensus & Kellermann, NRAO, 129

Patnaik, A.R., Porcas, R.W. & Browne, I.W.A, 1995, MNRAS, 274, 5p

Polatidis, A.G., Wilkinson, P.N., Xu, W., Readhead, A.C.S., Pearson, T.J., Taylor, G.B. & Vermeulen, R.C., 1995, ApJSup, 98, 1

Porcas, R.W. & Patnaik, A.R., 1995, in 10th working Meeting on European VLBI for Geodesy and Astrometry, Matera, in press

Refsdal, S., 1964, MNRAS, 128, 307

Shaffer, D.B. 1994, in Compact Extragalactic Sources, eds. Zensus & Kellermann, NRAO, 132

Xu, W., Readhead, A.C.S., Pearson, T.J., Polatidis, A.G., & Wilkinson, P.N., 1995, ApJSup, in press

1608+656: A QUADRUPLE LENS SYSTEM FOUND IN THE CLASS GRAVITATIONAL LENS SURVEY

STEVEN T. MYERS
California Institute of Technology
105-24, Pasadena CA 91125

Abstract. The first phase of a large gravitational lens survey using the Very Large Array at a wavelength of 3.6 cm has been completed, yielding images for 3258 radio sources. The Cosmic Lens All-Sky Survey, or CLASS, is designed to locate gravitational lens systems consisting of multiply-imaged compact components with separations $> 0\rlap{.}''2$. From this first phase has come the discovery of 1608+656, a quadruply-imaged object with maximum separation of $2\rlap{.}''1$. Images from the Palomar 5-m and Keck 10-m telescopes show the lensed images and the lensing galaxy. An optical spectrum obtained with the Palomar 5-m Telescope indicates a redshift of $z = 0.63$ for the lensing galaxy, and a newly-obtained Palomar spectrum indicates a redshift of $z = 1.39$ for the lensed source, which appears to be a galaxy. A simple single-galaxy lens model derived from the radio image reproduces the observed configuration and relative fluxes of the images, as well as the position, shape, and orientation of the lensing galaxy. Because a simple mass model is able to fit the observations, we argue that this lens system is promising for determining H_0. CLASS has also yielded the new double image lens system 1600+434. The second phase of the survey is scheduled for August and September 1995 on the VLA, and should yield images for an additional 5000+ targets, bringing the CLASS total to over 8000.

1. Introduction

The Cosmic Lens All-Sky Survey (CLASS) is a radio-based survey using the Very Large Array[1] (VLA) to image approximately 10000 flat-spectrum

---

[1]The National Radio Astronomy Observatory is operated by Associated Universities, Inc., under cooperative agreement with the National Science Foundation.

C. S. Kochanek and J. N. Hewitt (eds), Astrophysical Applications of Gravitational Lensing, 317–322.
© 1996 IAU

radio sources, with the aim of finding new gravitational lens systems suitable for cosmographic studies. CLASS is a collaboration between Caltech, Jodrell Bank and Leiden/Dwingeloo (see §5). In many respects, CLASS is an extension of previous VLA surveys such as the Jodrell-Bank/VLA astrometric survey (JVAS: Patnaik et al. 1992). Based on the statistics from JVAS, we expect to find about one lens per 500 sources observed, or 20 new lenses in the total CLASS sample.

Restricting the sample to flat-spectrum sources preferentially selects objects with compact and variable components well-suited to time delay measurements. The exclusion of complex extended steep-spectrum sources from the sample also simplifies the mapping and follow-up tasks. A total of 3258 targets has been observed in the first phase of the survey, CLASS-1. The survey has so far yielded two new lens systems: 1600+434, a double image system (Jackson et al. 1995) and the quadruple image lens 1608+656 (Myers et al. 1995), as well as other candidates yet to be followed up.

2. Observations

Targets were selected from the 4.85 GHz Green Bank Survey (87GB: Gregory & Condon 1991). At the time of the CLASS-1 VLA observations, we were able to select 683 sources with two-point spectral indices of $\alpha \geq -0.5$ between the 325 MHz Westerbork Northern Sky Survey (WENSS: de Bruyn et al. 1995) and 4.85 GHz, with 4.85 GHz flux densities above 25 mJy. The remaining 2575 sources were selected with $\alpha \geq -0.6$ between the 365 MHz Texas Survey (or upon the 4.85 GHz fluxes only in regions not covered by the Texas survey) and 4.85 GHz with 4.85 GHz flux densities above 50 mJy. Sources were restricted to galactic latitudes $|b| > 10°$. CLASS-1 covered the declination range from the northern limit of 87GB ($\delta \lesssim 75°$) down to $\delta \geq 45°20'$ ($\delta \geq 35°$ below the galactic plane).

The CLASS-1 observations took place during February through May 1994 in the A-configuration of the VLA at a frequency of 8.4 GHz ($0\overset{''}{.}2$ resolution). For the 2575 non-WENSS sources, two IFs of 25 MHz bandwidth were centered at 8.415 GHz and 8.465 GHz. An on-source dwell time of 30 seconds was used. For the WENSS selected sources, the full 50 MHz bandwidth was available due to the accurate WENSS source positions. A phase calibration source from the JVAS was observed every 13.5 minutes. The average target source observation rate was one per minute including overhead from calibration and slewing.

Following initial calibration using AIPS, automatic mapping of the 2575 87GB sources was performed using the DIFMAP package (Shepherd, Pearson & Taylor, 1994). From the resulting maps ~ 100 objects with multiple compact components have been selected for further investigation as poten-

tial gravitational lens candidates. Only a few of these are expected to be real
lensed systems. The WENSS-selected sources were analyzed separately (see
Jackson et al. 1995), but will be re-analyzed using the automatic mapping
for statistical completeness.

3. The Quad Lens 1608+656

The source 1608+656 (target source 87GB 16087+6540) was observed by
CLASS on 1994 March 1. In addition, 1608+656 was observed independently with the VLA in the B-configuration on 1994 July 23, with follow-up
observations with the WSRT and OVRO (Snellen et al. 1995). The CLASS
VLA data were well fit by a model consisting of four point sources, with
no missing extended flux in the residual image. The rms noise level in the
final image was 0.44 mJy. A total flux density of 73.2 ± 0.9 mJy was measured, with component flux densities of 35.6, 17.8, 15.2, and 4.6 mJy for
A, B, C and D respectively. The brightest component (A) is located at
$16^h 09^m 13^s.956 +65° 32' 28''.97$ (J2000), with estimated uncertainty $< 0''.05$.
The WSRT images indicate that 1608+656 is the lensed core of a large
(45") double radio source (Snellen et al. 1995).

Optical and infrared follow-up observations were undertaken in August
and September 1994. An optical image and spectrum of 1608+656 were
obtained with the Palomar Observatory 5-m Telescope. Approximate photometry on the optical images gives total magnitudes for images and lens of
Gunn $r = 19.4^m$ and $i = 19.2^m$. A 3000 second exposure using the Double
Spectrograph was taken and a redshift of $z = 0.6304$ for the lensing galaxy
was measured. In the galaxy spectrum, there are strong Balmer absorption
lines, indicating that there are A stars present and that the lens might
possibly be a post-starburst galaxy.

A $2.2\mu m$ (K band) infrared image of 1608+656 was obtained on the
W.M. Keck[2] 10-m Telescope using the NIRC, in 1" seeing. The three
brightest lens images are easily seen, as is the extended lensing galaxy.
The lensing galaxy has a K-band flux roughly four times greater than that
of the total emission from the lens components, and has an axial ratio of
$b/a = 0.56 \pm 0.10$ with its major axis in position angle 60°. Photometry
on the lensing galaxy yields a mean surface brightness of $60\,\mu Jy\,arcsec^{-2}$
within a circle of radius 1".1, giving a mean K surface luminosity of $L_E =
7.7 \times 10^8 L_\odot\,kpc^{-2}$.

The CLASS VLA and Keck images, and optical spectrum, are shown in
Myers et al.(1995), and are not reproduced here. Figure 1 shows a newly

---

[2]The W.M. Keck Observatory is operated as a partnership between the California
Institute of Technology and the University of California, and was made possible by the
generous gift of the W.M. Keck Foundation.

obtained VLA 5 GHz deep image of 1608+656. No extended emission is seen
to a level of 0.1 mJy at the full resolution of $0''.33$. Also shown is a MERLIN
5 GHz image. The four lensed images are found to be unresolved at the
resolution of 50 mas. VLBA observations are scheduled for August 1995,
and will show any milli-arcsecond structures that might be associated with
the core of a double radio source. Another 8.4 GHz VLA image of 1608+656
was obtained in July 1995. There is a significant change in relative fluxes of
the components since March 1994: the flux ratio A/B increased by 12%±3%
and the ratio C/B increased by 36% ± 5% (C became brighter than B).
There is clearly variability in this object, and therefore a time delay should
be measurable.

In late July 1995, a new spectrum of 1608+656 was obtained, again with
the Double Spectrograph on the Palomar 5-m Telescope. It was possible to
identify a feature near 9200Å previously seen in a Keck LRIS spectrum
as a Balmer break, along with the associated Balmer absorption series. A
redshift of $z_s = 1.394$ is derived for the lensed source. The spectrum is
indicative of a galaxy, not a quasar as was expected — if true, then the
images should be visibly extended in HST observations planned for Cycle
5.

A simple lens model for 1608+656 has been constructed using the positions and relative flux densities of the four images in the VLA 8.4 GHz radio
map. An oblate spheroidal mass model has been used for the lens, the details
of which are given in Myers et al.(1995). A Friedmann-Robertson-Walker
universe with a smoothed-out background matter distribution, $q_0 = 0.5$ and
$h = H_0/100\,\mathrm{km\,s^{-1}\,Mpc^{-1}}$, is assumed. For this model, and a background
source redshift of $z_s = 1.394$, the mass within the equivalent circular Ein-
stein ring radius $\theta_E = 1''.1$ ($r_E = 4.26$ kpc) is $M_E = 3.05 \times 10^{11}\,h^{-1}\,\mathrm{M_\odot}$.
Using the previously measured mean K surface luminosity L_E, we infer a
mean mass-to-light ratio of $M_E/L_E = 6.9\,h\,\mathrm{M_\odot/L_\odot}$ inside the cylinder r_E.

The position and orientation of the lensing galaxy visible in the Keck
infrared image agree with those predicted by the model, within the measurement uncertainties. The model axial ratio of $b/a = 0.28$ is somewhat
lower than the measured 0.56 from the Keck image, although this might be
accounted for by the effect of the seeing. The implied model magnifications
for images A–D plus missing image E are +2.58, +1.54, −1.36, −0.35, and
+0.04 respectively (signs denote parity), with a total magnification of 5.87.
According to the model, image B should vary first. The predicted time
delays (in h^{-1} days) relative to image B are 40 for A, 44 for C, and 115
for image D. The shortest relative pairwise delay is $4.2\,h^{-1}$ days between
images A and C.

The lensing galaxy in the 1608+656 system presents an interesting puz-
zle. Whether the lens is a single highly elliptical spheroid, disk galaxy, or a

pair of close galaxies has yet to be determined. The implied mass within the image radius is large, corresponding to a velocity dispersion of $500\,\mathrm{km\,s^{-1}}$ for an isothermal sphere of core radius 1.9 kpc, though the mass-to-light ratio ($7\,h$) is not atypical of other lensing galaxies. This mass is representative of first-ranked cluster ellipticals, while the spectrum and high ellipticity is more suggestive of an early-type spiral galaxy. It is possible that the lens is a close pair of less massive galaxies, explaining the high mass and high ellipticity. The HST observations scheduled for Cycle 5 should provide a better view of the lens and lensed galaxies. This system appears promising for time-delay measurements: it is measurably variable, with known redshifts, a simple mass model (unless it is two galaxies), and four images to provide more constraints on the geometry.

4. The Future

The CLASS survey is producing new lenses at around the expected rate of 0.2%. So far, with just over 3200 CLASS-1 targets observed, there are two confirmed lenses and two further very good candidates. CLASS-2 observations begin in mid-August 1995 and should survey an additional 5000 target sources, and yield another 10 new lenses. When completed, CLASS will yield a carefully selected uniform sample of radio lenses suitable for cosmographic studies. CLASS will also provide a large radio database for investigation of the AGN phenomenon.

Three of the CLASS-1 candidate lens systems are quads. It appears that just as in other radio-based surveys, quadruple image systems make up on the order of 50% of lenses. This points to the existence of highly elliptical potentials among the most massive lensing galaxies (eg. King et al. these proceedings).

5. The CLASS Team

CLASS is an ambitious undertaking, and is a collaboration between astronomers at Caltech, Jodrell Bank, Leiden, and Dwingeloo observatories. The CLASS team members are:

Caltech S.T. Myers, C.D. Fassnacht, T.J. Pearson, A.C.S. Readhead, M.C. Shepherd (DIFMAP), R.D. Blandford

Jodrell Bank I.W.A. Browne, N. Jackson, T. Muxlow (MERLIN), S. Nair, C. Sykes, P.N. Wilkinson.

Leiden/Dwingeloo A.G. de Bruyn, M. Bremer (WENSS), G.K. Miley, R.T. Schilizzi, I.A.G. Snellen (1608+656)

In addition, the optical/IR follow-up of 1608+656 involved:

Caltech/Palomar Observatory S.G. Djorgovski, K. Matthews, G. Neugebauer, J.D. Smith, D.J. Thompson, D.S. Womble

References

de Bruyn, A.G., et al., 1995, in preparation
Gregory, P.C. & Condon, J.J., 1991, ApJS, 75, 1011
Jackson, N., et al., 1995, MNRAS, 274, L25
King, L. et al., 1995, these proceedings
Myers, S.T., et al., 1995, ApJ (Letters), 447, L5
Patnaik, A.R., Browne, I.W.A., Wilkinson, P.N., & Wrobel, J.M., 1992, MNRAS, 254, 655
Shepherd, M.C., Pearson,T.J., & Taylor,G.B., 1994, BAAS, 26, 987
Snellen, I., de Bruyn, A.G., Schilizzi, R.T., Miley, G.K., & Myers, S.T., 1995, ApJL, 447, L9

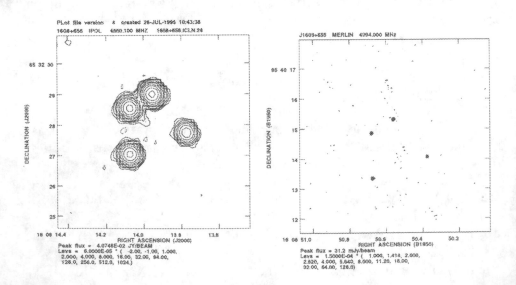

Figure 1. VLA 5 GHz (left), and MERLIN 5 GHz (right) images of 1608+656. Component A is at the top, B opposite A, C closest to A.

NEW "EINSTEIN CROSS" GRAVITATIONAL LENS CANDIDATES IN HST WFPC2 SURVEY IMAGES

K.U. RATNATUNGA, E.J. OSTRANDER, R.E. GRIFFITHS AND M. IM

Johns Hopkins University
Baltimore, MD 21218, USA

Abstract. We report the serendipitous discovery of quadruple gravitational lens candidates using the Hubble Space Telescope. We have so far discovered two good examples of such lenses, each in the form of four faint blue images located in a symmetric configuration around a red elliptical galaxy. The high resolution of HST has facilitated the discovery of this optically selected sample of faint lenses with small (~ 1") separations between the ($I \sim 25 - 27$) lensed components and the much brighter ($I \sim 19 - 22$) lensing galaxies. The sample has been discovered in the routine processing of HST fields through the Medium Deep Survey pipeline, which fits simple galaxy models to broad band filter images of all objects detected in random survey fields using WFPC2.

1. INTRODUCTION

Einstein (1936) computed the gravitational deflection of light by massive objects and showed that an image can be highly magnified if the observer, source and the deflector are sufficiently well aligned. However, the angular resolution available then to ground based optical telescopes made him remark that "there is no great chance of observing this phenomenon". Zwicky (1937) showed that "extragalactic *nebulae* offer a much better chance than *stars* for the observation of gravitational lens effects".

Over the last decade a number of lensed QSO candidates were located in radio surveys and subsequently the associated lensing galaxies were optically identified (See Schneider, Ehlers and Falco 1992 for review). Huchra et al. (1985) discovered the "Einstein cross" at the center of the bright

C. S. Kochanek and J. N. Hewitt (eds), Astrophysical Applications of Gravitational Lensing, 323–328.
© 1996 IAU

(V=14.6) galaxy 2237+0305, an object in the Center for Astrophysics red-shift survey: this lens is considered unique because of the very low probability of alignment of a QSO within 0″.3 of the center of a nearby (z=0.04) galaxy.

The Medium Deep Survey (MDS) is a Hubble Space Telescope (HST) key project, which relies exclusively on the efficient use of parallel observing time to take images of random fields which are several arcminutes away from the primary targets of other HST instruments (Griffiths et al. 1994). Similar observations have been made by the Guaranteed Time Observers (GTO) in parallel mode, in conjunction with primary GTO exposures. In addition to these two parallel surveys, a major 'strip' survey was performed (Groth et al. 1995) using the WFPC2 in primary mode. With the refurbished HST WFPC2 optics we have now been able to start an optical survey for gravitational lenses centered on field galaxies in the magnitude range I= 19 − 23, searching for background field galaxies which are lensed into components with magnitudes I= 23 − 27.

2. OBSERVATIONS

In cycle 4 of HST observations, from January 1994 to June 1995, the MDS and GTO parallel datasets have comprised about 35 and 15 independent high galactic latitude fields, respectively, with at least two WFPC2 exposures in each of the F606W(V) and F814W(I) filters for each field. The 42 arc min long Groth-Westphal strip (Groth et al. 1995) consists of 28 contiguous WFPC2 fields centered at $b = +60°25$ and $l = 96°35$. The observations, covering a total area about 120 square arc minutes, were taken between 7 March and 9 April 1994. After the one-year proprietary period, they were obtained from the HST archive and calibrated, stacked and processed in exactly the same way as fields obtained for the HST Medium Deep Survey. A root-mean-square(rms) error image reflecting both the excluded cosmic rays and the flat field was created and used in the object detection and subsequent image analysis algorithms.

In order to search for serendipitous objects and to correct any errors in the automated object detection process caused by confusion of overlapping or very bright images, the fields have been examined by eye. During this process, it was noticed that the I=19.7 elliptical galaxy (HST14176+5226) was flanked by four fainter images which were all at about the same magnitude and color and much bluer than the central elliptical which had a half light radius of 1″.2 (see Figure 1). The companion objects (V∼ 26) were about 1″.2 and 1″.6 distant from the center of the elliptical galaxy along the major and minor axes respectively. Furthermore, the objects on the cross appeared to be stretched in a direction at right angles to the line joining

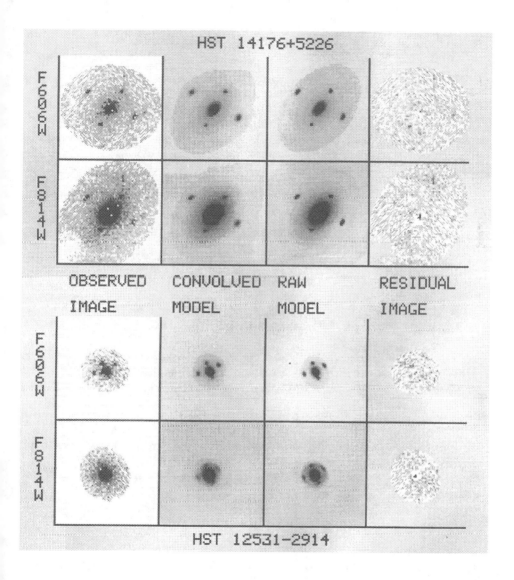

Figure 1. The maximum likelihood fits to the two gravitationally lensed images. The images are displayed as analyzed, without any interpolation over bad pixels. Each box is 6″4 square. The residuals show only a very faint trace of the subtracted images. Note that the convolution with the WFPC2 PSF does influence the appearance of the lensed images. The faint lensed images are blue and better resolved in F606W, even though the image exposure times in the F814W filter are 50% longer.

them to the center of the elliptical galaxy.

A second fainter (I=21.8) and smaller (half light radius 0″2) elliptical galaxy (HST12531−2914) was discovered when inspecting the residuals of the maximum likelihood model fits to galaxies in a deep MDS field urz00. Since the faint companion objects (V∼ 27) are only about 0″5 from the

TABLE 1. HST Quadruple Gravitational Lens Candidates

Name			HST14176+5226			
Equatorial(J2000)	14:17:36.3		+52:26:44		$V_3=32°93$	
HST WFPC2			Groth GTO:5090 11-Mar-1994			
Dataset[g][x,y]			U26X0801T[3][242,700]			
MDS Discovery			Eric J. Ostrander 02-May-1995			
HST WFPC2 Filter	F814W 4x1100s			F606W 4x700s		
Configuration	X	Y	V	±	V–I	±
Elliptical	0	0	21.68	0.04	1.97	0.04
A	−11	11	25.63	0.06	0.51	0.10
B	18	−4	25.77	0.07	0.39	0.11
C	10	12	25.99	0.08	0.52	0.13
D	−3	−9	25.97	0.08	0.42	0.14
Name			HST12531−2914			
Equatorial(J2000)	12:53:06.7		−29:14:30		$V_3=127°77$	
HST WFPC2			Griffiths GO:5369 15-Feb-1995			
Dataset[g][x,y]			U26K7G04T[3][755,326]			
MDS Discovery			Myungshin Im 12-Jun-1995			
HST WFPC2 Filter	F814W 4x2100s			F606W 3x1800s		
Configuration	X	Y	V	±	V–I	±
Elliptical	0	0	23.77	0.06	1.95	0.07
A	−6	−2	27.02	0.15	0.25	0.28
B	6	3	26.89	0.15	0.43	0.23
C	−2	4	26.72	0.11	0.34	0.21
D	3	−4	27.51	0.24	0.82	0.34

central elliptical, they had not been resolved as separate objects by the automated object detection algorithm. A "southern cross" just in time for presentation at this IAU symposium in Melbourne, Australia.

The observed image configuration, magnitudes and colors are given in Table 1. The magnitudes of the components were computed using a 0″3 square aperture, and corrected to total magnitudes assuming a point source. The offsets (X,Y) are in WFC 0″1 pixels from the centroid of the respective elliptical galaxy.

3. THE LENS MODEL

The details of the model fitting are given in Ratnatunga et al. (1995). Since we had already developed software for 2-dimensional 'disk + bulge' de-

composition of MDS galaxy images, we used the same procedure with a
slight modification to do 'bulge + gravitational lens' decomposition of the
observed light distribution. Given a set of model parameters, we generate 2-
dimensional images for the elliptical and the source galaxies. The expected
configuration of the lensed images is ray-traced by numerical integration.
The elliptical lens and the lensed source images are then convolved with the
adopted WFPC2 point spread function, and compared with the observed
galaxy image. The evaluated likelihood function (similar to a weighted χ^2)
is then minimized using a quasi-Newtonian method. This procedure iteratively converges simultaneously on the maximum likelihood model for the
lensing elliptical galaxy and the source, so as to produce the observed lensing galaxy and the configuration of images from the lensed source.

The inferred elliptical mass distribution is significantly flat (0.40). We
find that the lens configuration can be modeled using the gravitational field
potential of a singular isothermal ellipsoidal mass distribution (Kormann,
Schneider & Bertelmann 1994). We find that this potential is adequate to
obtain a fit better than any of the other potentials we tried, even using less
free parameters.

4. CONCLUSIONS

We have discovered two examples of quadruple gravitational lenses in HST
survey data, one in the MDS data and one in the archived Groth-Westphal
GTO survey.

The lensed image components are more distant from the centroid of
the lensing elliptical galaxy along the minor axis because the deflection is
proportional to the gradient of the potential. The ratio in separation is
equal to the inverse axis ratio of the potential. For the elliptical galaxy
HST14176+5226, the axis ratio (0.68) of the observed light distribution is
practically the same as that of the potential (0.74), and the orientation
is the same within 10 ± 2 degrees. A model independent inference is that
the stars are a trace population following the gravitational potential. The
Mass/Light distribution, increases radially outwards.

The critical radii of HST14176+5226 and HST12531+2914 are 1″5 and
0″6 , in each case larger than the half light radii of 1″2 and 0″2 for these
galaxies respectively. The distance of the intrinsic source from the centroid
of the lens needs to be less than about 0.15 of the critical radius for the
creation of a quadruple image. The impact parameters for these two objects
are about 0.08 and 0.09 of the critical radius.

These represent the first discoveries of lenses using the high resolution of
HST - indeed, apart from the exceptional original Einstein cross discovered
by Huchra et al. (1985), they represent the first discoveries of field-galaxy

gravitational lenses via the systematic study of optical images.

These objects would have been very difficult discoveries from the ground except under conditions of excellent seeing. We have not as yet observed these galaxies spectroscopically. The redshift of the lensed components in HST12531−2914 is probably a challenging observation for the Keck telescope in excellent seeing.

From the observed numbers of bright elliptical galaxies observed in the GTO survey strip (300 to I = 22), the numbers of faint objects in the fields (8000 to I = 26), and the expected cross-sections, we estimate that we should find one quadruple lens in every 20 − 30 WFPC2 fields surveyed. The number that has been discovered so far is therefore consistent with our expectations.

An on-going systematic and careful inspection, looking very specifically for possible gravitational lens candidates in the shallower MDS and GTO parallel fields is in progress in order to expand the sample. As further MDS data are taken in Cycle 5 and subsequent cycles, they will be examined for similar spectacular lenses, and also for more common lenses consisting of arcs or two or three components, to obtain a statistically representative sample of HST gravitational lens candidates, for statistical study.

These are a new class of gravitational lens candidates in which the cosmologically distant lens is a relatively bright elliptical galaxy with well understood properties. If a significant sample could be found and observed spectroscopically for redshifts, they will be very useful cosmological probes.

Acknowledgements: This paper is based on observations with the NASA/ ESA Hubble Space Telescope, obtained at the Space Telescope Science Institute, which is operated by the Association of Universities for Research in Astronomy, Inc., under NASA contract NAS5-26555. The Medium-Deep Survey is funded by STScI grant GO2684. We gratefully acknowledge Lyman Neuschaefer for help with the MDS pipeline and many helpful discussions, the anonymous referee for many useful suggestions, and *et tu* Broadhurst.

References

Einstein, A., 1936, Science, 84, 506

Groth, E. J., Kristian, J. A., Lynds, R., O'Neil E. J., Balsano, R., & Rhodes, J., 1994, BAAS, 26, 1403

Griffiths, R. E., et al., 1994, ApJ, 437, 67

Huchra, J., Gorenstein, M., Kent, S., Shapiro, I., Smith, G., Horine, E. & Perley, R., 1985, AJ, 90, 691

Kormann, R., Schneider, P., & Bertelmann, M., 1994, A&A, 284, 285

Ratnatunga, K. U., Griffiths, R. E., Casertano, S., 1995, ApJ, in preparation.

Ratnatunga, K. U., Ostrander, E. J., Griffiths, R. E., Im, M, 1995, ApJL submitted.

Schneider, P. Ehlers, J., Falco, E. E., 1992, Gravitational Lensing, (Berlin: Springer)

Zwicky, F., 1937, Phys Rev, 51, 290

IDENTIFYING OPTICAL EINSTEIN RINGS

S. J. WARREN
Imperial College of Science Technology and Medicine
Prince Consort Rd, London SW7 2BZ, United Kingdom

P. C. HEWETT AND G. F. LEWIS
Institute of Astronomy
Madingley Road, Cambridge CB3 0HA, United Kingdom

P. MØLLER
Space Telescope Science Institute
3700 San Martin Drive, Baltimore MD 21218, USA

A. IOVINO
Osservatorio Astronomico di Brera
Via Brera 28, I-20121 Milano, Italy

AND

P. A. SHAVER
European Southern Observatory
Karl-Schwarzschild-Strasse 2
D-85748 Garching bei München, Germany

1. Introduction

The discovery of gravitationally lensed radio–rings (Hewitt et al. 1988) opened up a new line of attack on the problem of dark matter in galaxies. High–resolution radio observations (Langston et al. 1989) resolve structure tangentially and radially within the rings, providing sophisticated analysis routines (Kochanek & Narayan 1992, Wallington, Narayan & Kochanek 1994) with enough constraints to compute realistic models of the mass distribution within the deflectors (Kochanek 1995). Five such systems are now known and extensive programs to identify further examples are underway.

Exploiting the full potential of radio–rings is hampered by two factors. Firstly, a familiar problem within the gravitational lensing field, examples of the phenomenon are rare and while further systems will be identified the

C. S. Kochanek and J. N. Hewitt (eds), Astrophysical Applications of Gravitational Lensing, 329–334.
© 1996 IAU

observational resources required are substantial. Secondly, once an example has been found it is in general very difficult to obtain unambiguous identifications of both the source and the deflector, since the optical counterparts of one or both can be extremely faint. Thus, of the five radio rings known, for only one, MG1654+134, are the redshifts of the source and deflector established. Without confirmed redshifts the basic geometry of the lens system remains unknown, severely limiting the quantitative information that can be deduced concerning the mass distribution within the deflector.

Miralda–Escudé and Lehár (1992) pointed out that the surface density of optical rings should greatly exceed that of radio rings. The deflector population, massive galaxies, is the same but the source counts of high–redshift, intrinsically faint, marginally extended galaxies in the optical far exceeds the number of extended radio sources. The availability of samples of optical rings would provide complementary information to that obtained from studies of the radio rings. Working at optical wavelengths holds the promise of identifying a much larger number of systems, allowing powerful statistical analyses to be undertaken. It is also more likely that the redshifts of lens and deflector will be obtained as part of the discovery process. All the same, the dynamic range and resolution of modern radio arrays, (*e.g.* the VLA and MERLIN) operating at high radio frequencies, provide far more information for an individual radio ring, although the availability of the Hubble Space Telescope for imaging means the situation in the optical is much improved.

2. THE OPTIMAL DEFLECTOR POPULATION

The key to establishing an efficient strategy for identifying optical rings lies in targeting the fraction of the galaxy population most effective at producing strong gravitational lensing distortions of the numerous, distant, $z > 1$, background source population. Confining attention to galaxies with the largest central velocity dispersions, σ_c, *i.e.* those with luminous massive bulges, maximizes the individual cross–sections, $\propto \sigma_c^4$, for strong lensing. A large surface density of lenses is also required to increase the total cross–section, effectively precluding the use of samples of nearby galaxies. The population of massive early–type galaxies in the redshift shell $z = 0.3 - 0.5$ satisfies both requirements while also providing desirable geometries for lensing, with values of $D_{LS}/D_{OS} \sim 0.5$ for sources with redshifts $z > 1.5$ (D_{LS} and D_{OS} are the angular diameter distances between the lens and source, and the observer and source respectively). The angular radius of the Einstein ring of a galaxy of redshift $z = 0.3$, modeled as a singular isothermal sphere with central velocity dispersion $\sigma_c = 250 \, \mathrm{km \, s^{-1}}$, for sources with redshift $z > 1.5$, is ~ 1.25 arcsec. Rings with diameters ~ 2.5 arcsec

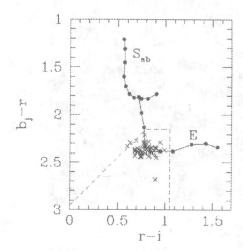

Figure 1. Two–color diagram, $(b_j - r, r - i)$, illustrating the predicted color–tracks for galaxies of type Sab and E over the redshift range $0.2 < z < 0.65$. The top of each locus corresponds to redshift $z = 0.2$. Symbols along each locus indicate the redshift at intervals of $\Delta z = 0.05$. The dashed line represents the color selection criteria for the galaxy sample, derived from APM scans of UK Schmidt plates, that forms the basis of our survey for optical rings. The crosses represent CCD photometry of 53 galaxies in the survey sample.

can be resolved from ground–based observatories in conditions of good atmospheric seeing and there is no reason in principle why they should not be detectable.

Early–type galaxies at distances of cosmological interest have been detectable for many years (Hamilton 1985). The relative ease with which such systems can be identified is due to the very strong spectral discontinuity arising from absorption line blanketing at $\sim 4000\text{Å}$, which, for nearby galaxies, produces large $U - B$ colors. The spectroscopic properties (determined from intermediate resolution observations of moderate signal–to–noise ratio) of the brightest early–type galaxies do not appear to evolve significantly out to redshifts $z > 0.5$. Thus, as the redshift of the galaxy increases the 4000Å spectral discontinuity shifts to longer wavelengths, resulting in the colors $B - V, V - R, ...$ attaining very large values within specific redshift ranges. This evolutionary behavior is in marked contrast to the blueward

Figure 2.   3600 s narrow–band image of the optical ring, obtained at the NTT during 1994 December, in 0.9 arcsec seeing, using a 60Å wide filter centered at the wavelength of the emission line. The image is 26 × 26 arcsec in extent, with a pixel size of 0.26 arcsec. North is up and East to the left.

trend in color exhibited by the bulk of the galaxy population,

Much attention has naturally focussed on the properties of the strongly evolving galaxy population and the physical processes responsible for the observed behavior. The far less dramatic behavior of luminous early–type galaxies has not featured as prominently. The reasons for this are two-fold; (a) the selection of galaxy samples using blue passbands – an L^* early–type galaxy at $z = 0.4$ has a B magnitude $m_B \sim 23$, and only the faintest spectroscopic surveys achieve the necessary depth, and (b) the low surface density on the sky – in the redshift shell $0.3 < z < 0.5$ only tens of objects per square degree are expected, and existing spectroscopic surveys contain very few examples. However, the use of wide–field photographic plate material in the R and I passbands enables a large sample of distant early–type galaxies to be defined.

3. THE FIRST CANDIDATE EINSTEIN RING

Our investigation employs APM scans of UK Schmidt Telescope (UKST) plates in B_J, R and I to identify the population of luminous early–type galaxies within the redshift shell $z = 0.3 - 0.5$. Galaxies are selected to lie in the color range $r - i < 1.05$, $b_j - i > 2.95$, and $b_j - r > 2.15$ (Figure 1), with magnitudes $16.4 < m_i < 18.85$. A galaxy at redshift $z = 0.4$ appearing at the sample limit has absolute magnitude $M_B \sim -21.5$ ($q_0 = 0.5$, $H_0 = 50\,\mathrm{kms^{-1}Mpc^{-1}}$, $\Lambda = 0.0$). The survey to date covers 5 UKST fields, $150\,\mathrm{deg^2}$, and the surface density of candidates is $\sim 40\,\mathrm{deg^{-2}}$. Spectroscopy of 162 objects within an area of $3\,\mathrm{deg^2}$ demonstrates the effectiveness of the selection procedures, producing 154 confirmed galaxies, only two of which have $z < 0.3$, and only 8 M–stars, a contamination rate of $< 5\%$.

Among the apparently normal galaxy spectra one object, a luminous early–type galaxy at redshift $z = 0.485$, possesses a strong emission feature at 5588Å. No other emission lines are evident over the wavelength range $\lambda\lambda 4456 - 7912$Å. Narrow–band imaging of the system at the emission–line wavelength reveals a semi–circular arc, radius $r = 1.35\,\mathrm{arcsec}$, centered on the peak of the galaxy surface–brightness distribution (Figure 2). The properties of the system can be reproduced by a gravitational lens model in which the deflector has a constant mass–to–light ratio (corrected for evolution) of $M/L_{B(0)} \sim 10$ ($H_0 = 50$) and the source is an intrinsically small but resolved (FWHM $\sim 0.2\,\mathrm{arcsec}$) object at high redshift, either at $z = 3.597$ (if the line is Lyα $\lambda1216$), or at $z = 2.607$ (if the line is CIV $\lambda1549$). Modeling of the system suggests an amplification of a factor ~ 20 with an unlensed magnitude for the source of $m_V \sim 27$.

Alternatives to the lensing explanation, in which the emission feature is associated with [OII] $\lambda3727$, with a velocity difference of $2900\,\mathrm{km\,s^{-1}}$ relative to the absorption–line redshift of the galaxy, appear to be extremely improbable. The system is almost certainly the first Einstein ring to be discovered in the optical, although unambiguous confirmation requires the detection of a second emission line. A full description of the observations and modeling is given by Warren et al. (1996).

4. FUTURE WORK

In common with most other examples of gravitational lensing phenomena, obtaining quantitative astrophysical information of interest on both the deflectors and the sources depends on the ability to identify a significant number of similar systems in a well–defined and efficient fashion. The multicolor selection technique described in §2 enables us to target distant, massive, early–type galaxies with relative ease. The availability of wide-field spectroscopic systems, particularly the 2dF on the Anglo–Australian Telescope means thousands of such galaxies can be identified with only a modest investment of time – the 2dF is capable of obtaining spectra of ~ 120 of our target galaxies in a single pointing of a few hours.

The discovery of further optical rings is predicated on the existence of an emission line(s) in the sources, thus producing a strong but readily quantifiable selection criterion, and, in general, allowing the measurement of the

Figure 3.   Simulation of 16 orbit observation with the HST PC of the optical ring. The simulation is based on the simple model for the system in which mass traces light in the deflecting galaxy, $z = 0.485$, and the source, $z = 3.597$, has a Gaussian profile with a FWHM= 0.2 arcsec. The image shows the field after subtraction of the galaxy. Contributions to the noise from the detector (read–out noise and dark current), and Poisson noise from the deflecting galaxy, the source, and from sky, are included. North is up and East to the left.

redshift of the source. Application of the statistical techniques developed by Kochanek (1993a, 1993b) to a sample of $10 - 20$ lenses, coupled with high–resolution imaging from HST of individual systems (Figure 3), would produce unique information concerning the surface density and properties of emission–line objects in the crucial redshift range $z = 2 - 4$ at magnitudes as faint as $m_V \sim 27$. Equally important, such a sample will provide quantitative constraints on the distribution of dark matter associated with early–type galaxies.

References

Hamilton, D., 1985, ApJ, 297, 371
Hewitt, J.N., Turner, E.L., Schneider, D.P., Burke, B.F., Langston, G.I. & Lawrence, C.R., 1988, Nature, 333, 537
Kochanek, C.S., 1993a, ApJ, 417, 438
Kochanek, C.S., 1993b, ApJ, 419, 12
Kochanek, C.S., 1995, ApJ, 445, 559
Kochanek, C.S. & Narayan, R., 1992, ApJ, 401, 461
Langston, G.I., et al., 1989, AJ, 97, 1283
Miralda–Escudé, J. & Lehár, J., 1992, MNRAS, 259, 31P
Wallington, S., Narayan, R. & Kochanek, C.S., 1994, ApJ, 426, 60
Warren, S.J., Hewett, P.C., Lewis, G.F., Møller, P., Iovino, A. & Shaver, P.A., 1996, MNRAS, in press

NEW OPTICAL AND MERLIN IMAGES OF THE QUADRUPLE GRAVITATIONAL LENS B1422+231

C. E. AKUJOR AND A. R. PATNAIK
Max-Planck-Institut für Radioastronomie, Bonn, Germany

AND

J. V. SMOKER AND S. T. GARRINGTON
Nuffield Radio Astronomy Labs, Jodrell Bank, U.K.

B1422+231 , a multiple component quasar at a redshift of 3.62, is believed to be a gravitationally lensed system (Patnaik et al. 1992). It has 4 components with maximum image separation of 1.3 arcsec. The three brighter components, A, B and C have similar polarization properties at 8.4 GHz. The radio spectra between 5 and 8.4 GHz of A, B and C are similar; since D is very weak its spectrum is not accurately determined.

We observed B1422+231 with the NOT; the seeing was 0.55 to 0.60 arcsec. The CCD camera has 520×520 pixels; each pixel is equivalent to 0.2 arcsec. A calibration field, F873–8 (Stobie et al. 1985) was observed in V, R and I bands. The data have been flat–fielded and sky–subtracted. B1422+231 was also observed at 1658MHz with MERLIN (Thomasson 1986) for 12 hrs in 'phase–reference' mode. The results are summarized in Table 1.

All the components including D are clearly detected at both radio and optical bands. These optical images are similar to those of Yee & Ellington (1994) who observed in r and g. The spectral indices of A, B, C and D are −0.42, −0.43, −0.41 and −0.15 respectively and confirm the turnover in the radio spectrum (Patnaik et al. 1992). No polarization is detected at 1658 MHz.

Patnaik et al. (1992) presented strong evidence for gravitational lensing but this was not as strong for D. Although the radio spectral index of D is slightly different at radio frequencies, the brightness ratio between the well-separated components C and D is similar at both radio and optical frequencies. This is one of the strongest arguments in favor of the lensing hypothesis for all the four components. These are amongst the earliest

C. S. Kochanek and J. N. Hewitt (eds), Astrophysical Applications of Gravitational Lensing, 335–336.
© 1996 IAU

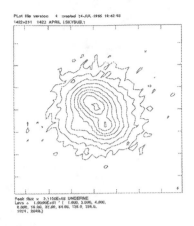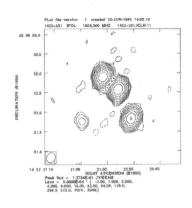

Figure 1. Images of B1422+231: I-band (1993, left); 1.7GHz MERLIN (right).

TABLE 1. The magnitudes of the whole source and magnitude and flux density ratios. The absolute magnitude calibration is not available for B-band; the typical error is ~10%

Filter	Obs date	Scan	mag	A	B	C	D
B	93april23	300s	-	1.0	1.49	0.61	≤0.025
V	92june30	300s	15.42±0.03	1.0	1.35	0.60	0.028
R	92june30	300s	15.59±0.06	1.0	1.52	0.73	0.037
I	92june30	300s	15.0±0.1	1.0	1.34	0.66	0.038
I	93april23	300s	15.2±0.1	1.0	1.20	0.52	0.027
1658MHz	93apr09	12hrs	350.4±5 mJy	1.0	1.01	0.54	0.03

observations of this lens system. So, they hopefully provide a standard for further monitoring of the lens images for possible variability.

Acknowledgements

We are grateful to Prof. A. Ardeberg, Dr. Jensen and NOT staff for support during the observations. CEA acknowledges Swedish NFR, and Av. Humboldt fellowships.

References

Patnaik, A.R. et al., 1992, MNRAS, 259, 1p
Stobie, R.S., Sagar, R. & Gilmore, G., 1985, AAS, 60, 503
Thomasson, P., 1986, QJRAS, 27, 413
Yee, H.K.C. & Ellingson, E., 1984, AJ, 107, 28

J03.13 A & B: A NEW MULTIPLY IMAGED QSO CANDIDATE

J.F. CLAESKENS

European Southern Observatory (Chile) and presently Aspirant au FNRS (Belgium) at the Institut d'Astrophysique, Université de Liège, 5 Avenue de Cointe, B-4000 Liège, Belgium

J. SURDEJ

STScI, 3700 San Martin Drive, Baltimore, MD 21218, U.S.A; also Directeur de Recherches au FNRS (Belgium)

AND

M. REMY

Institut d'Astrophysique, Université de Liège, 5 Avenue de Cointe, B-4000 Liège, Belgium

Abstract. We report the discovery of a new gravitational lens (GL) candidate for the quasar J03.13 (z = 2.55). The mean angular separation and magnitude difference between the A & B QSO images are found to be 0.84″ and 2.10 mag, respectively. A spatially unresolved medium resolution spectrum of J03.13 shows intervening absorption line systems at $z = 2.34$ and $z = 1.085$. The latter one is possibly associated with a $\sigma = 206$ km/s lens galaxy.

1. Introduction

The multiply imaged QSO candidate J03.13 was discovered in 1993, during the last campaign of an ESO Key-Program aimed at searching for gravitational lenses among highly luminous quasars (Surdej et al. 1987, 1989). These observations were conducted at ESO (La Silla, Chile) with a direct CCD camera attached to the Cassegrain-focus of the ESO/MPI 2.2m telescope. High angular resolution multicolor imaging and low resolution spectroscopic observations were carried out with the New Technology Telescope (NTT) and SUSI or EMMI, respectively, at La Silla in Chile, in 1994. These observations tend to confirm the gravitational lens hypothesis.

C. S. Kochanek and J. N. Hewitt (eds), Astrophysical Applications of Gravitational Lensing, 337–338.
© 1996 IAU

TABLE 1. Averaged photometry and relative astrometry.

Object	$B \pm \sigma_B$	$R \pm \sigma_R$	$i \pm \sigma_i$
J03.13 A	17.6±0.1	17.2±0.1	16.9±0.1
J03.13 B	19.7±0.1	19.3±0.1	18.9±0.1
J03.13 A-B	-2.16±0.05	-2.14±0.05	-1.99±0.05
$\Delta\theta_{AB} \pm \sigma$ (")	0.88±0.02	0.84±0.02	0.79±0.02

2. Results from the observations

Simultaneous fits of multiple point spread functions (PSFs) on the SUSI frames of J03.13 have been performed with a numerical profile fitting program, designed by Remy (1995). This allows us to derive the relative astrometric and photometric quantities of each individual component, reported in Table 1. Some residuals on the best R frame tend to show that J03.13 could possibly be better fitted with three point-like components.

A low resolution spectrum (3.5 Å per pixel) of J03.13 obtained with the NTT+EMMI shows that the QSO has a redshift $z = 2.55$ (instead of the value of 2.80 tabulated in Maza et al. 1993), and reveals two intervening absorption line systems: Lyα and CIV at $z = 2.34$ and MgII, MgI and FeII at $z = 1.085$. Unfortunately, because of the moderate angular resolution along the slit, this spectrum is spatially unresolved. More details about the reduction techniques and spectroscopic results may be found in Claeskens et al. 1995.

3. Conclusions

J03.13 is a new very promising GL having a small angular separation. The main arguments supporting this hypothesis are that i) both components A & B have approximately the same colors (contamination by a high z lens could account for the differences in the i filter reported in Table 1); ii) the absorption line system at z=1.085 could be due to the lens galaxy, whose velocity dispersion should be $\simeq$ 206 km/s to account for the observed angular separation between the two images.

HST observations of J03.13 should clear up the nature of the two (three?) images of the QSO, and eventually the presence of the lens.

References

Claeskens, J.-F., Surdej, J., Remy, M., 1995, submitted to A&A
Maza, J., Ruiz, M.T., González, L.E., et al., 1993, Rev. Mex. Astron. Astrof., 25, 51
Remy, M., 1995, PhD Thesis, Liège University, Belgium
Surdej, J., Magain, P., Swings, J.-P., et al., 1987, Nature, 329, 695
Surdej, J., Arnaud, J., Borgeest, U. et al., 1989, The Messenger, 55, 8

EVN-MERLIN OBSERVATIONS OF THE REMARKABLE LENS SYSTEM: 2016+112.

M.A. GARRETT, S. NAIR AND D.WALSH

NRAL, Jodrell Bank, Macclesfield, Cheshire SK11 9DL, UK.

AND

R.W. PORCAS AND A.R. PATNAIK

MPIfR, Auf dem Hügel 69,D-53121 Bonn, Germany.

1. Introduction & Analysis

2016+112 was observed simultaneously with the European VLBI Network (EVN) and MERLIN arrays during the May 1993 *joint* EVN-MERLIN session at λ18 cm. Common elements to both arrays included the Jodrell Bank 76-m Lovell and 32-m Cambridge telescopes. In order to simultaneously map the entire 4 arcsec^2 field of view, various wide-field mapping techniques were employed (see Garrett et al. 1994b).

2. Results & Conclusions

Earlier MERLIN λ6 cm observations by Garrett et al. (1994a) revealed C to be comprised of two components - C_1 and C_2, separated by $\sim$ 112 mas. Garrett et al. (1994a) suggested that C_2 was a third lensed radio image of the same background source that gives rise to images at A and B. The joint EVN-MERLIN map, with a resolution of 50 mas, is shown in Fig 1 (left). The EVN maps, with a resolution of 15 mas, show A and B to be partially resolved but C is well resolved. The MERLIN component C_2 is identified in the EVN map as an unresolved component but C_1 is further resolved into three components designated: C_{11}, C_{12}, C_{13}. Together with C_2 these three components form a remarkable chain which spans $\sim$ 200 mas. The resolved structure of C_1 seems at odds with its traditional identification with the galaxy at C, though this still remains a possibility. Taking into account the flat spectrum nature of C_1 and its bizarre resolved radio structure we

C. S. Kochanek and J. N. Hewitt (eds), Astrophysical Applications of Gravitational Lensing, 339–340.

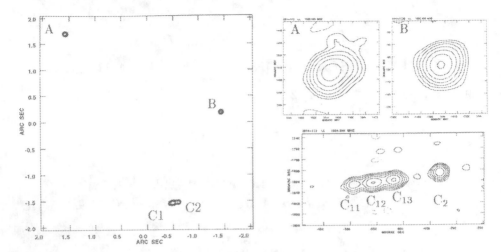

Figure 1. Left: the λ18 cm wide-field 50 mas resolution EVN-MERLIN maps of 2016+112 and Right: the λ18 cm 15 mas resolution EVN only maps of 2016+112 A,B,C1,C2.

are inclined to the view that part, if not all, of the emission at C_1 is in fact lensed in some way (see companion poster - Nair & Garrett these proceedings).

2.1. COMPARISON OF THE MERLIN 6CM AND EVN-MERLIN 18CM MAPS

One of the primary motivations for obtaining a joint λ18 cm EVN-MERLIN map is to allow the determination of an accurate flux ratio of C_2:B between λλ6 and 18cm. If C_2 is indeed a third image then the flux ratio C_2:B should be the same at both wavelengths, since gravitational lensing is achromatic. At λ18 cm MERLIN does not have the required resolution to resolve C_1 and C_2. However, it turns out that the joint EVN-MERLIN λ18 cm observations are well matched to the λ6 cm stand-alone MERLIN observations.

The ratio C_2:B is 0.41 from the joint λ18 cm map and 0.26 from the MERLIN λ6 cm map. At first sight this result argues against a "third image" interpretation for C_2 but see Nair & Garrett (these proceedings).

References

Garrett et al., 1994, MNRAS, 269, 902.
Garrett, M.A., Patnaik, A.R. & Porcas, R.W., 1994b, in Proc. EVN/JIVE Symposium, Torun, ed. Kus et al., TRAO, 73

MULTI-EPOCH, DUAL-FREQUENCY VLBI OBSERVATIONS OF PKS 1830-211 FROM JAPAN

YOSHIAKI HAGIWARA

The Graduate University for Advanced Studies,
Nobeyama Radio Observatory, Nagano 384-13, Japan

K. FUJISAWA, P. EDWARDS, H. HIRABAYASHI, Y. MURATA AND
H. KOBAYASHI

The Institute of Space and Astronautical Science,
3-1-1 Yoshinodai, Sagamihara, Kanagawa 229, Japan

AND

T. IWATA

Communications Research Laboratory,
Kashima Space Research Center, Ibaraki 314, Japan

1. Observations

PKS 1830-211 is a strong, flat-spectrum compact double source with a component spacing of 1 arcsecond. Observations of PKS 1830-211 were made with the Japanese domestic VLBI network at 2.3 GHz and 8.4 GHz bands in sessions between December 1991 and November 1994. The Usuda 64 m (ISAS) and Kashima 34 m (CRL) telescopes were used for all observations, and were used in conjunction with the Mizusawa 10 m (NAO) for observations in 1994. In addition, the total flux was measured with the Usuda 64 m at both bands. Data was recorded using K3 and K4 formatters and recorders, and correlated with NAOCO (the New Advanced One-unit COrrelator of the National Astronomical Observatory).

2. Analysis and Results

The model fitting results presented here were obtained from one 2 MHz bandwidth channel of the Usuda-Kashima baseline data. Results from other VLBI experiments enable us to assume values for the position angle and

341

C. S. Kochanek and J. N. Hewitt (eds), Astrophysical Applications of Gravitational Lensing, 341–342.
© 1996 IAU

TABLE 1. 2.3 GHz observations and Usuda-Kashima baseline results. Errors in fluxes are estimated to be of the order of 10%.

Date	Length [hr]	Total flux [Jy]	Corr. flux [Jy]	Flux ratio
1991–Dec–16	5.5	9.1	2.2	1.15 ± 0.06
1992–Mar–25/26	5.5	8.9	4.1	1.10 ± 0.06
1992–Jun–18	7	8.7	3.9	1.12 ± 0.06
1994–Mar–25/26	2	10.1	5.6	1.18 ± 0.02
1994–Nov–4/5	2.5	—	—	1.08 ± 0.02

component separation and thus to determine the flux ratio of the two components. The flux ratios show little variation in either band over 3 years but the visibilities show distinct variations during this period. The 2.3 GHz results are given in Table 1; similar results were obtained at 8.4 GHz. This indicates that the time-scale for the increase in the correlated flux density of both components is less than the separation between our observations.

3. Discussion

The 2.3 GHz correlated flux on the Usuda-Kashima baseline is a significantly smaller fraction of the total flux than at 8.4 GHz. This indicates that much of the broad ring feature, which contributes to the total flux at 2.3 GHz, is resolved out on this baseline (130 mas resolution at 2.3 GHz). The flatter spectrum core is thus more dominant at 8.4 GHz. In addition, the low galactic latitude and longitude of the source result in the source being increasingly broadened at lower frequencies by the effects of interstellar scintillation, which reinforces the expectation of higher correlated fluxes at higher frequencies (see also Jones et al. 1995).

The synchronization of the flux variation of both components is consistent with the effects of gravitational lensing. Recently, van Ommen et al. (1995) have derived a propagation delay of 44 ± 9 days for one component with respect to the other. Our observations were not designed for time resolution, so we are not able to observe this effect. Regular monitoring at radio wavelengths will enable the time delay to be constrained more tightly.

References

Fujisawa, K., 1992, M. Sc. Thesis, Tokyo University (unpublished).
Hagiwara, Y., 1995, M. Sc. Thesis, Nagoya University (unpublished).
Jones, D., et al., 1995, these proceedings.
van Ommen, T.D. et al., 1995, ApJ, 444, 561.

OPTICAL IMAGING OF B1422+231 –

PROSPECTS FOR DETERMINING THE HUBBLE CONSTANT

J. HJORTH

Institute of Astronomy, Madingley Rd., Cambridge CB3 0HA, UK

A.O. JAUNSEN

Nordic Optical Telescope, Ap. 474, S/C de La Palma, E-38700 Canarias, Spain

A.R. PATNAIK

Max Planck-Institut für Radioastronomie, Auf dem Hügel 69, D-53121 Bonn, Germany

AND

J.-P. KNEIB

Institute of Astronomy, Madingley Rd., Cambridge CB3 0HA, UK

Abstract. We report on observations indicating that the bright quadruple lens B1422+231 may be variable in the optical on a time scale of hours. We also find that a model can reproduce the optical positions and magnifications. Unfortunately, the system may be affected by possible microlensing and external shear. HST should provide a good model for the lens, the system should be monitored for several nights in superb seeing (0.3–0.4 arcsec FWHM), and the redshifts of the south-east galaxies should be determined. With this extra information, improved modeling will tell whether it is realistic to obtain H_0 from the time delays of B1422+231.

1. Introduction

B1422+231 (see Hammer, Rigaut, & Angonin-Willaime 1995 and references therein) is a prime candidate for a cosmological determination of the Hubble constant for several reasons. It is bright, quadruple, and has a short

C. S. Kochanek and J. N. Hewitt (eds), Astrophysical Applications of Gravitational Lensing, 343–344.
© 1996 IAU

predicted time delay. It also appears to have all parameters known, to be simple, variable, achromatic, and to be well modeled.

We here present preliminary results from a pilot study carried out at the Nordic Optical Telescope to check for optical variability. On the night of March 31 1995 we obtained 6 I images each separated by one hour. On the night of April 17 1995 we obtained 9 images, 8 of which were taken within one hour. The seeing ranged from 0.5 to 1.0 arcsec FWHM.

The point spread function (PSF) was determined from stars in the field and a multiple fit to the four components of the gravitational lens was performed using DAOPHOT II and ALLSTAR. Assuming that the positions determined for the brightest component (B) were correct, the best data were singled out by requiring the position of A to be correct within 0.01 arcsec. The positions of A, C, and D were then fixed according to (updated) relative radio positions and a new (constrained) multiple fit was performed.

2. Results

We find that the magnitudes for the A and B components are strongly anti-correlated. This is expected when $|AB| \leq$ FWHM (see Schechter 1993). Average values are A/B $= 0.82 \pm 0.01$, C/B $= 0.52 \pm 0.005$ and D/B $= 0.038 \pm 0.002$. These values seem to differ from those reported from the Spring 1993 where A/B $= 0.76 \pm 0.03$ and C/B $= 0.49 \pm 0.01$. This may indicate that the system is affected by microlensing, which could also explain why the optical intensity ratios differ from those in the radio.

We find that the combined light from A and B is constant within the 1 h time span on April 17 (variation less than 0.003 mag for the best images) whereas the images taken on March 31 during the 5–6 h time span show a monotonic increase in the brightness of A+B of 0.02 mag. We consider this as tentative evidence for intranight variability. It also shows that high-precision relative photometry can be achieved in relatively bad seeing.

On the combined images there is no evidence for the main deflector. The galaxies to the south-east appear to be early-type galaxies at a redshift of 0.3–0.4.

We have modeled the system and find that the configuration with the optical magnifications can be accounted for by an elliptical lens located very close to D, possibly with the added shear from the galaxies to the south-east.

References

Hammer, F., Rigaut, F., & Angonin-Willaime, M.-C. 1995, A&A, 298, 737

Schechter, P. L. 1993, in Proc. 31st Liège Int. Astrophys. Colloq., Gravitational Lenses in the Universe, eds. J. Surdej et al. (Liège: Université de Liège), 119

INTERSTELLAR SCATTERING AND THE
EINSTEIN RING PKS 1830-211

D.L. JONES, R.A. PRESTON, D.W. MURPHY AND D.L. MEIER
Jet Propulsion Laboratory, Mail Code 238-332,
4800 Oak Grove Drive, Pasadena, CA. 91109, USA

AND

D.L. JAUNCEY, J.E. REYNOLDS AND A.K. TZIOUMIS
Australia Telescope National Facility, CSIRO,
PO Box 76, Epping 2121, NSW, Australia

Abstract. The remarkably strong radio gravitational lens PKS 1830-211 consists of a one arcsecond diameter Einstein ring with two bright compact components located on opposite sides of the ring. We have obtained 22 GHz VLBA data on this source to determine the intrinsic angular sizes of the compact components. Previous VLBI observations at lower frequencies indicate that the brightness temperatures of these components are significantly lower than 10^{10} K (Jauncey et al. 1991), less than is typical for compact synchrotron radio sources and less than is implied by flux density variations. A possible explanation is that interstellar scattering is broadening the apparent angular size of the source and thereby reducing the observed brightness temperature. Our VLBA data support this hypothesis.

The position of PKS 1830-211 is only a few degrees away from the galactic center ($b=-5.7°$, $l=12.2°$). It is therefore plausible that significant interstellar scattering (ISS) within our galaxy may occur along this line of sight. The effects of ISS decrease rapidly with frequency, and should be small at 22 GHz.

We observed PKS 1830-221 with the full VLBA at 22 GHz in May 1994. The data were correlated twice using phase centers corresponding to the locations of the two compact components, which we designate the northeast (NE) and southwest (SW) components. After calibration and fringe fitting in AIPS, we used the Caltech program Difmap for editing,

C. S. Kochanek and J. N. Hewitt (eds), Astrophysical Applications of Gravitational Lensing, 345–346.
© 1996 IAU

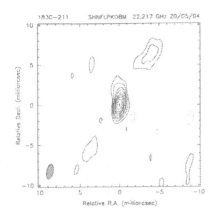 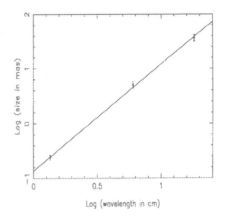

Figure 1.   **LEFT:** The contours are -2, 2, 5, 10, 15, 25, 50, 70, and 95%, and the restoring beam FWHM is 1.9 × 0.8 mas with the major axis along position angle -11.5°. The absolute flux density scale is not yet fully calibrated, but this does not affect the measured angular size of the core. **RIGHT:** The deconvolved minor axis angular sizes and their (formal) errors determined with the AIPS program IMFIT, using only the upper 50% or less of the brightness range to avoid bias by any extended low-level structure.

self-calibration, imaging, and deconvolution. This paper presents our image of the SW component; imaging of the NE component is in progress.

The two figures above show our VLBA image of the SW component of PKS 1830-211 and the deconvolved minor axis width of this component at three frequencies: 1.7 GHz (unpublished data from an ad-hoc VLBA experiment in 1990), 4.9 GHz (Jones 1994), and 22 GHz (new data).

The slope of the line fit to the angular size measurements is 1.95 ± 0.14, consistent with the λ^2 dependence expected for scattering. This suggests that angular size measurements made by VLBI at frequencies ≤ 22 GHz are indeed affected by angular broadening due to ISS. At 22 GHz the size of the SW component core is 0.6 × 0.2 mas and $T_b \sim 10^{11}$ K.

Acknowledgements: This research was carried out at the Jet Propulsion Laboratory, California Institute of Technology, under contract with NASA.

References

Jauncey, D.L., et al., 1991, Nature, 352, 132
Jones, D.L. 1994, in Compact Extragalactic Radio Sources, ed. J.A. Zensus & K.I. Kellermann (Socorro: NRAO), 135

FLUX DENSITY VARIATIONS OF PKS 1830-211

J.E.J. LOVELL, P.M. MCCULLOCH AND E.A. KING
Department of Physics, University of Tasmania, Australia

AND

D.L. JAUNCEY
Australia Telescope National Facility, CSIRO, Sydney, Australia

We have been observing the strong radio source and Einstein ring gravitational lens PKS 1830–211 as part of a flux density monitoring program at 2.3 and 8.4 GHz using the 26 m antenna at the Mt Pleasant Observatory. As can be seen from Figure 1 this source is variable at both frequencies. The 2.3 GHz flux density decreased from 10 Jy in 1990 November to a minimum of 9 Jy during 1992, and has since steadily increased to its present ∼12 Jy. Over the same interval, the flux density at 8.4 GHz has exhibited greater variability, more than doubling from an initial value of 5 Jy in late 1990 to a peak of ∼11 Jy early in 1992, before decreasing to a level of ∼7.5 Jy. VLA observations by van Ommen et al. (1995) show that the 15 GHz total flux density increased before the 8.4 GHz brightening seen at Mount Pleasant. Although the 15 GHz time series is not well sampled, it is clear from these data that the 15 GHz outburst precedes the 8.4 GHz feature by ∼400 days.

The radio morphology of PKS 1830–211 shows a steep spectrum ∼1 arcsecond-diameter ring upon the opposite sides of which lie two bright compact components (Jauncey et al. 1991). To establish the spatial properties of the variability, we augmented the single-dish flux density measurements with high-resolution observations using the SHEVE VLBI array (Preston et al. 1993). Data obtained in 1990 July and 1992 March clearly show that the flux density variability is dominated by changes in the two compact components (King 1994). Northern Hemisphere VLBI observations (Jones et al. 1993) have demonstrated variability in the compact structure over the period coincident with the total power variability.

The single peak clearly visible in the 8.4 GHz light-curve indicates that the time delay between the two components is either less than the temporal

C. S. Kochanek and J. N. Hewitt (eds), Astrophysical Applications of Gravitational Lensing, 347–348.
© 1996 IAU

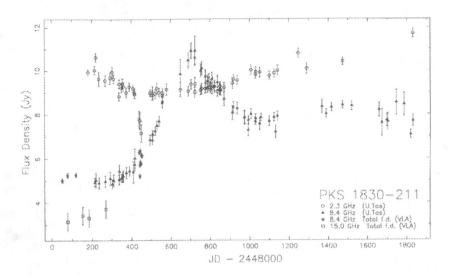

Figure 1. Complete 2.3 GHz and 8.4 GHz light-curves for PKS 1830–211 obtained with the Mt Pleasant 26 m antenna. VLA data from van Ommen et al. (1995) are also shown.

scale-size of the outburst, or very much greater than the time between outbursts. The high-resolution VLBI data show that the former is the case.

The confinement of variability to the compact components makes PKS 1830–211 an excellent target for relative time-delay studies. Such studies require frequent arcsecond-resolution observations. The steep spectrum of the ring structure means that the total flux density is dominated by the compact components at short wavelengths allowing straightforward modeling of the morphology.

We are about to begin a series of service observations with the Australia Telescope Compact Array over a period of approximately one year to monitor the relative intensities of the two compact components at 8.6 GHz as well as the total flux density at 4.8, 2.3 and 1.2 GHz. These observations are intended to provide a more direct and accurate measurement of the time delay as well as data on the propagation of outbursts as a function of wavelength.

References

Jauncey, D., Reynolds, J., Tzioumis, A., et al., 1991, Nature, 352, 132

Jones, D., Jauncey, D., Preston, R., et al., 1993, in Sub-Arcsecond Radio Astronomy, eds. R. Davis & R. Booth, (Cambridge: Cambridge Univ. Press) 150

King, E., 1994, Ph.D. Thesis, Physics Department, University of Tasmania

Preston, R., Jauncey, D., Reynolds, J., et al., 1993, in Sub-Arcsecond Radio Astronomy, eds. R. Davis & R. Booth, (Cambridge: Cambridge Univ. Press) 428

van Ommen, T., Jones, D., Preston, R., & Jauncey, D., 1995, ApJ, 444, 561

NEW RADIO OBSERVATIONS OF 'OLD FAITHFUL'

R.W. PORCAS AND A.R. PATNAIK
Max-Planck-Institut für Radioastronomie
Auf dem Hügel 69, D-53121 Bonn, Germany.

AND

T.W.B.MUXLOW, M.A.GARRETT AND D.WALSH
NRAL, Jodrell Bank, Macclesfield SK11 9DL, UK

We present new arcsecond-scale radio images of the gravitational lens system 0957+561 A, B. Observations at 1.6GHz were made in 1991 October with the VLA in A/B configuration with a resolution of 1.5 arcsec (Fig. 1, left). The lowest contour is 0.37 mJy/beam. In addition to the compact A and B image components, and the familiar NE/SW radio double-lobe structure surrounding image A, this map shows two interesting new features:

(a) a long, thin feature extending south and west from the NE lobe, some 10 arcsec in extent, confirming the detection by Avruch et al. (1994). It is reminiscent of the 'arc' features seen in optical images of lensing clusters, and hopefully can be used to constrain models of the cluster mass distribution. (For the superstitious, one can note that the position angle of the arc, 17°, is identical to that of the VLBI jet in image B, a few arcseconds to the west.)

(b) a low-level extension of the B image in the NW direction. We are not aware that this has been seen before. A possible interpretation of this feature is a continuation of the jet emission seen in VLBI images of B, whose counterpart in A is seen in higher resolution maps. The change of position angle between the VLBI and arcsec-scale jets would indeed have opposite parities in the A and B images.

Observations at 408 MHz with MERLIN were made in 1995 January. We present 2 maps; the first (Fig.1, right) uses natural weighting of the visibility function, and has a resolution of 1.5arcsec. This is a preliminary map, but we can see a 'hint' of emission from the arc, at the position of the brightest peak in the VLA 1.7GHz map. The lowest contour level is 1.96 mJy/beam. We deduce that the spectrum of the arc cannot be very steep (e.g. -1.0) or the arc would be more easily detected. The second MERLIN map (Fig.2)

C. S. Kochanek and J. N. Hewitt (eds), Astrophysical Applications of Gravitational Lensing, 349–350.

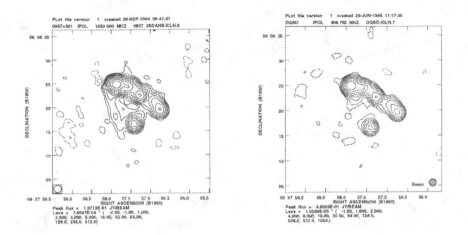

Figure 1.   1.5arcsec maps of 0957+561 at 1.7GHz (VLA, left) and 408MHz (MERLIN, right)

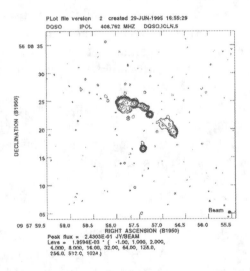

Figure 2.   MERLIN map of 0957+561 at 408MHz with 0.5arcsec resolution.

uses uniform visibility weighting, and has a resolution 0.5arcsec. The jet extension of image A, curving towards the NE lobe structure is clearly visible, although any counterpart in B is not seen, and must therefore be fainter at 408 MHz.

Acknowledgements We thank Prof. Bernie Burke for discussions regarding the 'arc' feature discovered by Avruch et al. (1993).

References

Avruch, I.M., et al., 1993, BAAS, 25, 1403

A GRAVITATIONAL LENS CANDIDATE BEHIND THE
FORNAX DWARF SPHEROIDAL GALAXY

C.G. TINNEY

Anglo-Australian Observatory
PO Box 296, Epping. NSW. 2121 Australia.

CCD astrometry makes possible milli-arcsecond precision relative astrometry at faint magnitudes (Tinney 1994a). The measurement of proper motions for the nearest galaxies is therefore a project requiring only years – not decades. A prerequisite for such measurements, however, is a set of unresolved, extragalactic reference objects – ie. QSOs. While searching for such objects behind the Fornax dSph galaxy, Tinney (1995b) discovered the QJ0240-343AB system as a pair of bright, UV-excess objects (U–B $= -1.0$, B $= 19.0$ & 19.8) with a separation of $6.1''$. Spectra obtained with the 3.9 m Anglo-Australian telescope and the RGO+FORS spectrographs (Fig. 1) have shown both components of the pair to be at a redshift of $z = 1.4$, with a rest velocity difference consistent with zero at a 1-σ limit of 180 km/s. The spectra also show a definite metal-line absorption system at $z = 0.543$ and a possible system at $z = 0.337$. No bright lens is seen, but the strong similarities in the spectra suggest a gravitational lens nature.

If the system is a lens, then the large difference in the line-to-continuum-ratio of the components suggests either strong variability, or microlensing of the continuum source. A search has been made of the literature and astronomical archives for images of the QJ0240-343AB pair – the data obtained to date are shown in Table 1. (The data from the NTT is in the R band, but non-photometric. Stars in the field have been used to create a uniform, but arbitrary, zero-point. The zero-point of the UKST data is very uncertain.) It seems clear from this data that the system is genuinely variable, on time scales of a few years (the time scale for micro-lensing should be ~30 years). The existing data does not enable statements about time delays to be made, but suggest that monitoring will straightforwardly provide much information on this system.

C. S. Kochanek and J. N. Hewitt (eds), Astrophysical Applications of Gravitational Lensing, 351–352.
© 1996 IAU

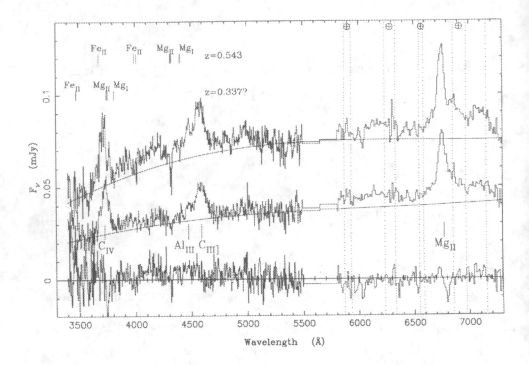

Figure 1.   Spectra obtained with the 3.9m AAT and RGO+FORS of QJ0240-343A *(upper trace)* and QJ0240-343B *(middle trace)*, and the difference between the line spectra *(lower trace)* after the subtraction of a continuum. Notice that the lines have roughly equal strength in the two components, while the continuum of A is twice that of B.

TABLE 1. QJ0240-343AB Photometry

Source	Epoch	A	B	Unc	A–B
Demers *et al.* 1994	20 Dec 1974	B_j=19.72	B_j=19.97	±0.1	ΔB_j=-0.38
	20 Dec 1974	V=19.71	V=20.43	±0.1	V=-0.72
UKST/COSMOS	16 Nov 1979	B_j=17.5	B_j=17.5	(±0.2)	ΔB_j=0.0
ESO/NTT Archive	5 Aug 1991	R_{91}=19.95	R_{91}=20.54	± 0.03	ΔR_{91}=-0.59
ESO/NTT Archive	24 Oct 1992	R_{91}=19.94	R_{91}=20.35	± 0.03	ΔR_{91}=-0.41
Tinney 1995	27 Oct 1994	B=19.00	B=19.77	± 0.05	ΔB=-0.77

References

Demers, S., Irwin, M. & Kunkel, W. 1995, AJ, 108, 1648
Tinney, C.G. 1995a, in Science with the VLT, eds. J.R. Walsh & I.J. Danziger, (Springer-Verlag: Heidelberg)
Tinney, C.G. 1995b, MNRAS, in press

MAPPING THE EXTINCTION IN DUSTY LENSES:
OPTICAL AND IR IMAGING OF MG J0414+0534

C. VANDERRIEST[1], M.-C. ANGONIN-WILLAIME[2] AND F. RIGAU
[1] *CFHT Corp., P.O. Box 1597, Kamuela 96743, Hawaii, USA*
[2] *Paris-Meudon Observatory, 92195 Meudon Cedex, France*

The source of the gravitational mirage MG J0414+0534, found in a radio survey (Hewitt et al. 1992), is an extremely reddened quasar at z $\simeq$ 2.63 (Lawrence et al. 1995) but the redshift of its lensing galaxy is still unknown ($\simeq$ 0.8 ?). Whether the extinction occurs mainly in the *lens galaxy* or in the *source* itself is a matter in debate. Also, a discrepancy is observed between the radio and optical flux ratios for images A_1 and A_2, which could be due either to differential magnification near a caustic or to differential extinction (Schechter et al. 1992, hereafter SM92; Angonin-Willaime et al. 1994, hereafter AW94). The new IR and optical data discussed here could throw some light on these obscure issues.

Data were obtained almost simultaneously at visible and IR wavelengths with the Canada-France-Hawaii telescope: in R band with the imaging mode of SIS (Subarcsecond Imaging Spectrograph) on the night of 15 August 1994; in K band the following night, with the Redeye camera (Simons et al. 1994). Near simultaneity eliminates any trouble from intrinsic variations and time delays for measuring the extinction. The pixel size was 0.173" for SIS and 0.201" for Redeye. Good seeing (see Table 1) allowed an accurate image decomposition. In R, a synthetic PSF was built from the surrounding standard stars (see AW94). The photometric accuracy for the decomposition is limited by photon noise. In K, the measurable stars were too faint and the PSF was determined iteratively from the object itself. The photometric accuracy, although limited by this process, is better than in the visible. In particular, the properties of the lensing galaxy are easily measured (this will be reported elsewhere). Table 1 summarizes the photometric data available from earlier studies and from these observations. The most striking feature is the dimming of A_1 by a factor 2 (0.75 mag) in R between 1992 and 1994. The other components, checked against the local photometric sequence, did not vary significantly.

C. S. Kochanek and J. N. Hewitt (eds), Astrophysical Applications of Gravitational Lensing, 353–354.
© 1996 IAU

TABLE 1. Summary of Measured Flux Ratios

color (date)	A/B	A_1/A_2	A_1/B	A_2/B	seeing (")	exposure (mn)
Radio (average)	5.0	1.15±0.05	2.7	2.3		
I (1991, cf. SM92)	3.2	2.5±0.2	2.2	0.9		
R (1992, cf. AW94)	3.2	3.0±0.4	2.4	0.8		
R (15/08/1994)	2.1	1.3±0.15	1.2	0.9	0.70	2×10
K (16/08/1994)	4.6	1.39±0.05	2.68	1.92	0.62	13×2

To understand this data set, we should note that: (1) the time delay $\Delta t(A_1-A_2)$ is too short to explain the A_1/A_2 variations; (2) in 1994, the flux ratios in the radio and in K are similar, while the R data agree with an extra extinction for A_1 and A_2, most likely due to the lens galaxy; (3) the reddening indices (R-K, or R-radio) are identical for A_1 and A_2 in 1994; (4) in 1991-92 the A_1/B ratio in R was close to the radio value.

Two different interpretations are then possible:
• In 1992, the extinction was comparable for A_1, B and C, but definitely larger for A_2. In 1994, A_1 was also more absorbed because its light beam was crossing a cloud in the lens galaxy. Gravitational magnification causes a "fast scanning" of highly amplified images across the galaxy (the proper motions can reach $\sim 10^4$km/s), which speeds up the observed variations.
• The fading of A_1 is simply the end of a microlensing event, the "normal" flux ratios being close to those measured in 1994. In 1992, the extinction of A_1 was compensated by microlensing amplification and it was just *by chance* that A_1/B in the visible was close to the radio value.

In any case, we can conclude that the large extinction of MG J0414+0534 is due **at least partly** to the lens galaxy. Since microlensing is achromatic, deciding between a microlensing event or variable extinction from "gravitational fast scanning" for image A_1 could have been done in 1992 with IR and visible photometry or, better yet, bidimensional spectroscopy. Such techniques are relevant for a future monitoring.

References

Angonin-Willaime M.-C., Vanderriest C., Hammer F. & Magain P., 1994, A&A, 281, 388 (AW94).
Hewitt J., Turner E., Lawrence C., Schneider D. & Brody J., 1992, AJ, 104, 968.
Lawrence C., Elston R., Jannuzi B. & Turner E., 1995, AJ, in press.
Schechter P. & Moore C., 1992, AJ, 105, 1 (SM92).
Simons D. et al., 1994, SPIE, 2198, 185.

GRAVITATIONAL TELESCOPES

R. D. BLANDFORD & D. W. HOGG
Caltech
Pasadena, CA 91125, USA

Abstract. Some ways in which gravitational lenses can act as crude tele scopes are reviewed. Magnification limits associated with finite source size, corrugation of the gravitational potential and finite wavelength are specified. It should be possible to obtain rotation curves for a small sample of ultra-faint galaxies imaged as giant arcs. There is an appreciable chance of eventually observing a radio jet component cross a caustic. The successful observation of microlensing in 2237+031 suggests that the continuum emission from this source at blue wavelengths is at least partly non-thermal. The maximum possible magnification observable, granted existing constraints is argued to be from γ-ray pulsars crossing caustic sheets formed by galaxies or a hypothetical population of intergalactic massive objects.

1. Magnification near caustics

1.1. FOLDS

It is convenient, though not obligatory, to illustrate large magnification effects using a weakly elliptical gravitational lens (*eg* Blandford & Kovner 1988, Schneider et al. 1992). Suppose that the scaled potential can be written in the form

$$\Phi(\vec{r}) = f(r) + \psi(r, \phi) \tag{1}$$

where we fix the angular radius of the unperturbed Einstein ring to be $r = b$ by setting $f'(b) = b$ and treat ψ as a perturbation. To first order, the source is located at

$$\vec{\beta} = b\, h_r\, \delta \vec{\hat{r}} - \nabla\psi \tag{2}$$

where $\delta = r - b$ is the radial displacement of the image from the unperturbed Einstein ring, $h_r = 1 - f''(b)$ and $\nabla\psi = \nabla\psi(b, \phi)$, etc. A small circular

C. S. Kochanek and J. N. Hewitt (eds), Astrophysical Applications of Gravitational Lensing, 355–364.

source of angular radius a will be magnified into an elliptical image with minor axis of length a/h_r and major axis a/h_t where $h_t = b^{-1}\delta h_r - b^{-1}\psi_{,r} - b^{-2}\psi_{,\phi\phi}$, rotated with respect to the radial direction by an angle $(b^{-2}\psi_{,\phi} - b^{-1}\psi_{,r\phi})/h_r$. The magnification becomes infinite on the critical curve where $h_t = 0$. This is displaced from the unperturbed Einstein ring by $\delta_c = (\psi_{,r} + b^{-1}\psi_{,\phi\phi})/h_r$. (The corresponding displacements of the equipotential and isodensity contour are $\delta_\Phi = -\psi/b$ and $\delta_\Sigma = (\psi_{,r} + b\psi_{,rr} + b^{-1}\psi_{,\phi\phi})/(h_r - bf''')$.) This critical curve maps onto the caustic in the source plane.

For the simple case of an isothermal potential with $f(r) = br$ and an external tidal perturbation, $\psi = \epsilon r^2 \cos(2\phi)$, the contours of constant surface density are circular. (This can occur if the tidal force cancels the centrifugal force, for example.) The radial magnification is unity and the critical curve has an ellipticity 4ϵ (in the epicyclic approximation). The tangential caustic is an astroid with four cusps located at $(\pm 4\epsilon b, 0), (0, \pm 4\epsilon b)$. The total cross section for magnification of a stationary source at a fold by more than $\mu \equiv 2/h_t$ (including both images) can be computed to be $2/\mu^2$ times the area of the Einstein ring (for $\mu \gg 1$), and if the source approaches a fold with a proper motion $\dot{\theta}$, then the observed magnification will vary with time according to $\mu = (\pm 3\epsilon\dot{\theta}|t|/b)^{-1/2}$. The total caustic length on one side of the caustic is $24\epsilon b$.

1.2. CUSPS

Three bright images can be observed when the source lies inside a cusp. In addition, as the contours $h_t = \text{constant}$ in the source plane encircle the cusp point and cross themselves, there is a small zone of dimensions $\sim h_t \times h_t^{3/2}$ lying just outside the cusp within which there will be one image formed with magnification $\mu > (h_r h_t)^{-1}$. The integral cross section for cusp magnification diminishes as $h_t^{5/2} \propto \mu^{-5/2}$ and so cusps are less likely to be involved in the highest magnification events of stationary sources than folds. At high enough magnification, the caustic can be treated locally like a fold except in the single image region, where the length scale over which a cusp magnification in excess of μ is created is $\sim 4/\mu$. Hence at high enough magnification ($\mu \gg \epsilon$), we expect that folds will dominate. Similar remarks apply a fortiori to the next simplest catastrophe, the hyperbolic umbilic, though here it is more useful to think of it as a singular point at the intersection of a pair of intersecting caustic sheets.

1.3. MAGNIFICATION LIMITS

A point source can be magnified by an arbitrarily large factor close to a caustic. In practice, three distinct limitations are relevant.

1.3.1. *Source size limit*

The fold magnification equation will break down when the source radius a is comparable with its distance from the caustic, *ie* when

$$\mu \sim \mu_{SS} \equiv \left(\frac{b}{\epsilon a}\right)^{1/2} \tag{3}$$

(where b is now introduced to be the Einstein radius). Note that the larger and more circular the lens, the greater the possible magnification. For a cusp, the peak magnification scales $\propto a^{-2/3}$ and so a greater magnification is possible for a given source.

1.3.2. *Corrugation limit*

This limiting magnification at a fold is inversely proportional to the third derivative of the potential along the direction of maximal linear magnification. If there are additional perturbations, for example those associated with individual galaxies in the case of clusters and individual stars in the case of galaxy lenses, the maximum magnification may be reduced. Suppose that a fraction $f(M)$ of the mass contained within the Einstein ring M_E, is contained in uncorrelated substructures of mass M. The closest such structure is expected to be at a distance $\sim (M/fM_E)^{1/2}b$ and it will dominate the third derivative at the new location of the critical curve and create $\sim f/f_{crit}$ extra images and increase the caustic length if $f \gtrsim f_{crit} \sim (\epsilon^2 M/M_E)^{1/3}$. In this case, fluctuations out to a distance $\sim (fM/\epsilon^2 M_E)^{1/4}b$ will also contribute and the potential will only be smooth on larger scales. This limits the magnification to

$$\mu \sim \mu_{CL} \equiv \mu_{SS}\left(\frac{f}{f_{crit}}\right)^{-3/4} ; \qquad a \lesssim \frac{f^{1/2}f_{crit}^{3/2}b}{\epsilon} \tag{4}$$

If there is substructure on more than one mass scale then this limitation to the magnitude must be applied hierarchically.

1.3.3. *Geometrical optics limit*

Geometrical optics can only be used within the diffraction limit, which depends upon the Fresnel length (*eg* Gould 1992, Ulmer & Goodman 1995). This limits the magnification to

$$\mu \sim \mu_{GO} \equiv \left(\frac{D}{\lambda}\right)^{1/3}\left(\frac{b}{\epsilon}\right)^{2/3} \tag{5}$$

2. Faint Galaxies

2.1. BASIC PROPERTIES

A remarkable discovery of recent years is the large excess of faint galaxies (Tyson 1988). So far, faint galaxy counts on the sky have reached 30 billion (Smail et al. 1995) and the number appears to be still rising at a rate of 2 per magnitude. Another way to express this is to note that the faint galaxies are, on average, about 4″ (or ∼ 5 kpc) apart, much smaller than a nominal galaxy size. It is found that the half power radii of these faint galaxies diminishes with decreasing flux to become ∼ 0.2″ for the faintest galaxies.

2.2. CLUSTER ARCS

Cluster arcs provide a natural way to image the faintest galaxies. The best example to date is A 2218 (Kneib et al. 1995) and here there are over ten well magnified structures that have been resolved in both directions by HST. Typically these arcs are observed over several thousand pixels. It should then be possible to use these observations to solve for the source structure. So far this program has only be carried out using model-fitting (*eg* Worthey et al.'s (1995, preprint) analysis of 0024+1654). However the HST data are sufficiently fine scale that it is worth developing techniques analogous to those used at radio frequencies to invert the images and obtain accurate source structure so that significant morphological information can be extracted.

The next step will be to perform one-dimensional imaging spectroscopy on these arcs (*cf* Pello et al. 1991) using large aperture, ground-based telescopes. In this way it might be possible to extract dynamical information and, given sufficient examples, determine the evolution of empirical relations like the Faber-Jackson relation. The limiting magnification is likely to be corrugation-limited by the individual cluster galaxies which comprise a fraction $f(10^{11}M_\odot) \sim 0.1$ of the total cluster mass within the Einstein ring. In this case, $f_{crit} \sim 0.02$ and we expect that the magnification will be limited to $\mu_{max} \sim 10 - 30$ as, indeed, is observed.

Cluster arcs can exhibit other high magnification effects such as triple imaging of supernovae (Kovner & Paczyński 1988). If the cluster potential were smooth, it would be possible to use the relative magnifications and time-delays to check fundamental scaling relations for cusps, (*eg* the central image flux should equal the sum of the fluxes from the outer images etc). However, it is more likely to provide information about the potential corrugation near the critical curve. Of most importance, though, for our topic is that as there can be notice of a supernova at high redshift which can be monitored intensively during its second or third appearance. In principle, such observations can provide a measurement of Ω_0 using a supernova that would ordinarily be too faint to observe if it were unmagnified.

3. Radio Jets

3.1. BASIC PROPERTIES

Active Galactic Nuclei (AGN) can be conveniently divided into the radio-quiet and radio-loud. The latter class are traditionally sub-divided into compact and extended sources, depending upon whether or not the dominant emission at an intermediate radio frequency comes from the nucleus of the galaxy (or quasar) or radio lobes external to the galaxy (eg Zensus & Pearson 1990). It appears that essentially all powerful radio sources are fueled by a pair of antiparallel, relativistic jets that beam their radio synchrotron emission along their directions of motion. According to the radio unification hypothesis, compact and extended sources belong to the same family with the compact sources being those that are beamed towards us. An extension of this hypothesis posits that the powerful radio galaxies are radio-loud quasars observed at latitudes $\lesssim 45°$ so that their broad emission line regions are obscured (eg Antonucci 1993). Radio jets in compact sources often exhibit core-jet structure and "superluminal expansion". The cores are commonly interpreted as the self-absorbed inner regions of the jets and, consequently have flat radio spectra. The jet emission is optically thin and is probably associated with outwardly propagating relativistic shock fronts.

In recent years, it has been discovered that many compact radio sources exhibit "intraday variability" at cm wavelengths. The significance of this observation is that if one takes the variability time scale as a measure of the light crossing time across the emitting region, then the derived radio brightness temperature can exceed $\sim 10^{18}$ K. This is six orders of magnitude larger than the so-called "inverse Compton" limit. As is well known this limit can be exceeded if there is relativistic expansion. However, the bulk Lorentz factors required are typically ~ 100 and the jet flows are radiatively inefficient. Alternatively, the variability may be due to refractive scintillation (eg Rickett et al. 1995). Several authors have proposed that the emission mechanism is coherent and, consequently, that there is no inverse Compton limit. Here, however, there is a new difficulty in that it is very difficult for radio emission with brightness temperatures as high as $\sim 10^{18}$ K to emerge from an AGN. A very small Thomson optical depth suffices to degrade the radio beam through a combination of induced Compton scattering and stimulated Raman scattering (eg Levinson & Blandford 1995).

3.2. MAGNIFICATION OF RADIO JETS BY INTERVENING GALAXIES

Using galaxy lenses as radio telescopes may clarify some of these issues. Large surveys like CLASS and WENSS should discover tens of multiply-imaged compact sources and the cores of the best cases will undoubtedly be monitored with the VLBA (Myers, these proceedings). Typical magnifications for cores are $\mu \sim 10$, so we ought to resolve them under the synchrotron hypothesis. If there is superluminal expansion with $V_{obs} \sim 10c$, then this will be seen as "hyperluminal expansion" with $V_{obs} \sim 100c$. However, we will not be able to measure higher brightness temperatures. For the brightest observed compact sources, with flux densities $S \sim 10$ Jy, these are limited by $T_{Bmax} \sim SR_{\oplus}^2/k \sim 10^{12}$ K (Levy et al. 1989). As observed core sizes appear to be wavelength-dependent, they will be subject to differential magnification (Connor et al. 1992, Nair, these proceedings).

It would be of great interest to use a galaxy lens to super-resolve a compact radio core. The most information would come from mm VLBI observations. For a source size $a \sim 10^{17}$ cm and $\epsilon \sim 0.1, b \sim 3$ kpc, the maximum observable linear magnification should be $\sim 10^3$. However, in order for a gravitational telescope to achieve this magnification requires that we find a source close enough to a fold caustic. If we adopt a transverse velocity ~ 1000 km s^{-1}, a mm core will only move its own diameter in ~ 30 yr. Therefore adopting the magnification cross section and allowing for selection effects, the maximum magnification of a mm core that we are likely to find is no more than several times the square root of the number of radio rings we discover, at most 30 unless we are very lucky. It is however possible that some of the observed intraday variability is attributable to microlensing variation associated with intervening stars. If there really are ultra-compact coherent sources within radio cores then the $\sim 10\mu s$ time delays associated with stellar microlensing may be detectable by autocorrelating the radio signal (Moore & Hewitt, these proceedings).

The probabilities are somewhat greater for moving features in a jet as these extend to much greater distances particularly at lower radio frequency, typically ~ 10 pc. As the characteristic linear size of a tangential caustic is ~ 300pc, there is a reasonable chance that in one of the radio rings to be discovered, a jet will straddle a fold caustic and features will be seen crossing it on a regular basis.

4. Accretion Disks

4.1. BASIC PROPERTIES

Radio-quiet quasars, and their lower luminosity cousins, the Seyfert galaxies, represent the majority of AGN that do not appear to possess prominent

relativistic jets. The popular supposition, for which there is some observational support, is that these are equatorial outflows of gas with speeds up
to $\sim 30,000$ km s^{-1}, at least in the case of the quasars. When the observer
is located in the equatorial plane a broad absorption line quasar (BALQ) is
seen; otherwise the object is classified as a radio-quiet quasar (*eg* Weymann
et al. 1991).

AGN in general are commonly interpreted in terms of the black hole -
accretion disk model. The circumstantial evidence in favor of this interpretation has accumulated and includes the large required efficiency, observed
rapid variability and dynamical estimates of the central mass, most recently
from observations of H$_2$O masers in M100 (Miyoshi et al. 1995), from X-ray
line profiles in MCG-6-30-15 (Tanaka et al.) and also from M87 (Harms et
al. 1994). Typical black holes masses are believed to lie in the $10^6 - 10^9$ M$_\odot$
range, so their Schwarzschild radii are $\sim 3 \times 10^{11} - 3 \times 10^{14}$ cm. We now
have better grounds to believe that the photoionizing UV and X-ray continuum originates from a very compact region close to the central black hole,
presumably the surface of an accretion disk. However, the failure to detect
predicted signatures of accretion disks such absorption edges, large linear
polarization and radial color variation is troubling (*eg* Antonucci 1993).

The broad emission line clouds, when present, are now thought to be
located over a broad range of radii dependent upon the ionization state,
but centered roughly on $\sim 0.1L_{46}^{1/2}$ pc. This can be estimated using the
technique of reverberation mapping (Peterson *et al* 1995). In the standard
model, a typical cloud has size $\sim 10^{13}$ cm and density $\sim 10^{10}$ cm^{-3}. The
filling factor of the emitting phase is generally very small $\sim 10^{-5}$. The
dynamical state and provenance of these clouds (moving with internal Mach
numbers $\gtrsim 1000$ is problematic, though there are several models (*eg* Netzer
1991).

The broad absorption line gas in quasars is believed to be located exterior to the broad emission line region at typical radii $\sim 10^{19}$ cm. Photoion-
ization arguments suggest densities $\sim 10^8$ cm^{-3} and individual cloud sizes
of only $\sim 10^9$ cm (*eg* Weymann et al. 1991).

4.2. MICROLENSING OBSERVATIONS

One of the most interesting uses of gravitational telescopes has been indi-
rectly in observing quasars that appear to be microlensing, most notably
2237+031. Here, the important deduction is that, in order to exhibit the
observed variable magnification, the size of the accretion disk, believed to
be responsible for the continuum emission must satisfy the source size limit
for an individual star ($\sim 1 - 3 \times 10^{15}$ cm; neither the emission line region
nor the compact radio sources are likely to be be microlensed and this is

one test that has been applied to validate claims of optical microlensing.) In the case of 2237+031, this must happen at the longest wavelength for which microlensing variation has been reported (*ie* 1μ or ~ 4000 Å in the quasar frame), and the black hole must be massive enough and the accretion disk consequently large enough to account for the total bolometric luminosity of the object, corrected for macrolensing by the intervening galaxy. This, in turn, suggests that some of the optical emission may be non-thermal. (See Rauch 1993 for a review of this topic.) In the future, it may be possible perform "microlens mapping" to measure the size of the emitting region as a function of wavelength (Wambsganss & Paczyński 1994). However, the simplest interpretation of the results of reverberation mapping of Seyferts suggests that the size of the emitting region is wavelength-independent (Peterson et al. 1995).

In a variant on this approach, Francis & Koratkar, (these proceedings), have reported that high redshift quasars have lower equivalent widths than low redshift quasars, an effect that they attribute to stellar microlensing preferentially magnifying the continuum. If this interpretation turns out to be correct, it introduces a much larger sample of microlensed sources, whose internal structure can be studied statistically.

4.3. PROBING THE BROAD ABSORPTION LINE REGION

Another interesting possibility is suggested by the observation that at the putative distance of the broad absorption line gas, the continuum source subtends an angular size ~ 1". Three gravitationally lensed quasars are reported to exhibit broad absorption lines UM425, 1115+080 (Michalitsianos & Oliverson, these proceedings) and HR 1413+117 (Kayser et al. 1990) and the typical rays are separated in angle by a comparable amount. Therefore, it is possible that by studying the absorption troughs associated with the individual images we can probe length scales $\sim 10^{14}$ cm in the absorbing region.

4.4. BLACK HOLE LENSES

The ultimate gravitational lens is undoubtedly a black hole and the caustics associated with the Kerr spacetime and a different observer have been calculated by Rauch & Blandford (1994). The best prospects for seeing high magnification are if there is a population of X-ray-emitting, accreting neutron stars orbiting the black hole at high speeds $\sim 0.03c$. Source size-limited magnifications of several thousand are possible, but the caustic crossing rate will be low and difficult to distinguish from regular X-ray variability.

5. Pushing the Envelope

Finally, I address the question of what is the largest magnification that we could imagine observing, granted the physical constraints listed above and observational limitations on sources and lenses (*cf* Miralde Escudé 1991). We are most interested in highly compact sources that satisfy the geometrical optics constraint. This points us towards γ-ray energies. The most compact, steady γ-ray sources that we know of are γ-ray pulsars, where the size of the emitting region is argued to be $a \sim 10^6$ cm. For the lenses, we can choose between clusters (with $M_E \sim 10^{14}$ $M_\odot$, $b \sim 100$ kpc, $\epsilon \sim 0.1$) and galaxies with ($M_E \sim 10^{11}$ $M_\odot$, $b \sim 3$ kpc, $\epsilon \sim 0.1$). In the former case we suppose that there is an intergalactic population of stars with $f \sim 0.01$ so that $f_{\rm crit} \sim 10^{-5}$. Typically the peak magnification is limited to $\mu_{\rm CL} \sim 3 \times 10^6$, while the total length of fold caustic has been lengthened roughly a thousand times to ~ 100 Mpc per cluster. In the latter case, we are in the strong microlensing limit with $f \sim 1$ and the stars are not perturbative. In the absence of a relevant simulation we retain the perturbative approach and estimate that $f_{\rm crit} \sim 10^{-4}$ and that the caustic length is ~ 30 Mpc per galaxy but that $\mu_{\rm CL} \sim 3 \times 10^5$. (A hypothetical population of intergalactic $\sim 10^{10}$ $M_\odot$ black holes would produce comparable effects with less caustic length but higher magnification.) Furthermore, the geometrical optics constraint requires that $\lambda \lesssim 1$ Å for maximal magnification and so the observed flux will be diminished at and below X-ray energies.

All of this is reminiscent of the observed properties of γ-ray bursts which occur at a rate of $R_b \sim 10^{-5}$ s^{-1} with a fluence $E_b \sim 10^{-7}$ erg cm^{-2}. It is possible that some observed bursts are actually persistent γ-ray sources crossing galaxy or cluster caustic surfaces. There are however, two important limitations. Firstly, the total γ-ray flux from the totality of unmagnified sources should not exceed the general γ-ray background flux, $F_{grb} \sim 4 \times 10^{-7}$ erg cm^{-2} s^{-1}. Secondly the intrinsic source luminosities must be large enough that we can see them at the observed fluences. In other words, if the burst duration is t, then $E_b \sim (L_\gamma/4\pi D^2)(b/\epsilon a)^{1/2}(f/f_{\rm crit})^{-3/4}t$. If the optical depth to multiple imaging is τ, and the number of bright γ-ray sources is N_γ, then the expect burst frequency satisfies

$$R_b \sim \frac{N_\gamma \tau}{\pi b^2} \frac{24\epsilon ab}{t} \frac{f}{f_{\rm crit}} \lesssim 24 \frac{F_{grb}}{E_b} \left(\frac{\epsilon a}{b}\right)^{1/2} \left(\frac{f}{f_{\rm crit}}\right)^{1/4} \tau \qquad (6)$$

On this rough basis, the galaxy-induced frequency is $R_b \lesssim 10^{-8}$ s^{-1} and the cluster-induced rate is somewhat smaller. Intergalactic black holes could lead to a larger burst frequency $R_b \lesssim 10^{-7}$ s^{-1}. No more than a minority of observed bursts are likely to be formed in this manner.

There is an interesting way to recognize caustic induced bursts. Interplanetary spacecraft locate bright bursts by timing a plane wave crossing three detectors presumably at the speed of light. Two solutions can be found and one of these is usually found to agree with other directional information. However if a caustic sheet passes through the spacecraft, the deduced wave normal should be orthogonal to the source location. (In a simple lens this caustic will move much slower than the speed of light. However, relative motions in the lens can lead to much faster caustic motion, (eg Wambsganss, these proceedings). A search for examples of caustic-induced bursts would be of interest.

Acknowledgements: We thank Rachel Webster and her colleagues for organizing an excellent meeting and for financial assistance. Helpful conversations with Ian Browne, Jordi Miralda-Escudé and Joachim Wambsganss are gratefully acknowledged. This research was supported by NSF contract AST93-23375.

References

Antonucci, R., 1993, ARAA, 31, 473

Blandford, R. D. & Kovner, I., 1988, Phys Rev A, 38, 4028

Connor, S., Léhar, J. & Burke, B., 1992, ApJ, 387, L61

Gould, A., 1992, ApJ, 386, L5

Harms, R. J. et al., 1994, ApJ, 435, L35

Kayser, R. et al. 1990, ApJ, 364, 15

Kneib, J.-P. et al., 1995, ApJ, in press

Kovner, I. & Paczyński, B., 1988, ApJ, 335, L9

Levinson, A. & Blandford, R. D., 1995, MNRAS, in press

Levy, G. S. et al., 1989, ApJ, 336, 1098

Miralda-Escudé, J., 1991, ApJ, 379, 94

Miyoshi, M. et al., 1995, Nature, 373, 127

Netzer, H., 1991, in Active Galactic Nuclei, eds. Blandford, R. D. Netzer, H. & Woltjer L, (Berlin: Springer-Verlag)

Pello, R. et al., 1991, ApJ, 366, 405

Peterson, B. M. et al., 1995, ApJ, 425, 622

Rauch, K. P., 1993, in Gravitational Lenses in the Universe, eds. J. Surdej et al. (Liège: Université de Liège) 385

Rauch, K. P. & Blandford, R. D., 1994, MNRAS, 421, 46

Rickett, B. J., et al., 1995, A&A, 293, 479

Schneider, P., Ehlers, J. & Falco, E. E., 1992, Gravitational Lensing (Berlin: Springer-Verlag)

Smail, I. R., Hogg, D. W., Yan, L. & Cohen, J. G., 1995, ApJ, in press

Tanaka, Y. et al., 1995, Nature, 375, 659

Tyson, J. A., 1988, AJ, 96, 1

Ulmer, A. & Goodman, J., 1995, ApJ, 442, 67

Wambsganss, J., Paczyński, B., 1994, AJ, 108, 1156

Weymann, R. J., Morris, S. L., Foltz, C. B. & Hewett, P. C., 1991, ApJ, 373, 23

Zensus, J. A. & Pearson, T. J., 1990, in Parsec Scale Radio Jets (Cambridge: Cambridge University Press)

PIXEL LENSING: THE KEY TO THE UNIVERSE

ANDREW GOULD

Dept of Astronomy

Ohio State University, Columbus, OH 43210

e-mail: gould@payne.mps.ohio-state.edu

Alfred P. Sloan Foundation Fellow

Abstract.

   Pixel lensing, the gravitational microlensing of unresolved stars, is potentially a powerful method for detecting and measuring microlensing events. Two groups are currently refining this method in observations toward M31. I show that the technique has wide application, from searching for intracluster Machos in the Virgo cluster, to improving the accuracy of follow-up observations of Galactic microlensing events, to measuring the star-formation history of the universe. I derive the equation for the pixel lensing event rate $\Gamma = (2/Q_{min}^2)\tau N_{res}\Gamma_{\bar{m}}\xi$ where Q_{min} is the minimum signal to noise for detection, τ is the optical depth, N_{res} is the number of telescope resolution elements in the field, $\Gamma_{\bar{m}}$ is the photon detection rate from a fluctuation magnitude star, and ξ is a suppression factor.

1. Introduction

We have heard talks which were described as unabashedly observational and unabashedly theoretical. This talk is unabashed advocacy. Pixel lensing is microlensing of unresolved stars. Or at any rate, it is usually thought of in these terms. Actually, as I will try to convince you, pixel lensing has wide-ranging applications that go far beyond this limited context.

   In classical lensing, such as proposed by Paczyński (1986) and now being carried out by MACHO (Alcock et al. 1993, 1995), EROS (Aubourg et al. 1993, 1995), OGLE (Udalski et al. 1994), and DUO (Allard 1995), one monitors the flux of some star with unmagnified flux F_0. If a microlensing event occurs, the flux is magnified by an amount $A(x) = (x^2 + 2)/\sqrt{x^2(x^2 + 4)}$

C. S. Kochanek and J. N. Hewitt (eds), Astrophysical Applications of Gravitational Lensing, 365–370.

© 1996 IAU

where x is the source–lens separation in units of the Einstein ring radius, θ_e. If the motion of the observer, source, and lens are all uniform, then $x(t; t_0, \beta, \omega) = \sqrt{\omega^2(t - t_0)^2 + \beta^2}$ where t_0 is the time of maximum, β is the impact parameter in units of θ_e and $\omega^{-1} \equiv t_e$ is the characteristic time of the event. From the measured light curve $F(t_i)$, one can hope to fit for the four parameters F_0, t_0, β, and ω according to

$$F(t_i) = F_0 A[x(t_i; t_0, \beta, \omega)]. \tag{1}$$

Crotts (1992) and Baillon et al. (1993) have begun to look for lensing toward M31. Here the potentially lensed stars are much fainter than the integrated light from all M31 stars within the point spread function (PSF) of the lensed star. Hence they look for the change in light within the PSF,

$$\Delta F(t_i) = F_0 \{ A[x(t_i; t_0, \beta, \omega)] - 1 \}, \tag{2}$$

in their ongoing observations toward M31.

Equations (1) and (2) look quite similar, but they have very different statistical properties. The correlation coefficients among the parameters are much larger in (2). The basic reason for this can be seen by focusing on the case (typical for pixel lensing toward M31) when ΔF can be detected only for $x \ll 1$. In this case $\Delta F \simeq F_0[\omega^2(t - t_0)^2 + \beta^2]^{-1/2}$ $= (F_0/\beta)[(\omega/\beta)^2(t - t_0)^2 + 1]^{-1/2}$. Clearly, the non-degenerate parameters which can be extracted are F_0/β, βt_e, and t_0.

2. Why Pixel Lensing Has A Bad Name

Thus, it seems clear that t_e, which is a key piece of information about individual events that can be extracted from classical lensing, cannot be directly measured in pixel lensing. This is very important because the optical depth can be determined simply by summing the observed time scales (appropriately weighted by the efficiencies). In addition, the time scales can be used to estimate the mass function of the lenses (Han & Gould 1995). Instead, in pixel lensing what one measures is the product of the time scale and the impact parameter, a random variable.

In fact, this difference between classical and pixel lensing is an illusion. In classical lensing one must take account of possible blended light B (from a binary companion to the source star, from the lens star, or from some random star in the field). Hence the flux should be written, $F = F_0 A + B$. When this additional parameter is included, the uncertainties in the time scale determination increase dramatically. For comparison, in pixel lensing one must also accurately measure the background flux B', so the true pixel lensing equation is $\Delta F = F_0(A - 1) + B'$. Since these two equations are formally identical, the only real difference between pixel lensing and classical lensing is signal to noise (S/N). If one compares pixel lensed stars in M31 with classically lensed stars in the LMC, then of course the classical lenses will have better signal to noise. But the real question is how do pixel lensing and classical lensing compare when applied to the same stars, or more fundamentally: what is the photon-noise limit of pixel lensing and how can it be achieved?

3. Achieving the Pixel Lensing Photon Limit

Crotts (1992) and AGAPE (Baillon et al. 1993) advocate two different methods of pixel lensing. Crotts convolves his best-seeing image with the current image and then subtracts the result from the current image. He then searches the difference image for PSFs (negative or positive) which would be a signature of a variable star. Note that these PSFs actually extend over several pixels. The variables are then classified into ordinary variables and lensing candidates. AGAPE follow individual pixels (or groups of "super pixels") and looks for variation. Melchior (1995) applied this technique to archival EROS I CCD data (Aubourg et al. 1995) and actually found variable stars that are 2 mag fainter than those followed by EROS in their original lensing study. I should mention, however, that in order to find these variables, she had to first take out the correlation between the pixel variation and the seeing. Seeing, it is clear, is the main problem to be overcome in pixel lensing.

Here I wish to propose a radically different approach to pixel lensing. In lensing studies, one usually accumulates over time dozens or even 100s of images of the same field. Classify these images according to the seeing. Take for example the (say 50) images which have seeing of $1.''30 \pm 0.''05$ and form their median. Then subtract this median from each of the individual images. The difference should contain only photon noise plus PSFs from any variable stars. Of course, in making this assertion I have implicitly assumed that the seeing is adequately characterized by only one number, the FWHM. In general, of course, the seeing disk will also have some ellipticity and may be variable in its radial profile as well. Nevertheless, I can report that I have performed the experiment of subtracting two images of M31 with the same FWHM taken by AGAPE. The seeing disks were not exactly the same, one was elongated vertically, the other horizontally. The result was very striking. The difference looked almost perfectly flat. There were two PSFs from variable stars that were very difficult to detect by eye simply by comparing the two images. Besides these, the remainder of the image looked almost perfectly flat except for photon noise and for an occasional "butterfly", a quadrupole residual where the foreground stars were imperfectly subtracted. This was a very minor problem in the M31 image.

It would be more severe in the LMC where there are more recognizable point sources. Nevertheless, the butterfly residuals are easily distinguished from PSFs and present no fundamental obstacle to finding the variables. I should emphasize that to apply this method the data should be oversampled. In the case of the AGAPE data, there was 1″.5 seeing and 0″.3 pixels, so the images could be very well aligned to a small fraction of a seeing disk using linear interpolation.

4. Applications of the Photon Limit I: M87

Once the photon limit is achieved, pixel lensing opens up many previously inaccessible regions of parameter space. I first consider pixel lensing of M87 or more generally, pixel lensing in the limit where the flux L from typical star is much fainter than the surface brightness S of the galaxy integrated over a resolution element Ω_{PSF}, $L \ll S\Omega_{PSF}$. Now, in this case the signal is $L * (A - 1)\alpha t_*$ where t_* is the integration time and α is the number of photons collected by the telescope per unit time per unit flux. The noise is $\sqrt{S\Omega_{PSF}\alpha t_*}$. Only events with $\beta \ll 1$ will be detectable. For these, the maximum signal will be $\sim L\alpha t_*/\beta$ while the width of the peak will be $\sim \beta/\omega$. Hence the total S/N of the event Q will be $Q^2 \sim \pi L^2 \alpha t_e/(\beta S\Omega_{PSF})$. Note that for fixed minimum S/N, $Q_{\min}$, the maximum impact parameter β for which the event can be detected $\propto L^2 t_e$. Since the event rate $\Gamma_0 = (2/\pi)\omega\tau$, where τ is the optical depth, the detectable event rate $\Gamma = \beta\Gamma_0$ is given by

$$\Gamma = \frac{2}{Q_{\min}^2}\tau\frac{L^2\alpha}{S}\frac{\Omega_{CCD}}{\Omega_{PSF}}, \tag{3}$$

where Ω_{CCD} is the area of the CCD. Of course, this formula is valid only for one class of star of flux L and we don't even know the flux of the star being lensed (although it can be estimated by measuring its color from the color of the lensing event). However, we can integrate equation (3) over the entire luminosity function, in which case $L^2/S \to \bar{L}$, where $\bar{L} = \int dL\phi(L)L^2/\int dL\phi(L)L$ is the "fluctuation flux", the same empirical quantity which is measured in surface-brightness fluctuation distance measurements (Tonry 1991). Hence the total rate from all stars is

$$\Gamma = \frac{2}{Q_{\min}^2}\tau N_{res}\Gamma_{\bar{m}}\xi, \tag{4}$$

where $N_{res} = \Omega_{CCD}/\Omega_{PSF}$ is the number of resolution elements on the CCD and $\Gamma_{\bar{m}}$ is the rate at which the telescope detects photons from a star at the "fluctuation magnitude" (Tonry 1991) (which is equivalent to $\bar{L}$). The correction factor ξ arises from the finite size of the source and the finite size of the Einstein ring and is explained in more detail elsewhere (Gould 1995b).

Equation (4) is important for two reasons. First it shows that the optical depth can be measured directly from measurable quantities. Recall that one criticism of pixel lensing is that because individual event times cannot be measured, it has been thought that the optical depth cannot be determined. Second it shows that lensing studies are potentially far more powerful than their present incarnations. Paczyński (1995) showed earlier this week that lensing events are currently being detected at one per 50–100 observations per τ^{-1}. Equation (4) shows that the limit is one per 50–100 photons.

Pixel lensing observations of M87 would be useful to search for intra-cluster Machos which might have formed in individual Milky-Way-like proto-galaxies in the proto-cluster and then dissolved into the cluster when the proto-galaxy was stripped of its gas (Gould 1995b). Such objects could be detected at a rate $\sim 5\,\mathrm{day}^{-1}$ if they exist. But pixel lensing can be put to less exotic applications.

5. Other Applications

For example, follow up observations of ongoing lensing events are now being made by two groups (Pratt 1995; Sackett 1995) with the aim of finding planetary systems, finding binaries, and measuring the proper motions and perhaps parallaxes of the Machos. At present the data are being analyzed using DoPhot, DAOPhot, and similar crowded-field photometry routines. As is well known, such techniques are generally limited to $\sim 1\%$ precision. But substantially more information could be extracted if more precise photometry could be done. Pixel lensing offers the prospect of photon-limited photometry (on the varying stars only – but that is all we are interested in) so it should be used here in place of traditional techniques. The same goes for the analysis of data from a proposed Macho Parallax Satellite (Refsdal 1966; Gould 1994, 1995a). In such an experiment, it is crucial that differential photometry be carried out in exactly the same way from the satellite and the ground. This is intrinsically very difficult using classical photometric techniques because the satellite PSF is diffraction limited (and hence color dependent) whereas the ground-based PSF is atmosphere limited. However, with pixel lensing it is straightforward.

The pixel lensing technique can be applied very widely. For example, Schechter (1995) has been monitoring close lensed quasar pairs for microlensing and time delays. Such measurements would likely be improved by the pixel technique. Even the basic microlensing searches of the LMC might well profit by applying the pixel technique. Here the argument is not quite so one-sided because the experimenters have generally tried to get critically sampled (not over-sampled) data in order to maximize their area coverage. It is not known how coarsely the data can be sampled before the linear interpolation required for pixel lensing breaks down. However, tests should be done to see if more events can be extracted using pixel lensing.

Finally, I want to mention another application: a search for microlensing among 10^6 quasars. Time prevents me from going into this in detail, but I refer you to my forthcoming paper (Gould 1995c).

Acknowledgements: This work was supported in part by grant NAG5-2864 from NASA.

References

Alcock, C., et al., 1993, Nature, 365, 621

Alcock, C., et al., 1995a, ApJ, 445, 133

Allard, C. 1995, IAU Symposium 173

Aubourg, E., et al., 1993, Nature, 365, 623

Aubourg, E., et al., 1995, A&A, in press

Baillon, P., Bouquet, A., Giraud-Héraud, Y., & Kaplan, J., 1993 A&A, 277, 1

Crotts, A. P. S., 1992, ApJ, 399, L4

Gould, A., 1994, ApJ, 421, L75

Gould, A., 1995a, ApJ, 441, L21

Gould, A., 1995b, ApJ, 455, 000

Gould, A., 1995c, ApJ, 455, 000

Han, C. & Gould, A., 1995, ApJ, submitted

Melchior, A.-L., 1995, Thèse, "Recherche de naines brunes par effet de microlentille gravitationnelle par la méthode des pixels. Analyse des données des collaborations AGAPE et EROS.", Collège de France, Paris

Paczyński, B., 1986, ApJ, 304, 1

Paczyński, B., 1995, IAU Symposium 173

Pratt, M., 1995, IAU Symposium 173

Refsdal, S., 1966, MNRAS, 134, 315

Sackett, P., 1995, IAU Symposium 173

Schechter, P., 1995, IAU Symposium 173

Tonry, J. L., 1991 ApJ, 373, L1

Udalski, A., et al., 1994, Acta Astron, 44, 165

THE STATISTICS OF NEARLY ON-AXIS
GRAVITATIONAL LENSING EVENTS

YUN WANG
NASA/Fermilab Astrophysics Center
Fermi National Accelerator Laboratory, Batavia, IL 60510-0500

Abstract. A small volume of space, nearly on-axis behind a gravitational lens with respect to a given source, will receive a greatly increased radiation flux. In the idealized case of a point mass lens acting on a point source in complete isolation, the volume will approach zero only as the flux tends to infinity; in fact, the volume weighted rms flux is divergent. In realistic cases, finite source size and the effects of other gravitational deflections (i.e., non-zero shear) limit the maximum flux and considerably complicate the physics, but very large fluxes are still produced in small volumes. We consider the physics and statistics of these Extreme Gravitational Lensing Events (EGLE) and present an initial examination of their possible astrophysical effects for various known and putative populations of lensing objects and sources, with particular attention to the case in which finite source size is important but shear is not.

1. Introduction

Nearly on-axis gravitational lensing events are extreme in magnification, and rare in occurrence for a given observer. However, every astrophysical object is a gravitational lens, as well as a receiver/observer of the light from sources lensed by other objects in its neighborhood. Statistically, extreme gravitational lensing events (EGLE) can have an effect on certain fragile objects in the sky, such as interstellar medium, molecular clouds, atoms, dust grains, etc.

An EGLE occurs when a moving observer crosses the line connecting a source and a lens. The maximum magnification of the source seen by the observer can be very large, limited only by the source size and the shear on

C. S. Kochanek and J. N. Hewitt (eds), Astrophysical Applications of Gravitational Lensing, 371–376.

the lens. As the observer moves away from the line connecting the source and the lens, the magnification of the source decreases. The duration of an EGLE depends on the velocity of the observer. A slowly moving observer in the neighborhood of a pair of small-size source and slightly-sheared lens can experience a strong burst of radiation due to the lensing of the source, which can be sufficient to affect the observer's properties.

If the observer moves a distance d away from the line connecting the source and the lens, it is equivalent to the source moving an angular distance of y (in units of the angular Einstein radius θ_E) from the optical axis (the line connecting the lens and the observer). For sufficiently distant lens or source, we have

$$y \simeq \left(\frac{D_{ds}}{D_d}\right)\frac{d}{D_s\theta_E}, \tag{1}$$

where D_{ds}, D_s, and D_d are angular diameter distances between the lens and source, observer and source, observer and lens respectively. The angular Einstein radius

$$\theta_E = \sqrt{\frac{2R_S D_{ds}}{D_d D_s}} = 10^{-6} \times \sqrt{\left(\frac{M_L}{5 \times 10^6 M_\odot}\right)\left(\frac{1\,\text{Mpc}}{D_d}\right)\frac{D_{ds}}{D_s}}, \tag{2}$$

where M_L is the mass of the lens. The dimensionless size of a source with physical radius ρ is defined as

$$R \equiv \frac{\rho}{D_s\theta_E} = \left(\frac{\rho}{3.09 \times 10^{18}\,\text{cm}}\right)\left(\frac{10^{-6}}{\theta_E}\right)\left(\frac{1\,\text{Mpc}}{D_s}\right). \tag{3}$$

For a given pair of lens and source, the shear on the lens is $\gamma \sim \tau$ (Wang & Turner 1995), where τ is the optical depth for microlensing, the probability that the source is lensed. The presence of shear considerably complicates the physics. Since $\tau = \Omega_L z_Q^2/4$ (Turner 1980, Turner, Ostriker & Gott 1984), where Ω_L is the density fraction in lenses and z_Q is the redshift of small sources, the shear γ is probably not important in the low redshift Universe. Here we briefly discuss the case in which the finite size of the source is important but shear on the lens is not, i.e., $\gamma < R$; a detailed discussion is presented in Wang & Turner (1995). For simplification, we only consider high magnification events (i.e., $R \leq 0.05$, $y \ll 1$).

2. Point source

Let us first consider a point source S with luminosity L_S, being lensed by a lens L with Schwartzschild radius R_S (mass M_L) at a distance D_{ds}. In a narrow tube-shaped volume $V_{SL}(f)$ behind the lens, which extends from the lens and tapers off to infinity, the flux from the source exceeds f. The cross section of the tube is $\sigma(f) = \pi d^2$. Hence

$$V_{SL}(f) = \int_0^{D_d(q)} dD_d \, \sigma(f). \tag{4}$$

Summing over S gives the total volume V_L in which the flux exceeds f for a given lens L; further summing over L gives the total volume $V_{tot}(> f)$. We use $D_s = D_d + D_{ds}$ for simplicity in our calculations.

In the absence of magnification, the flux from the source is $f_0 = L_S/(4\pi D_s^2)$. The magnified flux $f = \mu f_0$. We find

$$V_{SL}(f) = \frac{\pi R_S}{D_{ds}^2} \left(\frac{L_S}{4\pi f}\right)^2, \qquad V_L(f) = 4\pi^2 n_S R_S D_c \left(\frac{L_S}{4\pi f}\right)^2, \tag{5}$$

where n_S is the number density of sources, and D_c is the maximum separation between a lens and a source.

Let $\mathcal{F}_L(f)$ be the volume fraction of space in which the flux from the source exceeds f due to gravitational lensing, and $\mathcal{F}_S(f)$ the volume fraction of space in which the flux from the source exceeds f due to being close to the source. We have (Wang & Turner 1995)

$$\mathcal{F}_L(f) = n_L V_L(f) - \frac{3}{2} \tau N_S \left(\frac{f}{f_{min}}\right)^{-2},\qquad \mathcal{F}_S(f) = N_S \left(\frac{f}{f_{min}}\right)^{-3/2},$$
$$\frac{\mathcal{F}_L(f)}{\mathcal{F}_S(f)} = \frac{3}{2}\tau \left(\frac{f}{f_{min}}\right)^{-1/2}, \tag{6}$$

where τ is the optical depth, N_S is the total number of sources, and $f_{min} = L_S/(4\pi D_c^2)$. Note that the volume weighted rms flux due to lensing diverges logarithmically.

3. Finite source

Let us now consider a source S with physical radius ρ and luminosity L_S, being lensed by a lens L with Schwartzschild radius R_S (mass M_L) at a distance D_{ds}. The tube-shaped volume $V_{SL}(f, \rho)$ behind the lens in which the flux from the source exceeds f has finite length $D_d^m(f)$, because of the finite size of the source.

For a finite source with dimensionless size R, $\mu_{max} = 2/R$. Therefore $f = \mu f_0 \leq \mu_{max} f_0$. Let $f = \mu_{max} f_0$ at $D_d = D_d^m(f)$, i.e., the flux is equal to f on the line connecting the source and the lens. For given D_d, the flux decreases away from the line connecting the source and the lens, hence the volume in which the flux exceeds f converges to a point at $D_d = D_d^m(f)$,

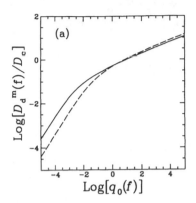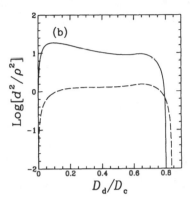

Figure 1. (a) Length of the tube volume $V_{\rm SL}(f, \rho)$, $D_{\rm d}^{\rm m}(f)/D_{\rm c}$; (b) Cross section of the tube volume $V_{\rm SL}(f, \rho)$.

i.e., $D_{\rm d}^{\rm m}(f)$ gives the length of the tube volume in which the flux exceeds f. $D_{\rm d}^{\rm m}(f)$ can be found analytically for arbitrary source size (Wang & Turner 1995).

Let us define a parameter $q_0(f)$ which measures the maximum magnification of the source relative to the minimum flux f,

$$q_0 \equiv \frac{8 R_{\rm S} D_{\rm c}}{\rho^2} \left(\frac{L_{\rm S}}{4\pi D_{\rm c}^2 f} \right)^2. \tag{7}$$

Figure 1(a) shows the length of the tube volume behind the lens in which the flux exceeds f, $D_{\rm d}^{\rm m}(f)/D_{\rm c}$, as function of $q_0(f)$, for $D_{\rm ds} = 0.2\, D_{\rm c}$ (solid line), $0.5\, D_{\rm c}$ (dashed line). The tube length is of order $D_{\rm c}$ for q_0 of order 1. Defining $x \equiv D_{\rm ds}/D_{\rm c}$, we have

$$\frac{D_{\rm d}^{\rm m}(f)}{D_{\rm c}} = \begin{cases} q_0(f)/x^2 & \text{for } q_0^{1/3} \ll x \\ [q_0(f)\, x]^{1/4} & \text{for } q_0^{1/3} \gg x \end{cases} \tag{8}$$

The tube volume $V_{\rm L}(f, \rho)$ has the cross-section $\sigma(f, \rho, D_{\rm d})$ which vanishes at $D_{\rm d} = 0$, $D_{\rm d}^{\rm m}$. We can write

$$\sigma(f, \rho, D_{\rm d}) = \pi \rho^2 \times d^2/\rho^2 = \pi \rho^2\, \bar{\sigma}(q_0, D_{\rm d}), \tag{9}$$

for given $D_{\rm ds}$. Figure 1(b) shows the cross-section $\bar{\sigma}(q_0, D_{\rm d}) = d^2/\rho^2$ with $q_0(f) = 4$, for $D_{\rm ds} = 0.2\, D_{\rm c}$ (solid line), $0.5\, D_{\rm c}$ (dashed line). The lens which is closest to the source has the thickest tube of high flux behind it.

The volume fraction in which the flux from the source exceeds f is

$$\mathcal{F}_{\rm L}(f, \rho) = 4\pi^2 n_{\rm L} n_{\rm S} R_{\rm S} D_{\rm c} \left(\frac{L_{\rm S}}{4\pi f} \right)^2 I(q_0), \tag{10}$$

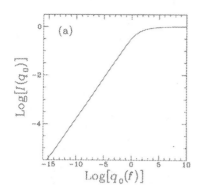 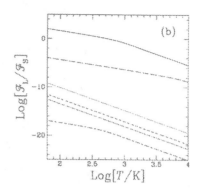

Figure 2. (a) Ratio of volume fractions, $I(q_0) = \mathcal{F}_L(f,\rho)/\mathcal{F}_L(f,\rho = 0)$; (b) Volume fractions for different sets of sources and lenses.

where $I(q_0) = \mathcal{F}_L(f,\rho)/\mathcal{F}_L(f,\rho = 0)$. To estimate $I(q_0)$, we can approximate the magnification of a finite source with the point source magnification cut off at $\mu_{max} = 2/R$; this leads to a slight underestimate of $I(q_0)$, but it is sufficiently accurate for all practical purposes. $I(q_0)$ is plotted in Figure 2(a). For $q_0(f) \ll 1$, $I(q_0) = 0.3583\, q_0^{1/3}$; for $q_0 \gg 1$, $I(q_0) \simeq 1$ (Wang & Turner 1995).

4. Possible astrophysical effects

In the context of EGLE, the relevant dimensional physical quantities are: lens mass M_L, size of the lens-source distribution D_c, source size ρ, source luminosity L_S, minimum lensed flux from the source f. All these collapse into a single dimensionless parameter $q_0(f)$ (see Eq.(7)), which can be written as

$$\log(q_0) = -5.62 + 0.8\, m_{bol} + 2\log\left(\frac{L_S}{L_\odot}\right) + \log\left(\frac{M_L}{M_\odot}\right) - 2\log\left(\frac{\rho}{R_\odot}\right)$$
$$-3\log\left(\frac{D_c}{1\,\text{kpc}}\right) \tag{11}$$

where the minimum flux f is measured by m_{bol}. For gamma-ray bursts, $\log(q_0) = 26 + 0.8\, m_{bol}$ (lensed by stars), and $\log(q_0) = 34 + 0.8\, m_{bol}$ (lensed by giant black holes).

Let us consider a population of sources (with number density n_S) lensed by a population of lenses (with number density n_L). For a given source, it induces one high flux tube behind each lens; for a given lens, it has one high flux tube coming out of it because of each source.

Let $\mathcal{F}_L(f)$ and $\mathcal{F}_S(f)$ denote the total volume fractions of space in which the flux from the source exceeds f due to gravitational lensing and due to being close to the source respectively. We have (Wang & Turner 1995)

$$\log \mathcal{F}_{\mathrm{I}} = -8 + \log I(q_0) + 0.8m_{\mathrm{bol}} + 2\log\left(\frac{L_{\mathrm{S}}}{L_\odot}\right) + \log\left(\frac{M_{\mathrm{L}}}{M_\odot}\right)$$
$$+ \log\left(\frac{D_{\mathrm{c}}}{1\mathrm{kpc}}\right) + \log\left(\frac{n_{\mathrm{S}}}{1\mathrm{pc}^{-3}}\right) + \log\left(\frac{n_{\mathrm{L}}}{1\mathrm{pc}^{-3}}\right)$$
$$\log\left(\frac{\mathcal{F}_{\mathrm{L}}}{\mathcal{F}_{\mathrm{S}}}\right) = -8.86 + \log I(q_0) + 0.2m_{\mathrm{bol}} + \frac{1}{2}\log\left(\frac{L_{\mathrm{S}}}{L_\odot}\right) + \log\left(\frac{M_{\mathrm{L}}}{M_\odot}\right)$$
$$+ \log\left(\frac{D_{\mathrm{c}}}{1\mathrm{kpc}}\right) + \log\left(\frac{n_{\mathrm{L}}}{1\mathrm{pc}^{-3}}\right) \tag{12}$$

where the minimum flux f is measured by $m_{\mathrm{bol}} = -10\log(T/0.56K)$; T is the minimum blackbody temperature of dust grains in the tube volume $V_{\mathrm{SL}}(> f)$. In Figure 2(b), we show $\mathcal{F}_{\mathrm{L}}/\mathcal{F}_{\mathrm{S}}$ versus T for gamma-ray bursts lensed by stars (solid line) and by giant black holes (long dashed line), QSO's (X-ray) lensed by stars (dotted line) and by giant black holes (dot-short dashed line), neutron stars lensed by stars (short dashed line) and by giant black holes (short-long dashed line). Only orders of magnitude are used for the sources and lenses in Figure 2(b). The largest effect comes from gamma-ray bursts lensed by stars in our simple model (zero shear).

Finally, we note that although the volume fractions of high flux due to lensing are small, the corresponding absolute volumes can be large. Further, since materials move across the high flux tubes constantly, the fraction of material affected by EGLE is much higher than the static volume fractions. We are still investigating various possible astrophysical applications.

Acknowledgements: This work is done in collaboration with Ed Turner. The author is supported by the DOE and NASA under Grant NAG5-2788.

References

Turner, E.L., 1980, ApJ, 242, L135
Turner, E.L., Ostriker, J.P. & Gott, J.R., 1984, ApJ, 284, 1
Wang, Y. & Turner, E.L., 1995, in preparation

GRAVITATIONAL LENSING BY CURVED COSMIC STRINGS

MALCOLM R. ANDERSON

Department of Mathematics
Edith Cowan University
2 Bradford St., Mt Lawley
WA 6050 Australia

Abstract. I briefly summarize all that is known about gravitational lensing by cosmic strings.

Cosmic strings are long thin filaments of Higgs field energy whose existence is predicted by a wide range of grand unified field theories (Vilenkin 1985). The earliest studies of cosmic string networks suggested that string loops could provide a simple and natural mechanism for the seeding of galaxy clusters (Zel'dovich 1980; Vilenkin 1981), but more recent numerical simulations indicate that networks have more small-scale power than first thought, leading to an excess of low-mass high-velocity loops which are unlikely to seed galaxy formation (Albrecht & Turok 1989, Bennett & Bouchet 1990).

Although cosmic strings have thus fallen into disfavor as cosmogonic agents, they may still have formed and played some role in the evolution of the early Universe. Apart from their characteristic Kaiser-Stebbins signature in the cosmic microwave background (Kaiser & Stebbins 1984) the best chance of detecting them seems to be through gravitational lensing.

The type of lensing behavior expected in the neighborhood of a cosmic string can be straightforward or extremely complex, depending on the geometric configuration of the string involved. The simplest possible model is the infinite straight string (Gott 1985; Hiscock 1985) which produces two unmagnified and undistorted images of any source whose line-of-sight passes sufficiently close to the string. The separation of the two images is proportional to the mass per unit length of the string. For a GUT string, the images would typically be separated by a few arcseconds.

C. S. Kochanek and J. N. Hewitt (eds), Astrophysical Applications of Gravitational Lensing, 377–378.
© 1996 IAU

However, numerical simulations suggest that strings have structure on all scales down to a characteristic radiative damping length. Realistic models need to take this small-scale structure into account. The first attempt to model lensing by string loops was made by Hogan and Narayan (1984), who assumed that the rays passing through the interior of a loop are undeflected while those passing outside it experience a spherical potential due to the mass of the string, plus a quadrupole moment. Although the model is very simplistic, it can give rise to one, two or three images, both magnified and unmagnified, plus caustics with often very complex geometry.

Unlike the metric outside a straight string – which is locally flat – the metric outside a curved horizon-sized string is characterized by the superposition of two distinct families of gravitational waves: traveling waves and curvature waves (Anderson 1995). Traveling waves, which are decoupled from the geometry of the string and can have arbitrary shape, were first discovered by Garfinkle (1990). Gravitational lensing by traveling waves has been examined numerically by Vollick and Unruh (1991), who found that they typically form double images whose separation varies as the wave components cross the line-of-sight.

Curvature waves are more interesting. Very difficult to treat mathematically, they are generated in response to high-curvature structure on the string. The only known exact solution containing curvature waves is the Aryal-Ford-Vilenkin (1986) metric, which describes a straight string passing through a Schwarzschild hole. It combines the lensing properties of the Schwarzschild metric and the straight string, and can produce up to four images. It is likely that the lensing effect of more general curvature waves would be similar to that of a series of black holes, each with mass proportional to the local curvature, threaded on a straight string. Their signature would thus be two pairs of two images, spaced symmetrically about the axis of the string, showing high variability in separation and magnitude.

References

Albrecht, A., & Turok, N., 1989, Phys Rev D, 40, 973
Anderson, M.R., 1995, Aust J Phys, in press
Aryal, M., Ford, L.H., & Vilenkin, A., 1986, Phys Rev D, 34, 2263
Bennett, D.P., & Bouchet, F.R., 1990, Phys Rev D, 41, 2408
Garfinkle, D., 1990, Phys Rev D, 41, 1112
Gott, J.R., 1985, ApJ, 288, 422
Hiscock, W., 1985, Phys Rev D, 31, 3288
Hogan, C., & Narayan, R., 1984, MNRAS, 211, 575
Kaiser, N., & Stebbins, A., 1984, Nature, 310, 391
Vilenkin, A., 1981, Phys Rev Lett, 46, 1169
Vilenkin, A., 1985, Phys Rep, 121, 263
Vollick, D.N., & Unruh, W.G., 1991, Phys Rev D, 44, 2388
Zel'dovich, Ya. B., 1980, MNRAS, 192, 663

COMPACT DOUBLES: TESTING THE LENSING HYPOTHESIS

GOPAL-KRISHNA AND KANDASWAMY SUBRAMANIAN

NCRA-TIFR, Poona University Campus, Pune 411007, India.

Abstract. If a compact double (CD) is caused by lensing, the orientation and/or flux ratio of its components can substantially change in ≈ 1 year.

1. Compact doubles as gravitationally lensed images

Major surveys using VLBA are underway to find lensed radio sources with separations $\approx 1 - 100$ milliarcseconds (e.g., Patnaik et al. 1995; Wilkinson 1995). Further, it has been suggested that some of the known CDs (Phillips & Mutel 1982) may in fact be such lensed images (e.g., Ostriker 1995). The required lensing agents of $\sim 10^6 - 10^7 M_\odot$ may be remnants of a generation of pregalactic stars or dwarf galaxies. Since the individual components of CDs are ≈ 1 mas, the lensed object is likely to be the radio core of a quasar. Such cores are identified with the base of a relativistic jet pointed roughly towards us (Blandford & Konigl 1979), which is itself often resolved by VLBI into one or more bright emission knots apparently separating at superluminal velocities, $v \sim 5 - 10c$, from a nucleus near the jet's origin (e.g., Vermeulen & Cohen 1994). We suggest that such superluminal motion of radio knot(s) in the lensed source can lead to significant structural variations in the small-separation images, even on time scales of ≈ 1 yr. VLBI monitoring can thus help in distinguishing any (*milli-*) lensed CD images from genuine CDs.

2. Temporal effects in the lensing scenario

For evaluating the temporal changes we concentrate on just the superluminal components (knots); they would dominate the core structure except in the strongly self-absorbed spectral regime. Consider a superluminal knot being multiply imaged by a point-like lens of mass M. The angular positions

C. S. Kochanek and J. N. Hewitt (eds), Astrophysical Applications of Gravitational Lensing, 379–380.

TABLE 1. Evolution of PA (Φ) and flux-ratio (R) for two ejection directions (ψ).

	$\psi = 60°$		$\psi = 120°$			$\psi = 60°$		$\psi = 120°$	
t (yr)	Φ (deg)	R	Φ (deg)	R	t (yr)	Φ (deg)	R	Φ (deg)	R
0	0	1.5	0	1.5	3	37	2.4	79	1.7
1	19	1.7	30	1.4	4	41	2.8	90	2.0
2	30	2.0	60	1.5	5	44	3.4	97	2.5

of the two images, α, measured from the lens (which is the origin of our co-ordinate system) are related to the source position β by : $\alpha^2 - \beta\alpha - \theta_L^2 = 0$. Here $\theta_L = 5(D_{LS}/D_L)^{1/2}(D_S/1000\text{Mpc})^{-1/2}(M/3 \times 10^6 M_\odot)^{1/2}$ mas is the Einstein ring radius, and D_S, D_L and D_{LS} are the source, lens and lens-source angular-diameter distances. For a $3 \times 10^6 M_\odot$ lens located roughly midway between us and a source ~ 1000 Mpc away the image separation $\Delta\alpha \sim 2\theta_L \sim 10$ mas, which is characteristic of CDs (Carvalho 1985). Another characteristic is the flux ratio of the components, $R \sim 1$ to 2 (note that $R \simeq 1.5$ if $\beta \simeq \theta_L/5$).

Now, suppose an emission knot initially located on the x-axis at $x = x_A$, moves with $v = \gamma c$ at an angle ψ to the x-axis. After a time t : $\beta^2(t) = x_A^2 + 2l x_A \cos\psi + l^2$, where $l = \dot\theta t = \gamma ct/D_S$ with $\dot\theta = 0.5(\gamma/8)(D_S/1000\text{Mpc})^{-1}$ mas/yr. The flux ratio is then: $R(t) = (2 + u(t)^{1/2} + u(t)^{-1/2})/(u^{1/2} + u^{-1/2} - 2)$, where $u = [1 + 4\theta_L^2/\beta^2]$. The position angle (or PA) $\Phi(t)$, of the line joining the two images relative to the x-axis varies as : $\tan\Phi(t) = (l \sin\psi)/(x_A + l\cos\psi)$. The evolution of these quantities is given in Table 1, assuming characteristic values $\theta_L = 5$ mas, $x_A = \theta_L/5 = 1$ mas and $\dot\theta = 0.5$ mas yr^{-1} (see above). Clearly, substantial variations can occur in the PA and R of the CD, if indeed it results from gravitational milli-lensing of a quasar core. Hence, by imaging a CD with VLBI arrays even ≈ 1 year apart one can verify if it is merely an illusion caused due to milli-lensing.

References

Blandford, R.D. & Konigl, A., 1979, ApJ, 232, 34.
Carvalho, J.C., 1985, MNRAS, 215, 463
Ostriker, J., 1995, in Quasars & AGN: High Resolution Radio Imaging, eds. K. Keller-mann & M. Cohen (Proc. Nat. Acad. Sci.), in press.
Patnaik,A.R., Garrett, M.A., Polatidis, A. & Bagri, D.S., 1995, these proceedings
Phillips, R. B.& Mutel, R. L., 1982, A&A, 106, 21
Vermeulen, R.C. & Cohen, M.H., 1994, ApJ, 430, 467
Wilkinson, P.N., 1995, in Quasars & AGN: High Resolution Radio Imaging, eds. K. Kellermann & M. Cohen (Proc. Nat. Acad. Sci.), in press

THE QUASAR LUMINOSITY FUNCTION

PAUL C. HEWETT

Institute of Astronomy

Madingley Road, Cambridge CB3 0HA, United Kingdom

Abstract. Recent results at high redshifts are briefly reviewed. An analysis of the evolution of the quasar luminosity function based on the recently published Hawkins & Véron (1995) variability–selected quasar sample is presented. The results support the conclusions of earlier work that suggested the simple two power law luminosity function, of invariant shape as a function of redshift, requires some modification in order to match the data. However, the contention of Hawkins & Véron that a single power law model provides an adequate fit to the variability–selected quasar sample and that there is no requirement for curvature in the form of the quasar luminosity function is not borne out by the analysis.

1. Introduction

The number of talks and posters at the 1995 Melbourne IAU Symposium 173, *Astrophysical applications of Gravitational Lensing*, focusing on the quasar population, whether in the context of strong lensing (multiple images) or weak lensing (quasar–galaxy associations), is significantly smaller than in the previous biennial gravitational lensing meetings. Their place has been taken by reports of new results in areas such as the weak lensing of background sources by galaxy clusters, complete with superb Hubble Space Telescope images, and the extraordinarily impressive results from the various consortia investigating the lensing of stars within our own Galaxy.

The reduced level of reporting that relates to quasar population statistics reflects what could be perceived by many as a lack of recent high–profile activity in the area. Published surveys containing 10^3 quasars now exist and large areas of the redshift–absolute magnitude plane are relatively well populated. Compiling new samples of $\sim 10^3$ objects to improve population

C. S. Kochanek and J. N. Hewitt (eds), Astrophysical Applications of Gravitational Lensing, 381–386.
© 1996 IAU

statistics, or undertaking campaigns to improve the quality of observational data (at all wavelengths), do not appear to be receiving high priority. This is perhaps due in part to the realization that observational capabilities are soon to increase by an order of magnitude or more. Planned facilities, notably the 2dF wide field multi–object spectrograph on the Anglo Australian Telescope, and the SLOAN project, will generate quasar samples with $10^4 - 10^5$ objects, dramatically improving the statistics for studies of the evolution, environments and clustering of the quasar population. These projects are likely to have most impact, initially at least, within the apparent magnitude range $16 < m_B < 22$ for objects that are, via their unusual broadband colors for example, relatively easy to detect.

Regions of the redshift–absolute magnitude parameter space not well covered remain so for good reason. In some cases observational campaigns of staggering size are required. For example, essentially all the know high–luminosity quasars at significant redshifts have been imaged to search for multiple imaging. However, while quasar surveys at the brightest magnitudes are underway, the very low surface densities and the consequent effort required to compile samples of even tens of quasars at $m < 16$, or hundreds at $m < 17.5$, means results are very slow to appear. Other portions of the redshift–absolute magnitude plane remain effectively inaccessible. The compilation of a large sample of intermediate luminosity quasars, $M_B \sim -24$, at redshifts $z = 2 - 3$, for example, requires the light gathering power and multi–object spectroscopic capability of the 8m class telescopes now under construction.

Notwithstanding this rather downbeat assessment of current activity there are important results appearing. Two of these, the recent progress at high–redshifts, and the apparent inconsistency between new results from a large variability–selected sample and the generally accepted picture of quasar evolution, are discussed in this short review.

2. The High–Redshift Quasar Luminosity Function

The most distant identifiable collapsed objects, and their use as direct or indirect probes of the early universe, continue to excite interest among astronomers and, notwithstanding the growing number of high–redshift radio galaxies, quasars remain the population identifiable to the highest redshifts. Activity in the 1980s was characterized by a long series of mutually incompatible claims concerning the behavior of the quasar luminosity function at redshifts $z > 2.5$ and in particular at redshifts $z > 4$. The latter years of the decade saw the instigation of major surveys for quasars at $z \geq 3$. The design of these experiments specifically included, in addition to the capability of simply detecting quasars, the ability to derive the intrinsic space density of quasars as a function of absolute magnitude and redshift.

The observational phase of the surveys of Schneider, Schmidt & Gunn (1994) and of Warren, Hewett & Osmer (1991) were completed some years ago. An indication of the scale and complexity of calculating a precisely specified selection function for experiments of this type comes from the length of the interval before the groups published their quantitative constraints on the form and normalization of the quasar luminosity function (Warren, Hewett & Osmer 1994, Schmidt, Schneider & Gunn 1995). Both groups find evidence for a significant decline in the space density of quasars by $z = 4$ although the preferred forms of parameterization are quantitatively different. The agreement between the two surveys is illustrated graphically in Figure 5 of Schmidt et al. (1995). The outcome of the high–redshift investigations has demonstrated that complex experiments of this nature can be quantitatively understood and characterized, allowing consistent conclusions to be obtained via entirely independent experimental approaches. Schmidt et al. (1995) summarize the situation: "The excellent agreement between our $z = 4$ luminosity function and that of Warren et al. (1994) is remarkable given the very different methods employed". Identical analysis techniques could now be applied using (some) existing as well as new quasar surveys to refine our understanding of the evolution of the quasar luminosity function. At lower redshifts, $z < 2$, utilizing the numerically larger samples of quasars available to begin investigating the evolution of the quasar population as a function of quasar spectral energy distribution is a key goal (see Hewett & Foltz 1994).

3. The Hawkins' Variability–Selected Quasar Sample

Hawkins has compiled a unique sample of quasars selected on the basis of their long–term (months to years) photometric variability in the observed–frame B_J band. Spectroscopically confirmed subsamples of quasars from this work have been available for some time, e.g. Hawkins & Véron (1993), however, the definition of the published samples employed a flux–limit related to the minimum magnitude achieved by objects over a specified time scale. Although, given the significant level of photometric variability exhibited by the quasars, such a definition has a number of advantages, it has not proved possible to perform a direct comparison of the variability–selected sample with other quasar samples defined using a flux limit at a single epoch. Recently Hawkins & Véron (1995) published an extensive sample of 315 variability–selected quasars from a single epoch flux–limited object catalogue, making a comparison with other samples possible.

Hawkins & Véron conclude that the new sample confirms their earlier claims (e.g. Hawkins & Véron 1993) that there is no evidence for the ex-

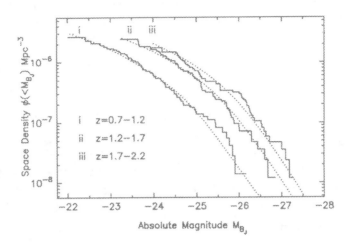

Figure 1. Cumulative luminosity functions in three redshift shells calculated using the Hawkins & Véron (1995) variability selected sample together with the predictions of a two–power law luminosity function model.

istence of any features in either the quasar number counts or the quasar luminosity function, and they model the quasar luminosity function, calculated for a number of redshift "shells", using a simple power–law form, $\phi = 10^{\beta(M-M^*)}$. Their conclusions contrast dramatically with the generally accepted view, based particularly on the data of Boyle et al. (1990), that the quasar luminosity function exhibits a distinct change in shape as a function of absolute magnitude, requiring a model with (for example) two power-laws of significantly different slope (see Boyle 1993 for a recent review). The potential relevance for gravitational lensing of Hawkins & Véron's claim is considerable, aside from the substantial implications for the evolution of the quasar population.

The very steep bright–end slope of the differential quasar luminosity function, $N_{qso}(L) \propto L^{-3.6}$, in the Boyle parameterization and the rapid transition to a much shallower slope, $N_{qso}(L) \propto L^{-1.5}$, means that strong amplification by gravitational lensing is potentially important at bright magnitudes but becomes essentially insignificant once close to the break luminosity. Hawkins & Véron's model, which corresponds to $N_{qso}(L) \propto L^{-2.58}$, would result in amplification by lensing proving to be only moderately important but present for essentially all luminosities. Wu (1994) illustrates the significantly different predictions for the number of quasar-galaxy associations that result from the two models.

Figure 1 shows a cumulative representation of the quasar luminosity function calculated using the Hawkins & Véron sample together with the predictions of a Boyle et al. (1990) two–power law luminosity function un-

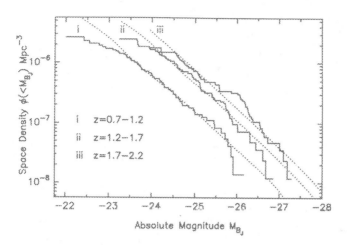

Figure 2. Cumulative luminosity functions in three redshift shells calculated using the Hawkins & Véron (1995) variability selected sample together with the predictions of a single power law luminosity function model.

dergoing pure luminosity evolution model. A cosmology with $q_0 = 0.5$, $H_0 = 50$ km s^{-1} Mpc^{-1} and $\Lambda = 0.0$ was employed. The quasar k–corrections were calculated using the composite quasar spectrum of Cristiani and Vio (1990). None of the conclusions below are sensitive to the adopted (conventional) cosmology or the details of the k–corrections.

Luminosity functions were calculated over three redshift shells $z = 0.7 - 1.2$ (57 quasars), $1.2 - 1.7$ (63 quasars) and $1.7 - 2.2$ (51 quasars). The solid lines in the figure are derived from calculating accessible volume for each quasar, ordering the quasars by absolute magnitude and plotting the contribution to the space density ($= 1/V_{acc}$). Such cumulative plots have significant advantages over the more prevalent differential representations, particularly when considering samples with small numbers of objects where differential representations necessitate binning over large intervals of absolute magnitude. Note that the contribution of individual objects to the space density is evident as a vertical step at the absolute magnitude of the object. The dashed lines are the predicted distributions of the run of $\Sigma 1/V_{acc}$ for a two–power law luminosity function model:

$$\phi\,(M_{B_J}, z)\,dM_{B_J}\,dz = \frac{\phi^*\,dM_{B_J}\,dz}{[10^{0.4(M_{B_J} - M_{B_J}(z))(\alpha+1)} + 10^{0.4(M_{B_J} - M_{B_J}(z))(\beta+1)}]}$$

with a pure–luminosity evolution of the characteristic magnitude, $M_{B_J}(z)$:

$$M_{B_J}(z) = M_{B_J}^* - 2.5 k_L \log(1 + z) \quad z < z_{max}$$

$$M_{B_J}(z) = M_{B_J}(z_{max}) \qquad\qquad z > z_{max}$$

with parameters: $\phi^* = 5.5 \times 10^{-7}\,\text{mag}^{-1}\,\text{Mpc}^{-3}$, $\alpha = -3.9$, $\beta = -1.5$, $M_{B_J}^* = -22.5$, $k_L = 3.45$ and $z_{max} = 1.7$.

No attempt to achieve a "best–fit" has been made. The two–power law model appears to represent the Hawkins & Véron data well, particularly the overall curvature. The data in the lowest redshift bin $z = 0.7 - 1.2$ appear to show some evidence for an excess of bright quasars and an improved model fit would be obtained by reducing the amplitude of the slope change in the two power law model. The sense of the deviations from the two–power law model are similar to those found by Hewett, Foltz & Chaffee (1993) whose analysis of the LBQS sample showed the quasar luminosity function steepening with increasing redshift.

Hawkins & Véron propose a model for the quasar luminosity function that corresponds to setting $\alpha = \beta = -2.58$. They do not specify what form of evolutionary model they favor and for purposes of illustration the pure luminosity evolution model has been retained. A slightly smaller exponent, $k_L = 3.20$, is necessary in order to match the observed space densities. The results are shown in Figure 2. The deviations between model and data appear to be larger than for the two–power law model predictions, although a single power law of the slope advocated by Hawkins & Véron does match substantial portions of the data.

The analysis of the Hawkins & Véron sample supports the results of earlier work that suggested the simple two power law luminosity function of invariant shape requires some modification. However, the contention that a single power law model provides an adequate fit to the Hawkins variability-selected quasar sample and that there is no requirement for curvature in the form of the quasar luminosity function is not borne out by the analysis presented here.

References

Boyle, B.J., 1993, in The Evolution of Galaxies and their Environment ed. H. Thronson and M. Shull (Dordrecht, Kluwer), 433

Boyle, B.J., Fong, R., Shanks, T. & Peterson, B.A., 1990, MNRAS, 243, 1

Cristiani, S. & Vio, R., 1990, A&A, 227, 385

Hawkins, M.R.S.H. & Véron, P., 1993, MNRAS, 260, 202

Hawkins, M.R.S.H. & Véron, P., 1995, MNRAS, 275, 1102

Hewett, P.C. & Foltz, C.B., 1994, PASP, 106,113

Hewett, P.C., Foltz, C.B. & Chaffee, F.H., 1993, ApJL, 406, L43

Hewett, P.C., Foltz, C.B. & Chaffee, F.H., 1995, AJ, 109, 1498

Schmidt, M., Schneider, D.P. & Gunn, J.E., 1995, AJ, 110, 68

Schneider, D.P., Schmidt, M. & Gunn, J.E., 1994, AJ, 107, 1245

Warren, S.J., Hewett, P.C. & Osmer, P.S., 1991, ApJS, 76, 23

Warren, S.J., Hewett, P.C. & Osmer, P.S., 1994, ApJ, 421, 412

Wu, X–P., 1994, A&A, 286, 748

GROUND-BASED AND HST DIRECT IMAGING OF HLQS

J. SURDEJ

STScI, 3700 San Martin Drive, Baltimore, MD 21218, USA,
Member of the Astrophysics Division,
Space Science Department of the European Space Agency,
and Directeur de Recherches au FNRS, Belgium

A.O. JAUNSEN

Nordic Optical Telescope, Ap. 474, S/C de La Palma,
E-38700 Canarias, Spain

J.-F. CLAESKENS

European Southern Observatory (Chile) and Aspirant au FNRS
Institut d'Astrophysique, Université de Liège,
Avenue de Cointe 5, B-4000 Liège, Belgium

S. GONZAGA AND A. POSPIESZALSKA-SURDEJ

STScI, 3700 San Martin Drive, Baltimore, MD 21218, USA

B. PIRENNE

ST-ECF, c/o ESO, Karl-Schwarzschild Str. 2,
D-85748 Gar-ching bei München, Germany

AND

A. PRIETO

MPI für Extraterrestrische Physik, Giessenbachstrasse,
D-85748 Garching bei München, Germany

Abstract. In the context of our studies on gravitational lensing effects among Highly Luminous Quasars (HLQs), we are presently compiling at STScI an archive of direct CCD frames for more than 1000 bright quasars observed with HST and ground-based telescopes. This archive will soon become publicly accessible through the Internet. On the basis of these observations, we are pursuing in a systematic way the analysis (subtraction of numerical PSFs and/or deconvolution) of the HLQ images in order to detect multiple QSO images and/or nearby foreground galaxies at very small angular separations. Residual images corresponding to several new possi-

C. S. Kochanek and J. N. Hewitt (eds), Astrophysical Applications of Gravitational Lensing, 387–392.

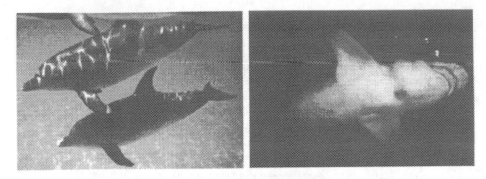

Figure 1. Sunlight caustics projected on dolphins (a) and a shark (b) observing multiple images of a background source (February 1994, Sea World in Orlando, Florida)

ble multiply imaged HLQs are presented here. From the observed number and image configuration of gravitational lens candidates identified in this large sample of HLQs, it is possible to infer realistic values for parameters characterizing the galaxy deflectors, the number counts of quasars, etc. (cf. Claeskens et al. 1995ab), and also to set interesting constraints on the cosmological density of compact objects in the mass range 10^{10} - 10^{12} $M_\odot$.

1. Preamble: dolphins, sharks, caustics and multiple images

Figure 1a illustrates very well defined caustics projected on the bodies of swimming dolphins at Sea World (Orlando, Florida). Of course, these caustics result from the complex lensing of sunlight at the wavy interface between air and water. In Fig. 1b, we have reproduced a photograph showing the multiple images of a background source as seen by a shark passing very near such a caustic. There is thus little doubt that dolphins, sharks and many other fishes had witnessed the formation of multiple images of distant background sources -namely the Sun and the Moon- well before 1979. Because of the similarity between the size of these caustics and the distance between the eyes of the big fishes, these animals are also probably aware of the parallax effect. However, it is highly unlikely that dolphins and sharks have the slightest idea of what a 'time delay' is, so that this very concept solely remains a pure astronomical invention (Refsdal 1964).

2. Introduction

Following several interesting discoveries of multiply imaged quasars (cf. UM673 AB, Surdej et al. 1987; H1413+117 A-D, Magain et al. 1988; 1208+1011 AB, Magain et al. 1992; 1009-025 AB, Surdej et al. 1993a) within a sample of HLQs (typically $M_V < -27$ for $H_0 = 50$ km s^{-1} Mpc^{-1} and $q_0 = 0.5$), our group has decided to systemize its search for gravitational lensing effects by carefully reanalyzing all direct CCD frames obtained with the ESO 2.2m, NTT, NOT, CFH, Pic-du-Midi 2m and HST telescopes. We presently have in our hands high angular resolution direct CCD frames for approximately 800 distinct quasars and we should obtain CCD frames for about 200 more HLQs during the coming year (planned observations at the Pic-du-Midi, NOT and also with HST during cycle 5). Let us note that most of the well known gravitational lens (GL) systems have been found from a direct visual examination of ground-based CCD frames. A first exception is the case of 1208+1011 A and B (angular separation of 0.45"), identified from ground-based direct images (seeing of 1.2", pixel size 0.41") following an efficient numerical PSF subtraction (Magain et al. 1992). Note that this GL has been independently detected in the HST quasar snapshot survey (Bahcall et al. 1992). Of course, by taking advantage of better angular resolution observations, the identification of sub-arcsec separation GLs among HLQs ought to significantly improve the statistical inference of parameters characterizing galaxy deflectors (cf. the velocity dispersion σ^*), counts of quasars and also the geometrical structure of the Universe (see Claeskens et al. 1995ab). Furthermore, because of the well known amplification bias affecting flux limited samples of quasars, one should detect a significant excess of bright (R < 22) foreground galaxies in the vicinity (typically < 3") of HLQs. Such an excess of galaxies has been reported by Van Drom et al. (1993), but its statistical significance needs to be improved further. Finally, the incidence of secondary lensed QSO images detected at very small angular separation from the primary ones may lead to interesting constraints on the cosmological density of compact objects (cf. Surdej et al. 1993b). All these arguments very clearly demonstrate the importance of reanalysing in a systematic way (subtraction of numerical PSF, image decomposition, deconvolution, etc. as opposed to a mere visual examination) the direct CCD frames of all HLQs imaged with ground-based telescopes and HST.

3. Search for very compact GLs among HLQs

Claeskens, Jaunsen and Surdej (in these proceedings) describe in detail how gravitational lensing statistics were applied to the direct imagery of a sample of more than 1000 HLQs to constrain the values of galactic parameters, the number counts of quasars and the cosmological constant. In order to efficiently perform such applications, it is not only important to have access to a large sample of quasars but also to make use of observations characterized by an optimal angular selection function (ASF; see

Surdej et al. (1993b) and Jaunsen et al. (1995) for a more detailed account
of the ASFs). In some way, the quality of an ASF does actually measure
our ability to resolve two QSO images on a CCD frame having a magnitude
difference (resp. an angular separation) that is as large (resp. as small) as
possible. An easy way to improve the ASF of a given observation consists
of course in subtracting from each HLQ image a numerical point spread
function (PSF), constructed from nearby stars selected on the CCD frame.
Provided that the PSF is spatially well sampled and that it is characterized by a good signal-to-noise ratio, this method is quite insensitive to the
image quality (cf. seeing and/or elongation due to telescope motions, etc.).
It is also possible to achieve very accurate relative photometry. If no such
comparison stars are available, the ASF will then not only be limited by the
seeing and the sampling of the CCD images but also by our visual ability
to separate nearby QSO components.

For approximately 200 quasars observed with the NOT, 250 quasars observed at ESO and 50 quasars observed within the HST snapshot survey, it
has been possible to define very good numerical PSFs (this work is still in
progress). We have proceeded as follows: in addition to a careful reduction
of all the ground-based data (through the determination of the bias, RON
and gain of the CCDs, test of CCD linearity, bias and flat field corrections),
we have determined on each CCD frame appropriate regions for sky estimates (void of objects and cosmic rays). Sky subtraction (modeled by a
2-D polynomial of 2nd degree in X and Y) has then been applied and stars
were selected to construct a good numerical PSF. Automatic batches perform accurate numerical PSF subtractions, including a weighting scheme
that takes into account the CCD RON, Poisson noise in the sky, stars and
the QSO, altogether with bi-quadratic rebinning and recentering of the individual PSF reference stars and HLQ images. This procedure is iterated
until we get an optimal PSF (i.e. excluding composite or extended stars,
etc.). Special decompositions of HLQ images into multiple components are
performed for the interesting cases. In the course of this project, we have
realized that all the resulting bias and flat field corrected CCD frames could
form a new public archive, since most of these ground-based CCD frames
are of good quality (good seeing, good sampling of the PSF, etc.) and could
be useful to the whole astronomical community for various other scientific
purposes (cf. relative photometry, astrometry, counts of galaxies, etc.). We
are thus presently checking, updating or creating FITS parameters, head-
ers, etc. for the CCD frames that will become part of the HLQs archive.
This archive should become accessible through the Internet by the end of
1996. We welcome very much observers to contact us in order to possibly
integrate their direct CCD frames of bright quasars in the HLQs archive.

From the HST snapshot observations, we have been able to construct

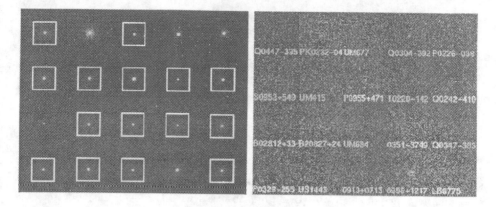

Figure 2. Direct images (**a, left**) and residuals (**b, right**) scaled in σ unit, for 20 selected HLQs observed with HST. These residuals were obtained by subtracting from each QSO image a scaled numerical PSF constructed from those quasar images surrounded by a large white square (see text)

accurate numerical PSFs for three sets of 23, 20 and 14 quasars imaged at the same location on the CCD planetary camera and observed during time intervals free from any telescope focus readjustment. Fig. 2 illustrates for the set of 20 quasars observed with HST the direct images (a: left) and the residuals (b: right; normalized in σ unit) left over after subtracting an average numerical PSF constructed from those quasars surrounded by a large white square. The 19th object in Fig. 2b, identified as 0956+1217, displays very significant residuals possibly due to the superposition of a foreground galaxy. Similar analyses of ESO (Fig. 3a) and NOT (Fig. 3b) observations have revealed several interesting gravitational lens candidates with very small angular separations. The residuals shown in Figs. 3ab are very symptomatic of double point-like images. These GL candidates will be further studied with HST during cycle 5.

4. New constraints on the density of compact objects

In order to significantly improve existing upper limits on the density Ω_L (measured in units of the critical density Ω_0) of putative lensing compact objects in the Universe with mass M_L, we have followed Surdej et al. (1993b) and made use of the observed number of multiply imaged quasars (5 confirmed GLs and 6 other candidates; cf. Figs. 3ab for two of the latter ones) in the sample of 1178 HLQs assembled by Claeskens et al. (1995ab) as well as of the observed angular separation (≤ 3") and magnitude difference (≤ 5) between their multiple images. At a 99.7% confidence level for $H_0 = 50$ km sec^{-1} Mpc^{-1}, $\Omega_0 = 1$ and $\Lambda = 0$, we find that $\Omega_L \leq 0.017, 0.005$ and 0.018 for $M_L = 10^{10}, 10^{11}$ and 10^{12} $M_\odot$, respectively. These are presently

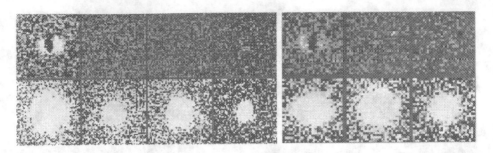

Figure 3. Direct images (lower row) and residuals (upper row) scaled in units of σ, for the new GL candidate J03.13 (**a, left**) identified by Claeskens et al. (1995c and d) at ESO and another GL candidate (**b, right**) observed with the NOT. In the latter case, the HLQ has been found to be compatible with two point-like components separated by 0.56" and a magnitude difference of 0.2. The direct images (lower row) and residuals (upper row) of the PSF reference stars are shown on the right side of their respective HLQ ones.

the best constraints existing on Ω_L in the mass range 10^{10} - 10^{12} $M_\odot$.

References

Bahcall, J.N., Hartig, G.F., Jannuzi, B.T., Maoz, D., & Schneider, D.P., 1992, ApJL, 400, L51

Claeskens, J.-F., Jaunsen, A., & Surdej, J., 1995a, in preparation

Claeskens, J.-F., Jaunsen, A., & Surdej, J., 1995b, in these proceedings

Claeskens, J.-F., Surdej, J., & Remy, M., 1995c, A&A, submitted

Claeskens, J.-F., Surdej, J., & Remy, M., 1995d, in these proceedings

Jaunsen, A.O., Jablonski, M., Pettersen, B.R., & Stabell, R., 1995, A&A, in press

Magain, P., Surdej, J., Swings, J.P., Borgeest, U., Kayser, R., Kühr, H., Refsdal, & S., Remy, M., 1988, Nature, 334, 325

Magain, P., Surdej, J., Vanderriest, C., Pirenne, B., & Hutsemékers, D., 1992, A&A, 253, L13, (Erratum in A&A, 272, 383)

Refsdal, S., 1964, MNRAS, 128, 307

Surdej, J., Magain, P., Swings, J.P., Borgeest, U., Courvoisier, T.J.-L., Kayser, R., Kellermann, K.I., Kühr, H., & Refsdal, S., 1987, Nature, 329, 695

Surdej, J., Remy, M., Smette, A., Claeskens, J.-F., Magain, P., Refsdal, S., Swings, J.P., & Véron, M., 1993a, in Gravitational Lenses in the Universe, eds. J. Surdej, D. Fraipont-Caro, E. Gosset, S. Refsdal and M.Remy, (Liège: Université de Liège) 153

Surdej, J., Claeskens, J.F., Crampton, D., Filippenko, A.V., Hutsemékers, D., Magain, P., Pirenne, B., Vanderriest, C., & Yee, H.K.C., 1993b, AJ, 105, 2064

Van Drom, E., Surdej, J., Magain, P., Hutsemékers, D., Gosset, E., Claeskens, J.-F., Shaver, P., & Melnick, J., 1993, in Gravitational Lenses in the Universe, eds. J. Surdej, D. Fraipont-Caro, E. Gosset, S. Refsdal and M.Remy, (Liège: Université de Liège) 301

THE PARKES LENS SURVEY

R. L. WEBSTER, P. J. FRANCIS, B. A. HOLMAN, F. J. MASCI
School of Physics,
University of Melbourne,
Parkville, Victoria, 3052, Australia

M. J. DRINKWATER
Anglo-Australian Observatory,
Coonabarabran, NSW, 2357, Australia

AND

B. A. PETERSON
Mount Stromlo and Siding Springs Observatory,
Private Bag, Weston Creek,
ACT, 2601, Australia

Abstract. We are undertaking an extensive survey of a sample of radio-selected flat-spectrum sources, which are predominantly quasars. The sample will be used for a range of gravitational lensing studies as well as studies of the generic properties of radio quasars. As yet we have not found any instances of multiply-imaged quasars, however we have found evidence for dust which we believe is reddening the observed spectra of many of the quasars. Samples which are selected to be complete in the radio are the best way to account for the effects of dust.

1. Introduction

Studies of the statistics of gravitational lensing require well-defined samples of source objects, such as quasars. The selection criteria must be specified carefully so that it is possible to determine which manifestations of gravitational lensing might have been detected. In the past, statistical samples have sometimes been chosen from heterogeneous collections of objects such as the Hewett-Burbidge catalogue of quasars. For such studies, it is almost impossible to assess the completeness of the detection limits.

C. S. Kochanek and J. N. Hewitt (eds), Astrophysical Applications of Gravitational Lensing, 393–398.
© 1996 IAU

Types of statistical studies include the numbers of lenses and their distribution with separation, observed magnitude in both optical and radio flux, and image configurations. In future more detailed studies might include other parameters such as variability and spectral characteristics, including evidence of reddening and emission-line strengths. Measurements of the statistics of quasar-galaxy associations require carefully chosen target source lists. For example, closely aligned quasars and galaxies are usually classified as extended galaxies using a simple automated selection procedure. More importantly, sources selected for UV-excess will be biased against quasars behind dusty galaxies. It is very difficult to model the effects of dust, as there is still controversy about the extent and distribution of dust in both spiral and elliptical galaxies. Finally, complete samples of multiply-imaged quasars provide a strong constraint on non-zero values of λ as discussed by other contributions in these proceedings.

2. The Quasar Sample

We are compiling a complete sample of radio-selected quasars. The sample (Parkes Half-Jansky Flat-Spectrum Sample (PHFS)) comprises 323 sources with fluxes > 0.5 Jy at 2.7 GHz. The sources have been chosen to have flat radio spectra: if the radio spectrum from 2.7 GHz to 5.0 GHz is defined by $S(\nu) \propto \nu^{\alpha}$, then sources with $\alpha > -0.5$ are selected. It is well-known that such sources are predominantly quasars, rather than radio galaxies, which are more common amongst the steep-spectrum sources. In addition, the sample was restricted to sources away from the galactic plane, with $|b| > 20 \deg$, and with declinations in the range $-45 \deg$ to $+10 \deg$. At present, a total of 258 of the sources have redshifts. Details of the sample will be published shortly (Drinkwater et al. 1996).

Figure 1 shows the redshift distribution of the PHFS compared to a sample of quasars selected by their optical spectra, the Large Bright QSO Survey (Hewett et al. 1995). The two redshift distributions are essentially similar.

Identification of the radio sources has required positions which are accurate to about one arcsecond. These have been obtained using the Australia Telescope Compact Array. The radio positions have then been mapped onto COSMOS maps of the sky. For 92% of the sources, an optical counterpart is located very close to the radio position; however the remaining sources are fainter than the optical limit of $b_J \sim 22.5$ (Drinkwater et al. 1996).

We have conjectured that these sources are reddened, and therefore their b_J magnitudes are fainter than the COSMOS limit. We are imaging all sources in the sample in the K-band, and are finding that many of the sources which are faint at optical magnitudes, are many magnitudes

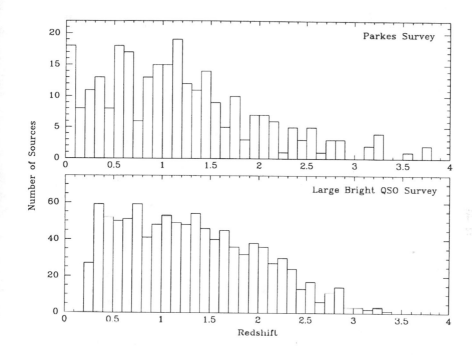

Figure 1. The top histogram shows the distribution of redshifts for the PHFS, and the lower panel shows the redshift histogram for the Large Bright QSO Survey.

brighter in the near-infrared. We measure $b_J - K$ colors ranging from 2 – 10 magnitudes. The redness might either be intrinsic, whereby the source actually has a red continuum, or the source may have been reddened by intervening dust. A strong argument in favor of the latter may be made by considering a plot of emission-line strength against reddening. If the quasar continuum is intrinsically reddened, then an anti-correlation would be predicted between redness and the emission-line strengths. The available data shows no such correlation (Webster et al. 1995).

We are able to compare the sample of radio-selected quasars, with a subsample of QSOs from the Large Bright QSO Survey, for which we have near-infrared data. The latter QSOs, which are largely selected for UV-excess, lie on the blue envelope of the $b_J - K$ distribution and have $b_J - K$ colors centered around 2.5 magnitudes. Our sample of radio-selected quasars are not biased against dust, and will thus reflect intrinsic quasar colors. Thus we are able to estimate the number of QSOs which would be missed from optically-selected samples, if they have the same reddening distribution as the radio-selected sample. We find that $\sim 80\%$ of quasars might be missing from these samples (Webster et al. 1995).

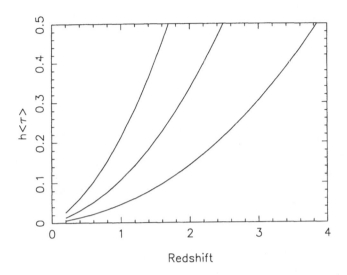

Figure 2.   Plot of the mean optical depth in dust (multiplied by the Hubble scaling factor) plotted as function of redshift, for three different galaxy scale lengths.

3. Dust-Effected Quasars

There are three possible locations for dust: a line-of-sight galaxy, the presumed host galaxy of the quasar and the quasar environs. We can model the effects of dust in intervening galaxies, and predict the dependence of color on redshift. This effect has been modeled by a number of researchers (see Masci & Webster (1995) and references therein), who predict a strong dependence with increasing redshift if the dust properties of the galaxies do not evolve with redshift. The actual probability of reddening, which is essentially the optical depth in dust, depends strongly on the assumed parameters for the dust distribution, in particular the spatial extent of the dust.

Generally we expect the lensing galaxies for multiply-images quasars to be strongly biased towards elliptical galaxies, as these have relatively higher velocity dispersions. We have therefore modeled the predicted mean optical depth in dust as a function of redshift for a population of elliptical galaxies. The elliptical galaxies are assumed to have a constant comoving number density of $0.006\,Mpc^{-3}$, and the dust is distributed like the stars, following a Hubble-type law: $\tau(r) = \tau_B(1 + r/r_0)^{-2}$, where the central optical depth in dust is taken to be $\tau_B = 2$, and the scale length r_0 is allowed to vary. Figure 2 shows the mean optical depth in dust as a function of redshift for $r_0 = 10, 20, 40$ kpc. For each model the distribution of optical depths in dust is essentially Poissonian, with a substantial tail to high optical depths.

We have constructed a mean quasar spectrum from $1000 - 12,000$ Å.

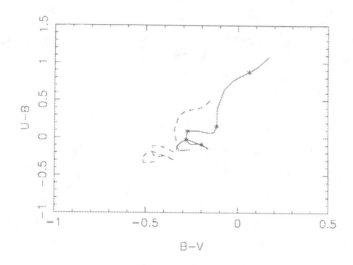

Figure 3. The dotted line shows the two color plot for a quasar with the generic spectrum, over the redshift range 0 − 2.8 (higher redshift quasars having more positive colors). The solid line shows the two color plot of the same quasar spectrum reddened at source by $\tau = 0.4$ of dust. Redshift intervals of 0.5 are marked. Stars occupy the top left corner of the plot.

The section of the spectrum from 1000 − 6000 Å is taken from Francis et al. (1991). The red-end of the spectrum has been constructed using a subsample of quasars from the Large Bright QSO Survey; these quasars have been selected principally by their UV-excess, and therefore are not appreciably reddened. The spectrum is a least squares fit to the broadband magnitudes of ∼ 30 of these quasars (Holman 1996). The generic spectrum is redshifted and convolved with the appropriate filters to construct the two-color plots. The quasar spectrum is then reddened using the mean reddening law derived by (Calzetti et al. 1994).

There are a number of techniques used to separate quasars from stars. In multicolor data, quasars candidates are those sources which lie in the low density parts of the multicolor plots. Thus if the quasars are reddened, and move into the heavily populated stellar locus, they are not detected. Such objects may still be detected by their emission lines.

Figure 3 shows the predicted colors of a generic quasar as a function of redshift. The dashed line is the normal *blue* quasar. The solid line shows the same quasar spectrum reddened by $\tau = 0.4$ in dust located at the same redshift as the source, though the calculation can be readily extended to dust located at intervening redshifts.

4. Conclusions

We have found no candidates gravitational lenses on scales of several arc-seconds in our radio-selected sample. However the data is not yet complete enough for us to be able to put strong limits on the lack of observed lensing.

We have found strong evidence for dust-effected quasars. In most cases, it is unlikely that the dust is in the line-of-sight to the quasars. However since the line-of-sight through most galaxies, including ellipticals, will be dusty, these lines-of-sight may be under-represented in QSO samples which are selected by UV-excess techniques. These quasars are also the ones which we expect to be either macro- or microlensed. Thus optically-selected quasar samples may be biased against lensed quasars. Quantification of this bias is difficult. However we note that lists of lensed quasars are over-represented by radio quasars, which are not biased against reddening by dust.

References

Holman, B.A., 1995, MSc Thesis, University of Melbourne

Calzetti, D., Kinney, A.L. & Storchi-Bergmann, T., 1994, ApJ, 429, 582

Drinkwater, M.J., Savage, A., Webster, R.L., Condon, J.J., Ellison, S.L., Francis, P.J., Jauncey, D.L., Lovell, J. & Peterson, B.A., 1996, MNRAS, submitted

Francis, P.J., Hewett,P.C., Foltz, C.B., Chaffee, F.H., Weymann, R.J. & Morris, S.L., 1991, ApJ, 373, 465

Hewett, P.C., Foltz, C.B. & Chaffee, F.H., 1995, ApJ, 109, 1498

Masci, F.J. & Webster, R.L., 1995, Proc Astron Soc Aust, in press

Webster, R.L., Francis, P.J., Peterson, B.A., Drinkwater, M.J. & Masci, F.J., 1995, Nature, 375, 469

A VLA/MERLIN/VLBA FOR INTERMEDIATE SCALE LENSES AND THE DISCOVERY OF A NEW LENS SYSTEM?

P. AUGUSTO, P.N. WILKINSON AND I.W.A. BROWNE

Nuffield Radio Astronomy Laboratories, Univ. of Manchester
Jodrell Bank, Macclesfield, Cheshire, SK11 9DL, U.K.

We are searching for small lens systems (50-250 mas or $10^8 - 10^9\ M_\odot$) in a sample of $\sim$ 1800 flat spectrum radio sources. This is the first time a systematic search has been made "between" the VLA and VLBI resolutions. Finding any would indicate the existence of other than the "conventional" spiral/elliptical lenses (only $\sim$ 0.01% chance - Turner et al. (1984)). For example, *faint galaxies* are numerous ($\sim 10^6$ gal/deg^2 - Lilly (1993), Glazebrook et al. 1995), compact (HST Medium Deep Survey (MDS) - Griffiths et al. 1994) and ideally placed for lensing ($< z > \sim$ 0.6 - MDS, Smail et al. (1994); *c.f.* Turner et al. 1984). Early-type *dwarf galaxies* (dE,N and cE), if extant at intermediate-z as favored by MDS are also obvious lens candidates. If no lenses are found, a limit 400 times better than the current one (Surdej et al. 1993), $\Omega_L < 0.001$, will be placed on the cosmological density of compact objects (*e.g.* black holes) for the above mass range.

From the JVAS (Patnaik et al. 1992) and the Cosmic Lens All Sky Survey (CLASS) (Browne et al. this volume) - we selected a parent sample of $\sim$ 1800 objects satisfying $\alpha_{1.4-5} < 0.50$ ($S_\nu \propto \nu^{-\alpha}$), $S_{8.4}^{total} \geq 100$ mJy and $|b^{II}| \geq 10°$. The 67 "candidates" were selected from the visibility plots (the sources >50 mas) and constitute an interesting population: two thirds of the candidates have $\alpha \sim 0.3 - 0.4$ whereas the parent sample has $< \alpha > = 0.0$; in addition, the candidate sample is clearly dominated by empty fields (40%) whereas the parent sample has 60% of QSOs.

Our 67 candidates were mapped with MERLIN "snapshots" at 5 GHz (4 $\times$ 15 min.). About 20 sources look very interesting for subsequent VLBA follow-up; the great majority of the remaining appear to be "core-jets".

One arcsecond scale source, 2114+022, looked promising enough for long track MERLIN observations (see Fig.1) and it is a good lens candidate despite its very strange configuration; there will be a multi-frequency VLBA follow-up.

C. S. Kochanek and J. N. Hewitt (eds), Astrophysical Applications of Gravitational Lensing, 399–400.

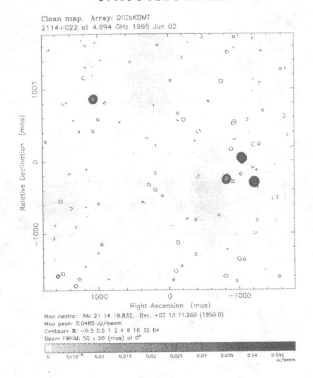

Figure 1. MERLIN map of 2114+022 at 5 GHz (11hrs. phase-referencing). The noise level is very low (< 0.1 mJy) and there are no signs of any extended structure; all the four images are unresolved. The (variable?) fluxes of the components are, starting from the east (left): 20 mJy (21 mJy - snapshot 10 days after), 13 mJy (10 mJy), 50 mJy (45 mJy), 66 mJy (60 mJy); the spectral indexes are: $\alpha_{1.6-5} = 0.32$ and $\alpha_{5-8.4} = 0.95$, $\alpha_{1.6-5} = -0.30$ and $\alpha_{5-8.4} = -0.87$, $\alpha_{1.6-5} = -0.65$ and $\alpha_{5-8.4} = 0.42$, $\alpha_{1.6-5} = 0.03$ and $\alpha_{5-8.4} = 0.78$. The optical identification (POSS) - 19^m elliptical red "blob" (optical counter-parts?) - is only 21° misaligned with the NE-SW radio-alignment.

Acknowledgements: P. Augusto is supported by a grant from JNICT/EC PRAXIS XXI.

References

Glazebrook, K., Ellis, R., Colless, M.M., Broadhurst, T.J., Allington-Smith, J., & Tanvir, N., 1995, MNRAS, 273, 157

Griffiths, R.E., Ratnatunga, K.U., Neuschaefer, L.W., Casertano, S., et al., 1994, ApJ, 437, 67

Lilly, S.J., 1993, ApJ, 411, 501

Patnaik,A.R., Browne,I.W.A., Wilkinson,P.N., & Wrobel,J.M., 1992, MNRAS, 254, 655

Smail, I., Ellis, R.S., & Fitchett, M.J., 1994, MNRAS, 270, 245

Surdej, J., Claeskens, J.F., Crampton, D., Fillipenko, A.V., et al., 1993, AJ, 105, 2064

Turner, E.L., Ostriker, J.P., Gott, J.R., 1984, ApJ, 284, 1

PRELIMINARY VLA SNAPSHOTS OF SOUTHERN RADIO SOURCES FROM THE PARKES-MIT-NRAO (PMN) SURVEY

A.FLETCHER, B.BURKE, S.CONNER, L.HEROLD, A.COORAY, D.HAARSMA, F.CRAWFORD AND J.CARTWRIGHT
Massachusetts Institute of Technology
26-344 M.I.T., 77 Mass. Ave., Cambridge, MA 02139, USA.
EMAIL: bfb@maggie.mit.edu

Abstract. Selection criteria for 1800 MIT-VLA snapshots of PMN radio sources are described, and 6 new MG & PMN lens candidates are presented.

1. Sample Selection

The Parkes-MIT-NRAO (PMN) Southern Hemisphere Sky Survey revealed $36,640$ sources over $\Omega = 4.51$ sr, and is $> 95\%$ complete and $> 90\%$ reliable down to $S_{4.85GHz} \approx 35$ mJy (Griffith et al. 1995). We have made 1800 $0.25''$-resolution 8.4 GHz VLA snapshots of PMN sources ($S_{4.85GHz} > 90$ mJy) in the $-30° < \delta < 0°$ strip ($|b| > 10°$). This sample is essentially complete down to $S_{4.85GHz} \approx 200$ mJy, and is divided roughly equally into a flat-spectrum sample ($S_\nu \propto \nu^{-\alpha}; \alpha < 0.5$), and a purely flux-limited one. Our Northern VLA campaign in the $0° < \delta < 37°$ strip of the MIT-Greenbank (MG) Surveys (Griffith et al. 1991) produced 5 confirmed lenses from $\approx$ 4000 snapshots: MG2016, MG1131, MG0414, MG1654 and MG1549.

2. New Results

Improvements in the MIT mapping pipeline (Conner et al. 1992) have uncovered several more good candidates, yielding a lensing frequency of $\approx 1/500$. Our initial candidate selection is by radio morphology. The 6 most promising cases from a new crop of 8.4 GHz MG & PMN snapshots are shown in Fig. 1. Optical R band imaging with the Michigan-Dartmouth-MIT 1.3m telescope has secured identifications ($R > 22.5$) for all 6 candi-

C. S. Kochanek and J. N. Hewitt (eds), Astrophysical Applications of Gravitational Lensing, 401–402.
© 1996 IAU.

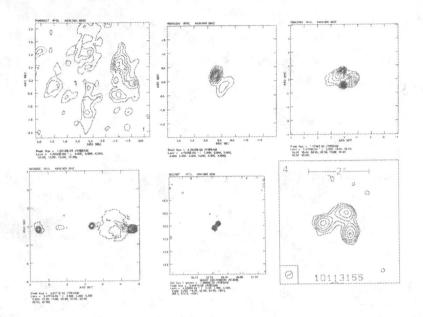

Figure 1.   VLA 8.4 GHz plots of 3 new PMN & 3 new MG gravitational lens candidates.

dates. MG0246 is almost certainly an Einstein Ring. MG1507 is just one of ≈ 130 'close doubles' $(0.25'' < \theta < 2.0'')$ in the 8.4 GHz database. It is not yet understood what these tiny radio galaxies are physically; it is likely that they are the lobes of distant, young or 'frustrated' classical doubles, but perhaps a few are doubly-imaged background radio sources. Theoretical predictions (Turner et al. 1984) that small angular size lenses should exist have been corroborated by several recent discoveries, e.g. the $0.33''$ ring B0218+35.7 (Patnaik et al. 1993). Further optical imaging and spectroscopy is required to investigate the lensing hypothesis for these new MIT candidates.

Acknowledgements: This work was supported in part by an NSF Grant. The VLA is run by the National Radio Astronomy Observatory, which is operated by Associated Universities, Inc., under cooperative agreement with the National Science Foundation.

References

Conner, S.R., Fletcher, A., Herold, L., & Burke, B.F., 1992, in Sub-arcsecond Radio Astronomy, eds. R.J. Davis & R.S. Booth (Cambridge: Cambridge Univ. Press), 154
Griffith, M.R., Langston, G., Heflin, M., Conner, S., & Burke, B., 1991, ApJS, 75, 801
Griffith, M.R., Wright, A.E., Burke, B.F., & Ekers, R.D., 1995, ApJS, 97, 347
Patnaik, A.R., Browne, I.W.A., King, L.J., Muxlow, T.W.B., et al., MNRAS, 261, 435
Turner, E., Ostriker, J., & Gott, R., 1984, ApJ, 284, 1

A RADIO SURVEY FOR GRAVITATIONAL LENSES IN THE SOUTHERN HEMISPHERE

J.E.J. LOVELL AND P.M. MCCULLOCH
Department of Physics, University of Tasmania, Australia

AND

D.L. JAUNCEY
Australia Telescope National Facility, CSIRO, Sydney, Australia

We are undertaking an imaging survey with the Australia Telescope Compact Array (ATCA) to find gravitational lens candidates in flat-spectrum Parkes Catalogue radio sources. Flat-spectrum radio sources typically possess a single high brightness temperature nucleus of milliarcsecond size. Such sources, if lensed, will show multiply imaged nuclei with separations that are large compared to their milliarcsecond sizes. Our flat-spectrum sample was selected using the criteria $\alpha_{2.7/5.0} > -0.5$ ($S(\nu) \propto \nu^{\alpha}$), $S_{2.7} > 0.34$Jy and $\delta \leq -20°$, and comprises a total of 461 sources.

Survey observations were made with the ATCA in "cuts" mode (with typically 7 "cuts" per source) at 3 and 6 cm simultaneously. The 3 cm observations allow images to be made at 1 arcsec resolution which, when combined with the 6 cm data, enable spectral index information to be obtained. Simulations show that a 1 min "cut" per source every 2 h over a ~12 h period is sufficient to detect a close (~1 arcsec) double. This method of observation allowed us to observe ~80 sources per day.

The data were edited and calibrated within AIPS and imaged using the Caltech Difmap program (Shepherd et al. 1995). The final self-calibrated images yielded typical dynamic ranges in excess of 100:1. The ATCA data provide source positions with sub-arcsec accuracy and so make the identification of optical counterparts feasible. The COSMOS/UKST Southern Sky Catalogue (see Drinkwater et al. (1995) for a description) has proven invaluable in this task.

To date we have identified three lens candidates, images of which are shown in Figure 1 and described below. Follow up work, including observa-

C. S. Kochanek and J. N. Hewitt (eds), Astrophysical Applications of Gravitational Lensing, 403–404.
© 1996 IAU

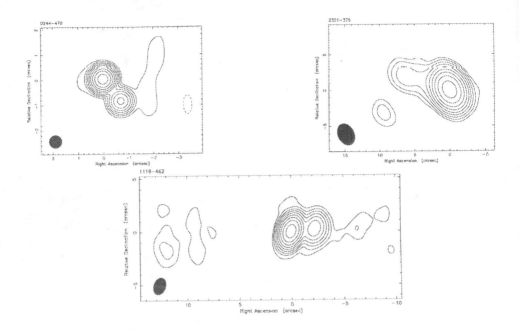

Figure 1. A 3 cm image of PKS 0244-470 super-resolved to reveal the compact nature of the two components (top left), PKS 2321-375 at 6 cm (top right) and PKS 1116-462 at 3 cm (bottom). Contour levels are 0.25,0.5,1,2,4,8,16,32,64 % of the peak for each source.

tions with the Australian Long Baseline Array and optical CCD imaging, is being carried out on these objects.

PKS 0244–470 is a 1 arcsec compact double. The flux density ratio of the two components is approximately 3:1 with a peak flux density of ∼1 Jy. Both components possess a flat spectrum, suggesting that this object may be a doubly imaged quasar.

PKS 1116–462 appears as a flat-spectrum double source at the ATCA, thus making it a radio lens candidate, and yet appears as a point source at I, V and B bands at the Anglo Australian Telescope. This may indicate that, if the source is a gravitational lens, then one component may be reddened by an intervening galaxy.

PKS 2321–375 consists of a compact flat-spectrum core with a steep spectrum "jet" to the east and a weak flat-spectrum component to the south-east of the "jet". This object could be in a lensing system where the compact core of a quasar has been doubly imaged.

References

Drinkwater, M., Barnes, & Ellison, 1995, PASA, in press
Shepherd, M.C., Pearson, T.J., & Taylor, G.B., 1994, BAAS, 26, 987

A VLBA 15 GHZ SMALL SEPARATION GRAVITATIONAL LENS SURVEY

A.R. PATNAIK
MPIfR, Bonn, Germany

M.A. GARRETT AND A. POLATIDIS
NRAL, Jodrell Bank, UK

AND

D. BAGRI
NRAO, Socorro, USA

1. Introduction

We have embarked on a 15 GHz VLBA survey of 1000 flat spectrum sources. We present the results from a 24 hour pilot observing run in which 72 sources were mapped. The primary aims of this project are:

- to search for small separation (1-150 mas) gravitational lens systems
- to identify targets for current *mm* and anticipated Space VLBI programs
- a morphological classification of compact radio sources at relatively high frequency with sub-mas resolution.

These observations are motivated by the fact that previous radio surveys, using the VLA and MERLIN (e.g. Patnaik et al. 1992), have only been sensitive to image separations > few hundred milliarcseconds (*i.e.* to lens masses > $10^9 M_\odot$). The aim of this survey is to search for multiply imaged VLBI radio cores with separations $\leq$ 150 mas.

2. Observations and Data Analysis

In 1995 January 20 we observed 72 sources in a single 24 hour period using 1 thin tape per station (a total of 12 hours of recording/observing time at 64Mb/s). Each source was observed for four scans of 2 minutes duration (including slew time), spread over $\sim$ 8 hrs in hour angle, using all 10 VLBA

C. S. Kochanek and J. N. Hewitt (eds), Astrophysical Applications of Gravitational Lensing, 405–406.

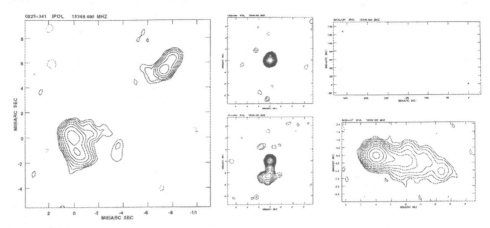

Figure 1. Left: A 15 GHz VLBA map of 0223+341, contours are spaced by factors of two in brightness, with the lowest at three times the rms noise level (0.36 mJy/beam). Right: Examples of (1) one of many excellent Space/mm-VLBI targets, (2) a wide-field map of the known lens system 0218+357, (3,4) sources with interesting resolved structure.

telescopes observing at frequency of 15 GHz. The data were correlated at the VLBA correlator in Socorro, NM, and were calibrated and mapped using the AIPS package. The map rms noise is typically $\sim$ 1 mJy/beam.

3. Results

Out of 72 sources we observed, 7 are points sources, 24 are point like (i.e. point sources with weak extensions), 24 exhibit typical core-jet structure, 10 are doubles and 4 are triple sources (i.e. 3 separate components, no evidence at the moment if they are one- or two-sided). Three sources were not detected on the intermediate and long baselines. The radio maps and a table of results are available via *ftp://jbss0.jb.man.ac.uk/pub/mikey/survey.ps*.

Wide-field maps of all the sources were also made in order to search for additional radio components up to 150 mas from the phase center of the map (see Fig 1).

4. Best Lens Candidate - 0223+341

While there are several potential candidates, there is one source which stands out from the rest of the sample – 0223+341 (see Fig 1). The double structure seen in both the north western and south eastern components is particularly suggestive. The source is identified optically as a "stellar object" but the redshift is unknown. Further observations are planned.

References

Patnaik A.R., Browne I.W.A., Wilkinson P.N., & Wrobel J.M., 1992, MNRAS, 254, 655

PROGNOSTICATING THE FUTURE
OF GRAVITATIONAL LENSES

WILLIAM H. PRESS
Harvard-Smithsonian Center for Astrophysics
60 Garden Street, Cambridge, MA 02138, USA

1. Introduction

About 20 years elapsed between my first and second papers on gravitational lenses (Press & Gunn 1973; Press, Rybicki & Hewitt 1992ab). Therefore, the conference organizers have asked me to prognosticate on the future of gravitational lenses. Their reasoning, if I understand it correctly, is that I will likely go to sleep for another 20 years immediately following this conference. So, if I simply look ahead to my very next paper on the subject, I will be in effect prognosticating over two decade's span.

2. Statistics on Lens Publications

In science, like so many other human activities, we march backward into the future. Let us examine, in the grossest of statistical terms, where we have been. Figure 1 shows the number of papers published per year on gravitational lensing. Note the logarithmic ordinate. Einstein's first paper on the subject is the one in 1911 (Einstein 1911). The "bulge" of papers in the mid 1960s is largely due to the prescient work of Refsdal (e.g., Refsdal 1964, 1966), and the fractious, but in some ways farsighted, papers of the Barnothys (e.g., Barnothy 1965, Barnothy & Barnothy 1968).

Looking at Figure 1, it is hard to spot where the first lens, 0957+561, was actually discovered (Walsh et al. 1979). I think that observers were *forced* to discover gravitational lenses by the rising tide of speculative theoretical papers, rather than the other way around!

Now turn, in Figure 2, to prognosticating the future. This is easily done, by the rigorous mathematical technique of linear extrapolation (on a log-linear plot, of course). One sees that by the year 2008, all astronomical work

C. S. Kochanek and J. N. Hewitt (eds), Astrophysical Applications of Gravitational Lensing, 407–414.

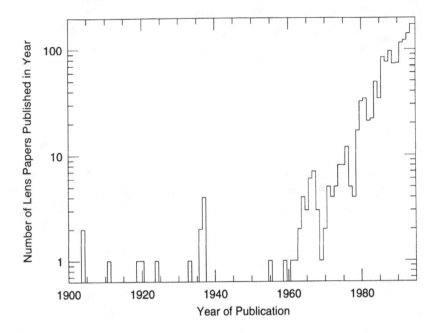

Figure 1. Number of published papers that relate to gravitational lensing, by year. Data from the bibliography of Pospieszalska-Surdej, Surdej & Véron (1994).

will be devoted to gravitational lenses; by the year 2046, gravitational lens work will take over all scientific research; and by the year 2100, all human activity will be devoted to our field.

Actually, the first of these conclusions (the year 2008) has perhaps more than a grain of truth in it. Gravitational lens studies are rapidly becoming mainstream astronomy. Gravitational lens effects, when we include weak lensing, are present along virtually every line of sight out to distances that we now call cosmological – but which will increasingly be the venue of extragalactic astronomy in the post-Keck era. In galactic astronomy, only a tiny fraction of lines of sight are lensed, but our technical capabilities for looking at multitudinous lines of sight are increasing explosively. Gravitational lensing may well disappear as a unique sub-specialty in astronomy, and instead become simply a ubiquitous observational technique. The index of refraction of space will be as much a fact of life to the 21st Century astronomer, as was the index of refraction of glass to his or her 19th Century counterpart.

To fulfill my charge as prognosticator, I must, however, make some more concrete predictions. I divide these into three categories: "New Heavy Industries," that is, large-scale observational efforts that are already gathering force; "Not Whether, But When," that is, advances that seem fully within the reach of foreseeable technology, but that will require dedicated commit-

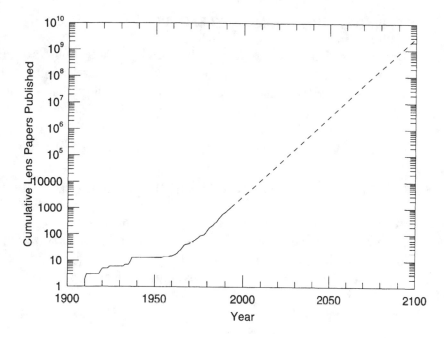

Figure 2. Cumulative number of published papers that relate to gravitational lensing, by year. Data from the bibliography of Pospieszalska-Surdej, Surdej & Véron (1994). The dashed part of the curve is extrapolation to the future.

ment to complete; and "Not When, But Whether," that is, advances that may or may not prove to be possible at all. I have only a few words to say about each item in each category, so my contribution becomes, from this point on, little more than an outline. (Prognosticators, from the Oracle at Delphi to the present, know the wisdom of brevity.)

3. New Heavy Industries

3.1. MAPPING THE SKY IN WEAK LENSING

The number of faint blue galaxies that can be imaged by Keck-class telescopes on the ground (Lawrence in this volume), and HST-class instruments in space, is so numerous that statistical detection of weak gravitational lensing can be done along virtually every line of sight in the sky (see Tyson, and Kneib & Soucail in this volume). It is becoming possible literally to map the distribution of intervening dark (and luminous) matter (Kaiser & Squires (1993), also Miralda-Escudé and Schneider in this volume). Tyson and others have discussed the feasibility of a dedicated dark matter telescope. See also Stebbins, McKay & Frieman (in this volume).

3.2. EXPLOITING GRAVITATIONAL TELESCOPES

Highly distorted, and therefore highly magnified, arclets can be found around virtually every cluster of galaxies at moderate redshift. These images will be increasingly exploited not just to tell us about the lensing cluster, but also (in some cases primarily) to increase the magnification of our terrestrial telescopes for studying the internal structure of faint high-redshift galaxies. See, e.g., Blandford & Hogg, Kneib & Soucail, Smail et al. in this volume.

3.3. LENS INVERSION AS A LARGE-SCALE INVERSE PROBLEM

Instead of using a handful of point-image locations and magnifications to solve for a handful of lens parameters, the lens inversion problem is rapidly becoming one of simultaneous solution, in both the source and lens planes, of millions of equivalent pixels. A recent example is Chen, Kochanek & Hewitt (1995).

3.4. MICROLENS SEARCH FOR PLANETS

NASA Administrator Dan Goldin wants to find extra-solar planets, and he is not a man to be trifled with. We can debate whether this is a scientific or a technological goal; but NASA is in the business of doing both. Personally, I find this a very exciting project in the spirit of Galactic exploration. See Mao & Paczynski (1991), Gould & Loeb (1992) and, more recently, Gould (1994, 1995) and Albrow et al. (in this volume).

4. Not Whether, But When

4.1. MEASURE H_0 TO TWO SIGNIFICANT FIGURES

I am confident that there will be such a measurement, *some day*. Whether it will come from lenses, as opposed to (e.g.) Type II supernovae using the expanding photosphere method, is more problematic. It depends on whether the "perfect" lens comes along: perhaps something like 0218+357, with a delay that can be measured to 2 hour accuracy (see Browne et al. in this volume). One would need the lensed object to be an extended source, to get a handle on the radial mass profile of the lens.

4.2. SHOW THAT $\Lambda < 0.5$ AT 99% CONFIDENCE LEVEL

This will emerge from the continued careful analysis of the statistics of known lenses, not just their number, but also their distributions with redshift, etc. See Rix, Kochanek, and Claeskens et al. (in this volume).

4.3. MEASURE Ω OR Q_0 TO ONE SIGNIFICANT FIGURE

Lens studies will contribute in two ways. First, statistical lens studies will put constraints on the geometry of spacetime at redshifts > 1. Second, dark matter mapping studies will "weigh" the clustered matter. Lens methods are competing with both cosmic microwave background fluctuation measurements on the 1 degree scale, and also with the dynamical data that will be mined from the Sloan Digital Sky Survey and other similar large-scale surveys.

4.4. QUANTIFY THE BINARY POPULATION IN OUR GALAXY

Although it is a messy statistical inverse problem, it is clear that once a significant number of binary microlensing events are in hand, we will know a lot about the population of binary stars in our Galaxy.

4.5. MAP OUR GALAXY'S BAR AND HALO WITH MICROLENSING

The prediscoveries and rediscoveries of the Galactic bar make for an interesting case study in astronomy's "blind spots". See Paczynski (in this volume).

4.6. DETERMINE THE MASS DISTRIBUTION IN GALAXIES AND CLUSTERS GENERALLY

Weak lensing studies will ultimately say a great deal about this. See (e.g.) Schneider, Kneib & Soucail, and Brainerd, Blandford & Smail all in this volume. Kochanek (in this volume) discusses what can come from an approach based on the statistics of strong lenses. Bartelmann & Narayan (in this volume) discuss the exploitation of a subtle effect that allows simultaneous determination of the redshift distribution of faint blue galaxies, and the mass distributions of foreground clusters of galaxies.

5. Not When, But Whether

5.1. MEASURE H_0 TO THREE SIGNIFICANT FIGURES

Am I kidding? It depends on whether there is any possibility of measuring "interferometrically accurate" lens delays, by observing coherence effects between different lens images. (Of course, it also depends on exquisite knowledge of lens mass profiles.)

5.2. MEASURE Ω OR Q_0 TO TWO SIGNIFICANT FIGURES

Now, am I kidding? Probably.

5.3. DISCOVER AN INCONTESTABLE DARK MATTER OR SHADOW SECTOR GALAXY

It will be a breathtaking moment when a future Tony Tyson (or even the present one) shows us a picture with the classic circular pattern of highly distorted arclets, but with *nothing in the middle*. In CDM and related scenarios, baryons and dark matter are not necessarily tied together in all cases, because baryons can be stopped by shocks that allow the dark matter to coast through. Two-stream or other instabilities might then perturb the dark matter, allowing dissipationless gravitational collapse to form completely dark galaxies.

5.4. RULE OUT ALL MACHO MASS RANGES

The implication would be that the dark matter is in distributed form, lending credibility to the belief that it is composed of exotic particles. See Paczynski, Pratt et al., and Alard in this volume.

5.5. MAP STELLAR SURFACES BY LENS CAUSTICS

Jaroszynski & Paczynski (1996) have pointed out that lens caustics are, in a sense, better-than-perfect occulting edges. Where the edge of a physical object (the Moon, say) has a transmission coefficient that varies sharply from 1 to 0, the edge of a lens caustic jumps from an integrable infinity to zero. This causes a new scale – not the Fresnel scale – to enter, with the possibility of sub-picoarcsecond resolution.

5.6. DISCOVER AN $\Omega \sim 1$ POPULATION OF OBJECTS RESPONSIBLE FOR (SOME) QUASAR VARIABILITY

Hawkins (1993, 1995) puts forth the unconventional hypothesis that a part of the variability in quasars not intrinsic to the quasar, but rather due to omnipresent microlensing from a hypothetical population of compact objects with $\Omega \sim 1$.

5.7. DETECT A MICROLENS BY DIRECT COHERENCY OF IMAGES

We first learned to find gravitational lenses in cases where they have multiple images. Later, we learned to find them in microlens surveys when they cause time-varying magnifications. We are now learning to find them statistically in the small shears of faint blue galaxies.

There are at least two other ways that lenses might, in certain circumstances, be found: (1) time autocorrelation on unresolved, time-varying, objects; and (2) direct detection of time-lagged coherency on non-varying sources. As I discussed in my talk, and won't repeat here, the latter idea contains conceptual pitfalls that are clearly avoided only in the case of *extremely* small source sizes.

Still, I have a nagging feeling that there is a pony (or kangaroo) somewhere in the pile of papers that have been written on one or another aspect of coherency effects in lensing. See Schneider & Schmid-Burgk (1985), Mandzhos (1981), Deguchi & Watson (1986), Peterson & Falk (1991), Krauss & Small (1991), and Spillar (1993), as well as the previously mentioned Jaroszynski & Paczynski (1995). For the microscopic case ("femtolensing"), see Gould (1992), and Ulmer & Goodman (1994).

References

Alard, C., 1996, this volume
Albrow, M., et al., 1996, this volume
Barnothy, J.M., 1965, AJ, 70, 666
Barnothy, J.M., & Barnothy, M.F., 1968, Science, 162, 348
Bartlemann, M.L. & Narayan, R., 1996, this volume
Blandford, R.G., & Hogg, D.W., 1996, this volume
Brainerd, T.G., Blandford, R.B., & Smail, I., 1996, this volume
Browne, I.W.A., Corbett, E.A., Wilkinson, P.N., & Patnaik, A.R., 1996, this volume
Chen, G.H., Kochanek, C.S., & Hewitt, J.N., 1995, ApJ, 447 62
Claeskens, J.F., Jaunsen, A.O., & Surdej, J., 1996, this volume
Deguchi, S., & Watson, W.D., 1986, ApJ, 307, 30
Einstein, A. 1911, Annalen der Physik, 35, 898
Gould, A. 1992, ApJL, 386, L5
Gould, A. 1994, ApJL, 421, L75.
Gould, A. 1995, ApJL, 441, L21.
Gould, A., & Loeb, A. 1992, ApJ, 396, 104
Hawkins, M.R.S. 1993, Nature, 366, 242
Hawkins, M.R.S. 1995, preprint (submitted to MNRAS)
Jaroszynski, M, & Paczynski, B., 1995, submitted to ApJ (preprint available as ftp://astro.princeton.edu/library/preprints/pop615.ps.Z)
Kaiser, N. & Squires, G., 1993, ApJ, 404, 441
Kochanek, C.S., 1996, this volume
Kneib, J.P., & Soucail, G., 1996, this volume
Krauss, L.M., & Small, T.A., 1991, ApJ, 378, 22
Lawrence, C.R., 1996, this volume
Mandzhos, A.V. 1982, Sov Astron Lett, 7, 213
Mao, S., & Paczynski, B., 1991, ApJL, 374, L37
Miralda-Escudé, J., 1996, this volume
Paczynski, B., 1996, this volume
Petterson, J.B., & Falk, T., 1991, ApJL, 374, L5
Pospieszalska-Surdej, A., Surdej, J., & Véron, P. 1994, document available as http://www.stsci.edu/ftp/stsci/library/grav_lens/grav_lens.html
Pratt, M.R., et al. (14 authors), 1996, this volume
Press, W.H., & Gunn, J.E., 1973, ApJ, 185, 397
Press, W.H., Rybicki, G.B., & Hewitt, J.N., 1992a, ApJ, 385, 404
Press, W.H., Rybicki, G.B., & Hewitt, J.N., 1992b, ApJ, 385, 416
Refsdal, S., 1964, MNRAS, 128, 295
Refsdal, S., 1966, MNRAS, 132, 101

Rix, H.-W., 1996, this volume

Schneider, P., 1996, this volume

Schneider, P., & Schmid-Burgk, J., 1985, A&A, 148, 369

Smail, I., Dressler, A., Kneib, J.-P., et al., 1996, this volume

Spillar, E.J., 1993, ApJ, 403, 20

Stebbins, A., McKay, T., & Frieman, J.A., 1996, this volume

Tyson, J.A., 1996, this volume

Ulmer, A., & Goodman, J. 1995, ApJ, 442, 67

Walsh, D., Carswell, R.F., Weymann, R.J., 1979, Nature, 279, 381

THE 'GRAVITATIONAL LENSING' BIBLIOGRAPHY

A. POSPIESZALSKA-SURDEJ
STScI, 3700 San Martin Drive, Baltimore, MD 21218

J. SURDEJ
STScI, 3700 San Martin Drive, Baltimore, MD 21218,
presently member of the Astrophysics Division, Space Science
Department of the European Space Agency and also Research
Director, FNRS, Belgium

AND

P. VERON
Observatoire de Haute-Provence,
F-04870 Saint-Michel l'Obser-vatoire, France

Abstract. We present a non-exhaustive bibliography on "Gravitational Lensing" (GL), totalizing nearly 1500 titles (see Pospieszalska-Surdej et al. 1993 for an earlier version of this bibliography). It also includes recent abstracts of papers dealing with gravitational lensing and submitted to astro-ph@babbage.sissa.it. The GL bibliography is accessible through a World-Wide-Web page at the URL (cf. next page)
http://www.stsci.edu/ftp/stsci/library/grav_lens/grav_lens.html
It is also possible to retrieve -via FTP- the latex file 'grav_bib1.tex' or the postscript file 'grav_bib1.ps' containing the most recent version of this bibliography. Please proceed as follows: ftp stsci.edu, Name: anonymous, Password: your E-mail address, cd stsci/library/grav_lens, get grav_bib1.tex or get grav_bib1.ps, and finally bye).

We are aware that the present compilation is not complete, not always uniform and that there are still errors, typos, etc. We would appreciate very much if you could communicate to us (via E-mail: surdej@stsci.edu) missing or new titles (published or accepted for publication), errors, changes and suggestions. It would also be nice for us to receive in the future your preprints or reprints related to 'Gravitational Lensing' studies. Please send your material to: A. Pospieszalska-Surdej, STScI, 3700 San Martin Drive, Baltimore MD 21218, USA.

C. S. Kochanek and J. N. Hewitt (eds), Astrophysical Applications of Gravitational Lensing, 415–416.

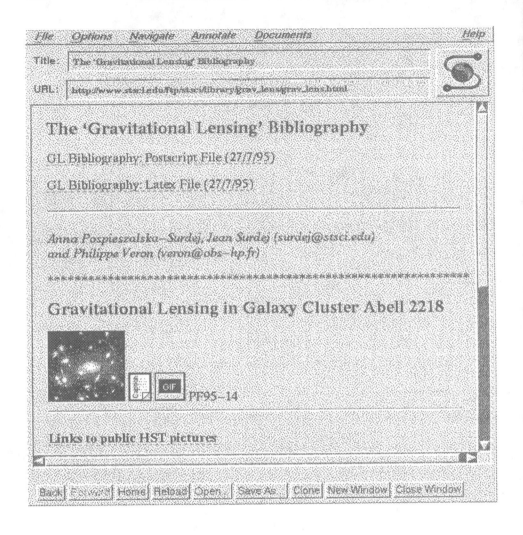

Figure 1. World-Wide-Web home page for the GL bibliography accessible at the URL http://www.stsci.edu/ftp/stsci/library/grav_lens/grav_lens.html

References

Pospieszalska-Surdej, A., Surdej, J., & Véron, P., 1993, in Gravitational Lenses in the Universe, eds. J. Surdej, D. Fraipont-Caro, E. Gosset, S. Refsdal & M. Remy (Liège: Université de Liège), 671

FORMATION OF GIANT LUMINOUS ARCS AND ARCLETS USING AN OPTICAL GRAVITATIONAL LENS EXPERIMENT

J. SURDEJ
STScI, 3700 San Martin Drive, Baltimore, MD 21218, USA,
Member of the Astrophysics Division,
Space Science Department of the European Space Agency,
and Directeur de Recherches au FNRS, Belgium

S. REFSDAL
Hamburg Observatory, Gojenbergsweg 112,
D-21029 Hamburg, Germany

AND

A. POSPIESZALSKA-SURDEJ
STScI, 3700 San Martin Drive, Baltimore, MD 21218, USA

Abstract. It is well known that simple optical lenses simulating light deflection due to a foreground object may easily reproduce all types of image configurations observed among known multiply imaged QSOs and AGN (cf. simulations in Refsdal and Surdej 1992, 1994, Surdej et al. 1993). Such an optical lens experiment can also be used to simulate the formation of multiple giant luminous arcs and arclets seen near massive foreground galaxy clusters. The optical setup used to make this experiment is shown in Fig. 1 (a similar setup was presented during the Melbourne conference). A compact light source is located on the right side (not very clearly seen), then comes to the left a 'spiral galaxy disk' optical lens (manufactured by the authors) and a cardboard perforated by multiple pinholes. In the absence of light deflection (i.e. by simply removing the optical lens), direct (non distorted) images of the background galaxies are seen by an observer, like those projected on the distant white screen (see Fig. 2a). When the optical lens is inserted between the light source and the multiple pinholes screen (cf. Fig. 1), images of the background galaxies get distorted (arclets) and/or transformed into multiple images, including giant luminous arcs (see Figs. 2bc for examples). We showed during the conference that by covering the pinholes with various transparent colored filters and translating the multi-

C. S. Kochanek and J. N. Hewitt (eds), Astrophysical Applications of Gravitational Lensing, 417–418.

Figure 1. Setup of the optical gravitational lens experiment used to simulate distorted images (cf. Figs. 2bc) of background galaxies (Fig. 2a) by a foreground cluster (see text)

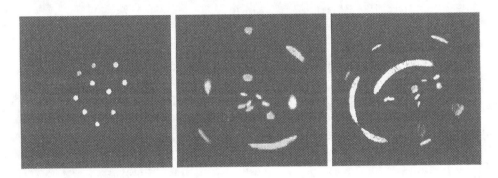

Figure 2. Giant luminous arcs and arclets (Figs. 2bc) resulting from the gravitational lens distortion of background galaxies (Fig. 2a) by a foreground cluster

ple pinholes screen, it was possible to reproduce, in a very inexpensive way, distorted images of background galaxies like the very impressive strong lensing artifacts seen behind the rich cluster Abell 1689 (cf. the color videotape of simulations by Tony Tyson).

References

Refsdal, S., & Surdej, J., 1992, in Highlights of Astronomy, ed. J. Bergeron, 9, 3

Refsdal, S., & Surdej, J., 1994, Rep Prog Phys, 56, 117

Surdej, J., Refsdal, S., & Pospieszalska-Surdej, A., 1993, in Gravitational Lenses in the Universe, eds. J. Surdej, D. Fraipont-Caro, E. Gosset, S. Refsdal & M. Remy, (Liège: Université de Liège) 199

SUMMARY OF DATA ON
SECURE MULTIPLY IMAGED SYSTEMS

C.R. KEETON II AND C.S. KOCHANEK
Harvard-Smithsonian Center for Astrophysics
60 Garden Street, Cambridge, MA 02138, USA

Abstract. We present an extensive summary of the position, redshift, and flux data on the secure, multiply imaged, quasar and radio lenses. It includes neither the less-promising (in the opinion of the author) lens candidates nor the cluster lenses (which are too hard to reduce to short tables of numbers). A broader listing of suggested lenses and also of the cluster lenses is available at (see Pospieszalskia-Surdej et al. in this volume): http://www.stsci.edu/ftp/stsci/library/grav_lens/grav_lens.html

1. Introduction

Table 1 lists the lenses and lens candidates discussed in this summary. The objects are sorted by a letter grade (A="I'd bet my life this is a lens," B="I'd bet your life this is a lens," and C="You should worry if I'm betting your life.") The B objects are almost certainly lenses, but are either very new candidates or double images missing a few too many observational checks. Some objects in C are certainly *not* lenses, because the C list contains all of the problematic wide separation double quasars, some of which are simply correlated quasars rather than lenses. Within each grade, the lenses are sorted in order of decreasing image separation or tangential critical radius.

Following Table 1 we present short data summaries for the individual objects, schematic diagrams or radio images, and time series of the flux ratios between the images where there is both data and space. We summarize only the data for the lenses (not models), and we do not attempt to give a complete list of all observational work on the object. References are only given for the data physically presented in the summaries, and

419

the summaries include only the most recent or most accurate observations. Whenever possible, we included error estimates for the data. *Please do not use this summary as your only reference for data on these lenses – the authors of the original papers deserve the credit for their work!*

Acknowledgements: We would like to thank E. Falco, J. Hewitt, D. Jones, J. Lehár, A. Patnaik, R. Porcas, K. Ratnatunga, and P. Schechter for their assistance in compiling this data and for commenting on the format.

Summary of Multiply Imaged Systems									
Name	G	z_s	z_l	m_s mag	m_l mag	F_{GHz} mJy	N_{im}	$\Delta\theta$ "	Page
0957+561	A	1.41	0.36	16.7	18.5	$F_5 = 65.6$	(3)E	6.1	426
2016+112	A	3.27	(1.01)/(?)	$i=22.1$	$i=21.9$	$F_5 = 84.6$	(3)E	3.8	427
0142−100=UM673	A	2.72	0.49	16.8	19		2	2.2	424
PG 1115+080	A	1.72	0.29	15.8	20.2		4	2.2	428
MG 0414+0534	A	2.64		$I'=19.3$	$I'=21.1$	$F_5 = 977$	4E	2.1	429
CLASS 1608+656	A	1.39	0.63	(~20)	(~20)	$F_8 = 73.2$	4	2.1	430
MG 1131+045	A				(22)	$F_5 = 205$	2R	2.1	436
MG 1654+1346	A	1.74	0.25	$r=20.9$	$r=18.7$	$F_5 = 130.0$	R	2.1	436
B 1938+666	A				(23)	$F_5 = 316$	R	1.8	437
2237+0305	A	1.69	0.039	$B=16.8$	$B=15.6$	$F_8 = 0.336$	4	1.7	431
MG 1549+3047	A		0.11	23.3	16.7	$F_5 = 185$	R	1.7	437
B 1422+231	A	3.62	(0.65)	$r=15.6$	$r=21.8$	$F_5 = 557$	4E	1.3	432
H 1413+117	A	2.55		17.0		$F_8 \sim 0.1$	4	1.2	433
PKS 1830−211	A					$F_2 \sim 10^4$	2ER	1.0	438
MG 0751+2716	A		0.35		19	$F_5 = 202$	R	0.9	438
B 0218+357	A	(0.96)	0.68		$r=20.0$	$F_5 = 1209$	2ER	0.34	439
LBQS 1009−0252	B	2.74		$B=18.1$			2	1.5	424
CLASS 1600+434	B	1.61		~20		$F_8 = 132$	2	1.4	424
HST 14176+5226	B			$V=24.3$	$V=21.7$		4	1.4	434
HST 12531−2914	B			$V=25.5$	$V=23.8$		4	1.2	435
BRI 0952−0115	B	4.5					2	0.95	425
J03.13	B	2.55		17.1			2	0.84	425
1208+1011	B	3.80		$V=18.1$			2	0.48	425
2345+007	C	2.15		18.5	24.5		2	7.1	422
1120+019=UM425	C	1.47	(~0.6)	15.7			2	6.5	422
QJ 0240−343	C	1.4		$B=18.6$			2	6.1	422
1429−008	C	2.08		17.7			2	5.1	423
1634+267	C	1.96	(~0.6)	~19			2	3.8	423
HE 1104−1805	C	2.32		$B=16.9$			2	3.0	423

TABLE 1. The source and lens redshifts are z_s and z_l, the source and lens magnitudes are m_s and m_l, and the image separation or the diameter of the tangential critical line is $\Delta\theta$. The f GHz radio flux is given by F_f. Blank entries are unknown, and entries in parenthesis have low accuracy or substantial uncertainty. The magnitudes are R magnitudes unless otherwise specified. The number of compact images is N_{im}, an E indicates the images are extended, and an R indicates a ring.

2-image systems

Lens	Separation ($''$)	Brightness	Redshift

2345+007

$\Delta\theta_{AB} = 7.06 \pm 0.01$ $z_s = 2.15$

$(x,y)_A = (\ \ 5.08, \ \ 2.90) \quad m_R(A) = 18.78 \pm 0.05$

$(x,y)_B = (-0.78, -1.04) \quad m_R(B) = 19.96 \pm 0.03$

$(x,y)_{G1} = (\ \ 0.00, \ \ 0.00) \quad m_R(G1) = 24.5 \pm 0.5$

Coordinates: A, $23^h45^m45.90^s$, $+00°40'40.4''$ (B1950)

Comments: The galaxy G1 is considered to be a lens candidate (F94). S91 find significant metal absorption systems at redshifts 1.4828 and 1.4912, as well as systems at 0.7545 (in component B only), 1.7717 (stronger in B than in A), 1.7977(A) and 1.7998(B), and 1.9832 (in A only).

References: F94 (Fischer et al., 1994, ApJL, 431, L71); S91 (Steidel & Sargent, 1991, AJ, 102, 1610); W82 (Weedman et al., 1992, ApJL, 255, L5).

1120+019 =UM425

$\Delta\alpha_{A-B} = \ \ 3.1 \pm 0.1 \qquad m_R(A) = 15.7 \qquad\qquad\qquad z_s = 1.465$

$\Delta\delta_{A-B} = -5.7 \pm 0.1 \qquad m_R(B) = 20.1$

$\qquad\qquad\qquad\qquad\qquad m_R(B) - m_R(A) = 4.42 \pm 0.12$

$\qquad\qquad\qquad\qquad\qquad\qquad\qquad\qquad\qquad\qquad\qquad (z_l \sim 0.6)$

Coordinates: $11^h20^m46.7^s$, $+01°54'13.2''$ (B1950)

Comments: The absolute magnitudes of A and B are uncertain by a couple of tenths of a magnitude because of the poorly determined zero point. z_l was estimated by subtracting a scaled spectrum of A from the spectrum of B and interpreting the residuals as a galaxy.

References: Meylan & Djorgovski, 1989, ApJL, 338, L1; Michalitsianos & Oliversen, 1995, ApJ, 439, 599.

QJ0240−343

$\Delta\theta = 6.1 \qquad\qquad\qquad m_B(A) = 19.00 \pm 0.05 \qquad\qquad z_s = 1.4$

$\qquad\qquad\qquad\qquad\qquad m_B(B) = 19.77 \pm 0.05$

Comments: There is a metal-line absorption system at $z = 0.543$, and a possible system at $z = 0.337$.

References: Tinney, these proceedings.

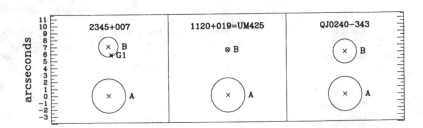

Figure 1. Schematic diagrams of 2-image systems. The orientation is arbitrary. The ratios of the circle areas are equal to the flux ratios. (The scale varies between Figs. 1-4.)

2-image systems

Lens	Separation ($''$)		Brightness	Redshift

1429−008

$\Delta\theta = 5.14 \pm 0.10$ $m_R(A) = 17.74 \pm 0.05$ $z_s = 2.076$

$m_R(B) = 20.77 \pm 0.10$

Coordinates: A, $14^h 29^m 54.5^s$, $-00°53'04''$ (B1950)

Comments: H89 find Mg II absorption systems at $z = 1.55$ and $z = 1.62$. S95 find a Ly-α absorption line at $z = 1.662$, as well as a C IV doublet at $z = 1.42$ with a velocity difference of 580 km/s between A and B.

References: H89 (Hewett et al., 1989, ApJL, 346, L61); S95 (Smette et al., these proceedings).

1634+267[†]

$\Delta\alpha_{A-B} = 0.63$ $m(A) = 19.15$ $z_s = 1.961$

$\Delta\delta_{A-B} = 3.72$ $m(B) = 20.75$

$(z_l \sim 0.57)$

Coordinates: A, $16^h 34^m 59.1^s$, $+26°42'04''$ (B1950)

Comments: ([†]St91 suggest that this object's correct designation is 1634+267, not 1635+267.) In the spectrum of A, St91 find a metal absorption system at $z = 1.1262$ and a less-definite system at $z = 1.8389$. T88 estimate z_l by subtracting a scaled spectrum of B from the spectrum of A and interpreting the "excess flux" as a galaxy. St91 do not find evidence for "excess flux," but note that their $1''$ slit was smaller than T88's $2.5''$ slit.

References: D84 (Djorgovski & Spinrad, 1984, ApJL, 282, L1); T88 (Turner et al., 1988, AJ, 96, 1682); Sr78 (Sramek & Weedman, 1978, ApJ, 221, 468); St91 (Steidel & Sargent, 1991, AJ, 102, 1610).

HE 1104−1805

$\Delta\alpha_{A-B} = -2.7 \pm 0.1$ $m_B(A) = 17.06 \pm 0.01$ $z_s = 2.319$

$\Delta\delta_{A-B} = 1.4 \pm 0.1$ $m_B(B) = 18.91 \pm 0.01$

Coordinates: A, $11^h 06^m 33.45^s$, $-18°21'24.2''$ (J2000)

Comments: A damped Ly-α plus metal absorption system is seen at $z = 1.66$, and a Mg II absorption system is seen at $z = 1.32$ (only in A).

References: Wisotzki et al., 1993, A&A, 278, L15; Wisotzki et al., 1995, A&A, 297, L59.

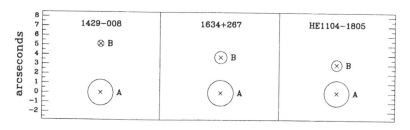

Figure 2. Schematic diagrams of 2-image systems. The orientation is arbitrary. The ratios of the circle areas are equal to the flux ratios. (The scale varies between Figs. 1-4.)

2-image systems

Lens	Separation (″)	Brightness	Redshift
0142−100 =UM673	$\Delta\theta_{AB} = 2.22 \pm 0.03$	$m_R(A) = 16.9 \pm 0.2$ $m_R(B) = 19.1 \pm 0.2$	$z_s = 2.719$
	$\Delta\theta_{BG} = 0.8 \pm 0.2$	$m_R(G) \approx 19$	$z_l = 0.49$

Coordinates: $01^h42^m48.6^s$, $-10°00'13''$ (B1950)
Comments: G lies very nearly along the line joining A and B. It was identified by subtracting normalized spectra of A and B and matching the residual spectrum to the spectra of galaxies, and by noting the presence of faint absorption lines due to Ca II H and K. There is a high ionization absorption line system at $z = 2.3564$, a suspected Lyman absorption system at $z = 2.7363$, and a possible absorption line system at $z = 1.8987$.
References: MacAlpine & Feldman, 1982, ApJ, 261, 412; Surdej et al., 1987, Nature, 329, 695; Surdej et al., 1988, A&A, 198, 49.

LBQS 1009−0252	$\Delta\theta = 1.53 \pm 0.01$	$m_B(A) = 18.2$	$z_s = 2.739$
	$\Delta\alpha_{A-B} = 0.61$	$m_B(B) = 20.8$	
	$\Delta\delta_{A-B} = 1.41$	$m_B(B) - m_B(A) = 2.52 \pm 0.20$	

Coordinates: A, $10^h09^m43.97^s$, $-02°52'12.0''$ (B1950)
Comments: The relative coordinates ($\Delta\alpha_{A-B}$, $\Delta\delta_{A-B}$) are taken from a figure caption in Hewett et al.; values given in a table are inconsistent with values given in the figure caption. A second quasar (labeled C) lies $4.62''$ from image A and has $m_B = 19.3$ and $z = 1.627$. Mg II absorption lines are seen at $z = 0.8688$ and $z = 1.6266$ in A and B; the $z = 0.8688$ line is also seen weakly in C.
References: Hewett et al., 1994, AJ, 108, 1534; Surdej et al., 1993, Liège proceedings, p. 153.

CLASS 1600+434	$\Delta\theta = 1.39 \pm 0.01$	$F_{8.4GHz}(A) = 73$ mJy/beam	$z_s = 1.61$
	$\Delta\alpha_{A-B} = -0.72$	$F_{8.4GHz}(B) = 56$ mJy/beam	
	$\Delta\delta_{A-B} = 1.19$	$A/B = 1.30 \pm 0.04$	

Coordinates: A, $16^h01^m40.446^s$, $+43°16'47.76''$ (J2000)
References: Jackson et al., 1995, MNRAS, 274, L25.

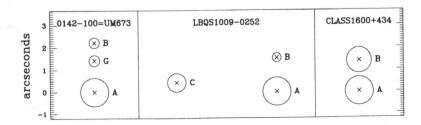

Figure 3.   Schematic diagrams of 2-image systems. The orientation is arbitrary. The ratios of the circle areas are equal to the flux ratios. (The scale varies between Figs. 1-4.)

2-image systems

Lens	Separation (")	Brightness	Redshift
BRI 0952−0115	$\Delta\theta = 0.95$	$\Delta m = 1.35$	$z_s = 4.5$

Coordinates: $09^h 52^m 27.20^s$, $-01°15'53.0''$ (B1950)
References: McMahon et al., 1992, Gemini, 36, 1.

J03.13	$\Delta\theta = 0.84 \pm 0.02$	$m_R(A) = 17.2 \pm 0.1$	$z_s = 2.55$

$m_R(B) = 19.3 \pm 0.1$
$m_R(B) - m_R(A) = 2.14 \pm 0.05$
Comments: There are absorption line systems at $z = 2.34$ (Ly-α and C IV) and $z = 1.085$ (metal).
References: Claeskens et al., these proceedings.

1208	1011	$\Delta\theta = 0.476 \pm 0.004$	$m_V(A) \approx 18.3$	$z_s = 3.803$

$m_V(B) \approx 19.8$
$A/B = 4.2 \pm 0.1$
Coordinates: $12^h 08^m 23.73^s$, $+10°11'07.9''$ (B1950)
Comments: Magnitude uncertainties are not stated. The flux ratio is from aperture photometry in the F555W (V) band; the uncertainty represents scatter in the measurements. Measurements in other bands (F439W, F702W, F785LP) also yield a flux ratio of approximately 4:1.
References: Bahcall et al., 1992, ApJL, 392, L1; Bahcall et al., 1992, ApJL, 400, L51; Hazard et al., 1987, Nature, 322, 38; Magain et al., 1992, A&A, 253, L13.

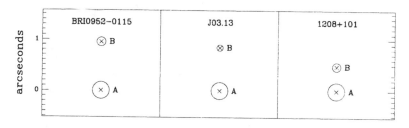

Figure 4. Schematic diagrams of 2-image systems. The orientation is arbitrary. The ratios of the circle areas are equal to the flux ratios. (The scale varies between Figs. 1-4.)

0957+561 (A: $09^h 57^m 57.324^s$, $+56°08'22.344''$ [B1950; Go84])

		$-\Delta\alpha$ ($''$)	$\Delta\delta$ ($''$)	F_{GHz} (mJy)	Jet Length (mas)	Jet PA ($°$)
	A	1.25252 ± 0.00003	6.04662 ± 0.00004	$F_5 = 35.9 \pm 0.2$	48.3 ± 0.1	19.9 ± 0.1
	B	0.0	0.0	$F_5 = 27.2 \pm 0.1$	58.8 ± 0.1	17.8 ± 0.1
	G'	-0.181 ± 0.001	1.029 ± 0.001	$F_2 = 0.6 \pm 0.1$		
	G/G1	-0.19 ± 0.03	1.00 ± 0.03	$F_5 = 2.50 \pm 0.09$		

Positions and fluxes (left): The position of A is from Go84. The position of the third VLBI component G' is from Go83. The position of the optical galaxy G1 is from S80. The fluxes of A, B, and G are from Ro85, and the flux of G' is from Go83.

Jet orientation (right; Ga94): The core-jet configuration is modeled with 6 Gaussian components; the lengths and position angles given here are for the most luminous components (A_5, B_5). Data from VLBI observations, 1989 Nov 8.

Transformation matrix (Ga94): The transformation matrix M^{BA} is specified by its eigenvalues (M_i), the position angles of its eigenvectors (ϕ_i), and the spatial derivative of the eigenvalues along the direction of the jet ($\dot{M}_i$). ($\dot{\phi}_i \equiv 0$ by assumption.)

M_1	$=$	1.23 ± 0.04	$M_2 =$	-0.50 ± 0.03
ϕ_1	$=$	$18.6° \pm 0.1°$	$\phi_2 =$	$118° \pm 6°$
$\dot{M}_1$	$=$	$(0.5 \pm 1.5) \times 10^{-3}$ mas^{-1}	$\dot{M}_2 =$	$(2.6 \pm 0.8) \times 10^{-3}$ mas^{-1}

Lens velocity dispersion: $\sigma = 303 \pm 50$ km/s (Rh91).

References: Ga94 (Garrett et al., 1994, MNRAS, 270, 457), Go83 (Gorenstein et al., 1983, Science, 219, 54), Go84 (Gorenstein et al., 1984, ApJ, 287, 538), P95 (Porcas et al., these proceedings), Rh91 (Rhee, 1991, Nature, 350, 211), Ro85 (Roberts et al., 1985, ApJ, 293, 356), and S80 (Stockton, 1980, ApJL, 242, L141).

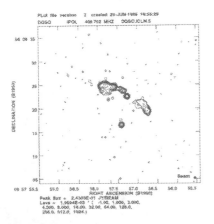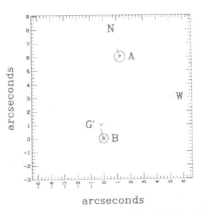

Figure 5.   Left: 408 MHz MERLIN map of 0957+561, from P95. Right: Schematic diagram of 0957+561. (The scale differs from the MERLIN map.) The ratio of the circle areas is equal to the core flux ratio (0.75 ± 0.02, Ga94). The lines give the direction of the jets, and the ratio of the line lengths is equal to the ratio of the jet lengths.

2016+112 (A: $20^h 16^m 55.4790^s$, $+11°17'46.537''$ [B1950; G94])

ID	$-\Delta\alpha$ "	$\Delta\delta$ "	F_{5GHz} mJy	$-\Delta\alpha$ "	$\Delta\delta$ "	$\lambda 5180$ mag($\pm$mag)	g mag($\pm$mag)	Object
A	0.000	0.000	15.8	0.00	0.00	20.95(0.02)	22.72(0.03)	QSO image
B	3.009	−1.489	17.2	3.02	−1.49	21.48(0.02)	23.09(0.03)	QSO image
C_2/C'	2.188	−3.205	4.4	2.03	−3.37	22.85(0.08)	24.39(0.12)	QSO image
C_1/C	2.077	−3.220	47.2	2.09	−3.21			Radio galaxy?
D				1.68	−1.95			Galaxy
A1				2.9	2.0	22.8 (0.2)		Gas cloud
B1				5.8	−1.2	22.8 (0.2)		Gas cloud

G94 data (left): C_1 and C_2 denote two radio sources resolved in this map. The uncertainties in the relative positions are $\leq$ 5 mas. The map has an rms noise of 75 μJy/beam. Data from 5GHz MERLIN observations, 1992 July 11-12.

S86 data (right): C denotes a radio source, and C′ denotes a source identified via a Ly-α emission line. $\lambda 5180$ denotes a filter centered at 5180 Å with FWIIM 100 Å. The uncertainties in the positions are $0.1''$ for A, B, and C′; $0.3''$ for A1 and B1; $0.01''$ for C (from a VLA map); and $0.1''$ for D (from i-band observation in S85). The uncertainty given for the magnitudes is photon noise. Data from the Hale telescope, 1985 Oct 13.

Miscellaneous: A, B, C′, A1, and B1 are identified via a Ly-α emission line at 5193 Å (S86); A1 and B1 are thought to be two gas clouds located near the lensed QSO and are thought to be magnified but not multiply imaged (S87). Though the radio source C/C_1 was thought to be a galaxy, new observations that resolve it into three subcomponents question this interpretation (G95).

References: G94 (Garrett et al., 1994, MNRAS, 269, 902), G95 (Garrett et al., these proceedings), S85 (Schneider et al., 1985, ApJ, 294, 66), S86 (Schneider et al., 1986, AJ, 91, 991), and S87 (Schneider et al., 1987, AJ, 94, 12).

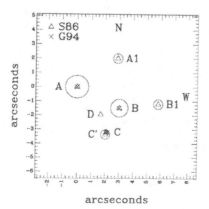

Figure 6. Positions from G94 ($\times$) and S86 ($\triangle$). The ratios of the circle areas are equal to the $\lambda 5180$ flux ratios from S86.

PG 1115+080 ($11^h15^m41.5^s$, $+08°02'24''$ [B1950; W80])

ID	x	y	V	I	x	y	V	B	R
	$''$	$''$	mag	mag	$''$	$''$	mag	mag	mag
A1	−1.294	−2.036	16.90	16.12	−1.27	−2.08	16.99	17.48	16.71
A2	−1.448	−1.582	17.35	16.51	−1.44	−1.62	17.27	17.74	16.95
B	0.362	−1.949	18.87	18.08	0.39	−1.95	18.74	19.19	18.46
C	0.000	0.000	18.37	17.58	0.00	0.00	18.26	18.71	17.97
G	−0.355	−1.322		18.36	−0.33	−1.35	20.89	> 21.6	20.20

K93 data (left): x is approximately west, y is approximately north. The internal position uncertainties are 5 mas for the quasar images and 50 mas for the galaxy. The relative fluxes are uncertain by 1.5% in I (F785LP) and 3% in V (F555W), but the zero-point for magnitudes is uncertain by 0.3 mag. Data from HST WFPC, 1991 Mar 3.

C87 data (right): x is approximately west, y is approximately north. The position uncertainties are not clearly stated, but separate reductions of B, V, and R frames agreed within an rms difference of 4 mas. Magnitude uncertainties for the quasar images are 0.03 in V and 0.05 in $B - V$ and $V - R$; magnitude uncertainties for the galaxy are 0.05 in R and 0.1 in $V - R$. Galaxy magnitudes determined with a 1.6'' diameter aperture. Data from CFHT, 1986 Feb 19.

References: C87 (Christian et al., 1987, ApJ, 312, 45), K93 (Kristian et al., 1993, AJ, 106, 1330), S93 (Schechter, 1993, Liège Proceedings, p. 119), V86 (Vanderriest et al., 1986, A&A, 158, L5), and W80 (Weymann et al., 1980, Nature, 285, 641).

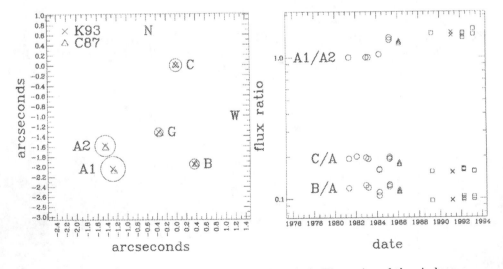

Figure 7. Left: Positions from K93 (×) and C87 (△). The ratios of the circle areas are equal to the I-band flux ratios from K93. Right: Flux ratios from C87 (△), K93 (×), S93 (□), and V86 (o).

MG 0414+0534 (B: $04^h14^m37.7321^s$, $+05°34'44.270''$ [J2000; K93])

ID	$-\Delta\alpha$ $''$	$\Delta\delta$ $''$	Peak mJy/beam	$-\Delta\alpha$ $''$	$\Delta\delta$ $''$	F_{5GHz} mJy	$-\Delta\alpha$ $''$	$\Delta\delta$ $''$	I' mag
A1	−0.577	−1.945	185.9	−0.588	−1.934	401	−0.59	−1.94	20.07
	−0.594	−1.939	39.0						
A2	−0.712	−1.543	122.3	−0.721	−1.530	362	−0.75	−1.54	20.95
	−0.728	−1.516	28.7						
B	0.000	0.000	38.6	0.000	0.000	156	0.00	0.00	20.95
	−0.048	−0.023	13.8						
C	1.345	−1.454		1.361	−1.635	58	1.40	−1.69	21.83
G							0.42	−1.36	21.08
X?							0.98	−0.03	23.32

E95 data (left): 5GHz VLBI data, with A1, A2, and B resolved into two subcomponents each. Position uncertainties are 11-14 mas for A1 and A2, and 22 mas for B. The flux uncertainty is estimated to be 10-15 mJy/beam; in addition, the VLBI observations are thought to have resolved out the low surface brightness emission. The VLBI detection of C is probably spurious. Data from 1992 June 7.

K95 data (center): 5GHz VLA data. The uncertainties in the relative positions are $0.01''$. The uncertainties in the peak fluxes are ∼ 4%. Data from Jan 1993.

S93 data (right): The position uncertainties are $0.01''$-$0.04''$. Magnitude uncertainties are 0.03-0.11, and 0.16 for X. Data from Hiltner 2.4m telescope, 1991 Nov 2-4.

References: A93 (Annis & Luppino, 1993, ApJL, 407, L69), E95 (Ellithorpe 1995, Ph.D. thesis, MIT), H92 (Hewitt et al., 1992, AJ, 104, 968), K93 (Katz & Hewitt 1993, ApJL, 409, L9), K95 (Katz et al., 1995, in preparation, given in E95), S93 (Schechter & Moore, 1993, AJ, 105, 1), and V95 (Vanderriest et al., these proceedings).

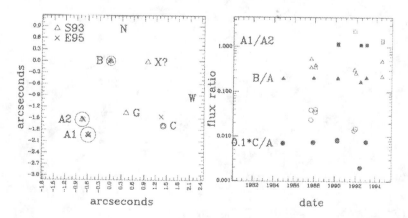

Figure 8. Left: Positions from E95 (×) and S93 (△). The ratios of the circle areas are equal to the flux ratios from K95. Right: Flux ratios A1/A2 (□), B/A (△), and C/A (○). Filled symbols denote radio data (H92, K93, E95, K95), while other symbols denote optical (H92, S93, V95) or infrared (A93) data.

CLASS 1608+656 (A: $16^h09^m13.956^s$, $+65°32'28.97''$ [J2000; M95])

ID	$-\Delta\alpha$ $''$	$\Delta\delta$ $''$	Flux Ratio (8.4 GHz)
A	0.00	0.00	2.06 ± 0.06
B	-0.74	-1.96	$\equiv 1.00$
C	-0.75	-0.46	0.85 ± 0.03
D	1.13	-1.24	0.26 ± 0.03
G	0.56	-1.16	

QSO images: The data for A, B, C, and D are from M95. The uncertainties in the positions are not stated. The flux of component B is 17.80 ± 0.44 mJy; the total flux is 73.2 ± 0.9 mJy. Data from VLA (A configuration), 8.4 GHz, 1994 Mar 1.

Galaxy: The data for G are from S95. The uncertainty in the position of the galaxy is correlated with the shape of the galaxy; it is estimated to be ~ 10 mas. Data from Hiltner 2.4m telescope, Apr 1995.

Miscellaneous: (from M95) Optical observations from Palomar Observatory 5m telescope (1994 Aug 9) give magnitudes for the entire system (including lensing galaxy) of $r = 19.4$, $i = 19.2$, uncertain by a few tenths of a magnitude. Observations with Keck 10m telescope (1994 Aug 22) indicate that in the K-band the flux of the lensing galaxy is ~ 4 times greater than total flux from the 4 images, and that the galaxy image has an axial ratio of $b/a = 0.56 \pm 0.10$ with its major axis having a position angle of $60°$. (The galaxy is likely to be more elliptical, because the image is circularized by the seeing).

References: M95 (Myers et al., 1995, ApJL, 447, L5; also these proceedings), and S95 (Schechter, 1995, private communication).

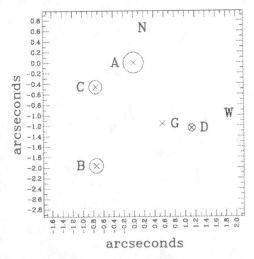

Figure 9. Positions from M95 (A, B, C, D) and S95 (G). The ratios of the circle areas are equal to the flux ratios from M95.

2237+0305 (22^{h}37^{h}57.3^s, +03°05'49" [B1950; Hu85])

ID	$-\Delta\alpha$ "	$\Delta\delta$ "	g mag	B mag	$-\Delta\alpha$ "	$\Delta\delta$ "	R mag	U mag	Peak μJy/beam
A	0.000	0.000	17.74	17.96	0.000	0.000	17.42	16.63	65.5
B	0.672	1.673	17.60	17.82	0.676	1.686	17.29	16.53	64.2
C	−0.626	1.202	18.41	18.66	−0.625	1.200	18.11	17.56	26.5
D	0.854	0.517	18.62	18.98	0.869	0.520	18.34	17.97	59.4
G	0.093	0.936			0.083	0.918			

Cr91 data (left): The relative positions have a formal error of 5 mas, largely due to limitations in the correction for detector distortion. The magnitude of B is 17.60 ± 0.10 in g (F502M), and 17.82±0.07 in B (F430W). The relative magnitudes have uncertainties of 0.05 mag for A and 0.10 mag for C and D. Data from HST FOC, 1990 Aug 27.

Ri92 data (center): The relative positions have an uncertainty of 15 mas. The relative magnitudes are accurate to ±0.04 mag in R (F702W) and ±0.06 mag in U (F336W). The R band was calibrated by comparing photometry of a star 9" from the center of the galaxy with r-band measurements by Ra91; the U band was not calibrated, so U brightnesses are given in instrumental magnitudes, subject to an arbitrary zero point. Data from HST WFPC, 1990 Dec 18.

Fa95 data (right): The uncertainty in the peak fluxes in 8.3 μJy/beam. The total flux is 336 ± 60 μJy. Data from 8 GHz VLA observations, 25 June 1995.

Miscellaneous: The central velocity dispersion of the galaxy is $\sigma_p = 215\pm30$ km/s (Fo92). The bulge of the galaxy has a position angle PA = 68° on the sky (Ri92).

References: Co91 (Corrigan et al., 1991, AJ, 102, 34), Cr91 (Crane et al., 1991, ApJL, 369, L59), Fa95 (Falco 1995, private communication), Fo92 (Foltz et al., 1992, ApJL, 386, L43), Ho94 (Houde & Racine, 1994, AJ, 107, 466), Hu85 (Huchra et al., 1985, AJ, 90, 691), N91 (Nadeau et al., 1991, ApJ, 376, 430), Ra91 (Racine, 1991, AJ, 102, 454), Ra92 (Racine, 1992, ApJL, 395, L65), Ri92 (Rix et al., 1992, AJ, 104, 959), and Y88 (Yee, 1988, AJ, 95, 1331).

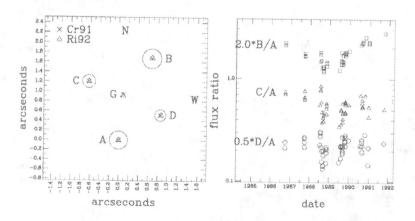

Figure 10. Left: Positions from Cr91 (×) and Ri92 (△). The ratios of the circle areas are equal to the B-band flux ratios from Cr91. Right: flux ratios from Ho94, which includes data reported originally in Co91, Cr91, N91, Ra91, Ra92, Ri92, and Y88.

B1422+231 (B: $14^h24^m38.094^s$, +22°56'00.59" [J2000; P92])

ID	$-\Delta\alpha$ "	$\Delta\delta$ "	F_{5GHz} mJy	$F_{8.4GHz}$ mJy	$-\Delta\alpha$ "	$\Delta\delta$ "	r mag	g mag
A	−0.3908	0.3194	216	148	−0.39	0.32	16.77	16.92
B	0.0000	0.0000	221	153	0.00	0.00	16.45	16.64
C	0.3357	−0.7457	115	79	0.34	−0.77	17.25	17.44
D	−0.9449	−0.8059	4.5	3.9	−0.96	−0.80	20.40	20.56
G					−0.68	−0.58	21.8	> 22.4

P92 data (left): The relative positions are from the 5 GHz MERLIN map and are uncertain by a few mas. Data from 8.4 GHz VLA map, 1991 June 16, and from 5 GHz MERLIN map, 1991 Aug 31.

Y94 data (right): The relative positions have an uncertainty of 0.01" for A, B, and C, and 0.02" for D. The galaxy position has an uncertainty 0.05". The relative magnitudes have an uncertainty of ~ 0.02 mag for A, B, and C, and ~ 0.1 mag for D; the absolute magnitudes have a systematic uncertainty of 0.07 mag. Data from CFHT, 1993 Apr 26.

Miscellaneous: The lensing galaxy is observed with a redshift $z_l = 0.647 \pm 0.001$ (Ha95). Two bright galaxies several arcseconds away from the lens are thought to perturb the lensing potential (Ha95, Ho94); the positions relative to B are as follows:

ID	$-\Delta\alpha$ (")	$\Delta\delta$ (")
G2	−9.0	−5.2
G3	−3.6	−7.3

References: A95 (Akujor et al., these proceedings), Ha95 (Hammer et al., 1995, A&A, 298, 737), Ho94 (Hogg & Blandford, 1994, MNRAS, 268, 889), L92 (Lawrence et al., 1992, MNRAS, 259, 5P), P92 (Patnaik et al., 1992, MNRAS, 259, 1P; also private communication), R93 (Remy et al., 1993, A&A, 278, L19), and Y94 (Yee & Ellingson, 1994, AJ, 107, 28).

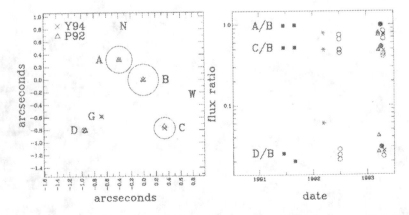

Figure 11. Left: Positions from Y94 (×) and P92 (△). The ratios of the circle areas are equal to the r-band flux ratios from Y94. Right: Flux ratios from L92 (∗), P92 (■), R93 (△), Y94 (×), and A95 (o, ●). Filled symbols denote radio data.

H 1413+117 ($14^h 13^m 20.11^s$, $+11°43'37.8''$ [B1950; K90])

| ID | $-\Delta\alpha$ | $\Delta\delta$ | Flux Ratio | | |
	''	''	R	I	B
A	0.000	0.000	1.00	1.00	1.00
B	−0.741	0.168	0.85	0.90	0.84
C	0.497	0.715	0.74	0.76	0.83
D	−0.351	1.046	0.61	0.59	0.61

S95 data (left): The uncertainty in the relative positions is about 4 mas, and the uncertainty in the alignment of the CCD is about 0.1°. Data from Hiltner 2.4m telescope, Apr 1992.

K90 data (right): The uncertainties in the flux ratios are not stated. The R magnitude for A is assumed to be 18.30. The system is detected at radio wavelengths, and the images have peak fluxes $\sim 0.1 - 0.2$ mJy/beam. Optical data from Danish 1.54m telescope at ESO, 1988 Apr 27. Radio data from VLA (A configuration), 8.415 GHz, 1989 Jan 13.

References: A90 (Angonin et al., 1990, A&A, 233, L5), K90 (Kayser et al., 1990, ApJ, 364, 15), M88 (Magain et al., 1988, Nature, 334, 325), R95 (Remy et al., these proceedings), and S95 (Schechter, 1995, private communication).

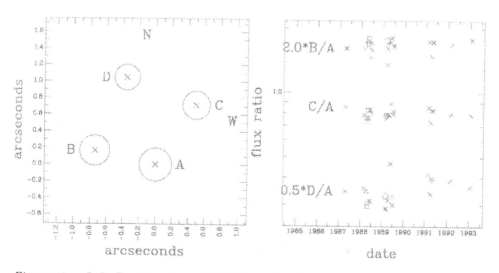

Figure 12. Left: Positions from S95. The ratios of the circle areas are equal to the R-band flux ratios from K90. Right: Flux ratios from A90 (o), K90 (□), M88 (△), and R95 (×).

HST 14176+5226 ($14^h 17^m 36.3^s$, $+52°26'44''$ [J2000])

ID	x $''$	y $''$	$-\Delta\alpha$ $''$	$\Delta\delta$ $''$	V mag	I mag
A	−1.1	1.1	−1.31	−0.85	25.63	25.12
B	1.8	−0.4	0.77	1.68	25.77	25.38
C	1.0	1.2	−0.96	1.23	25.99	25.47
D	−0.3	−0.9	0.82	−0.48	25.97	25.55
G	0.0	0.0	0.00	0.00	21.68	19.71

R95 data: The (x, y) positions are in the HST observation frame, with the x-axis 12.07° clockwise from North; they were given (in R95) in units of WFC 0.1″ pixels. The $(\Delta\alpha, \Delta\delta)$ positions were computed by rotating the (x, y) positions and retaining an extra digit. The position uncertainties are estimated to be about 0.03″. The bands are V (F606W) and I (F814W). The magnitudes of the quasar images were computed using a 0.3″ square aperture and were corrected to total magnitudes assuming a point source; the magnitude uncertainties for components A, B, C, D (respectively) are 0.06, 0.07, 0.08, 0.08 in V and 0.10, 0.11, 0.13, 0.14 in $V - I$. The magnitude of the galaxy was determined by a fit to the light distribution; the magnitude uncertainties of the galaxy are 0.04 in V and 0.04 in $V - I$. Data from HST WFPC2, 1994 Mar 11.

References: R95 (Ratnatunga et al., these proceedings; also private communication).

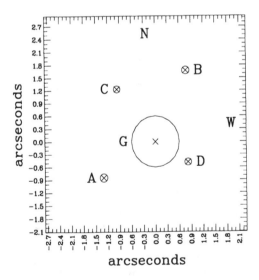

Figure 13. Positions from R95 (×). The ratios of the circle areas are equal to the V-band flux ratios.

HST 12531−2914 ($12^h53^m06.7^s$, $-29°14'30''$ [J2000])

ID	x $''$	y $''$	$-\Delta\alpha$ $''$	$\Delta\delta$ $''$	V mag	I mag
A	−0.6	−0.2	0.62	0.12	27.02	26.77
B	0.6	0.3	−0.63	−0.22	26.89	26.46
C	−0.2	0.4	0.15	−0.42	26.72	26.38
D	0.3	−0.4	−0.25	0.43	27.51	26.69
G	0.0	0.0	0.00	0.00	23.77	21.82

R95 data: The (x, y) positions are in the HST observation frame, with the x-axis 82.77° counter-clockwise from North; they were given (in R95) in units of WFC 0.1″ pixels. The $(\Delta\alpha, \Delta\delta)$ positions were computed by rotating the (x, y) positions and retaining an extra digit. The position uncertainties are estimated to be about 0.03″. The bands are V (F606W) and I (F814W). The magnitudes of the quasar images were computed using a 0.3″ square aperture and were corrected to total magnitudes assuming a point source; the magnitude uncertainties for components A, B, C, D (respectively) are 0.15, 0.15, 0.11, 0.24 in V and 0.28, 0.23, 0.21, 0.34 in $V - I$. The magnitude of the galaxy was determined by a fit to the light distribution; the magnitude uncertainties of the galaxy are 0.06 in V and 0.07 in $V - I$. Data from HST WFPC2, 1995 Feb 15.

References: R95 (Ratnatunga et al., these proceedings; also private communication).

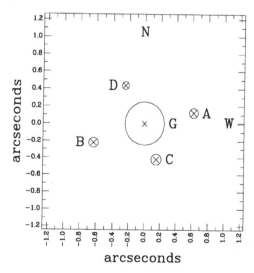

Figure 14. Positions from R95 (×). The ratios of the circle areas are equal to the V-band flux ratios.

Extended-emission systems

Lens		RA	Dec	Flux
MG 1131+0456	A	$11^h 31^m 56.48^s \pm 0.1''$	$+04°55'49.8'' \pm 0.1''$	$F_{15GHz}(A) = 3.18 \pm 0.10$ mJy
	B–A	$-1.725'' \pm 0.010''$	$-1.087'' \pm 0.010''$	$F_{15GHz}(B) = 3.90 \pm 0.12$ mJy
	D–A	$-0.4'' \quad \pm 0.1''$	$-0.4'' \quad \pm 0.1''$	$F_{8GHz}(D) = 0.32 \pm 0.06$ mJy
	C–D	$-1.710'' \pm 0.010''$	$-1.299'' \pm 0.010''$	$F_{8GHz}(C) = 1.16 \pm 0.03$ mJy
				$F_{5GHz}(\text{total}) = 205$ mJy

Comments: Positions are J2000. Component A is resolved into two components, A1 and A2. The compact components A2 and B appear to be variable.
References: Chen & Hewitt, 1993, AJ, 106, 1719; Hewitt et al., 1995, AJ, 109, 1956.

MG 1654+1346	A	$16^h 54^m 41.83^s$	$+13°46'22.0''$	
	G	$\Delta\alpha = -0.7''$	$\Delta\delta = -0.5''$	$m_r = 18.66 \pm 0.03$
	Q	$1.5''$	$1.3''$	$m_r = 20.93 \pm 0.03$
	Lensed Radio Lobe			$F_{5GHz} = 130.0$ mJy

Comments: Positions are J2000. Component A is a peak in the ring, but is not a compact component. The uncertainty in the absolute position of A is $0.1''$, and the uncertainty in the absolute positions of G and Q are $0.05''$. The quasar (Q) has $z = 1.74$, and the galaxy (G) has $z = 0.254$.
References: Langston et al., 1989, AJ, 97, 1283; Langston et al., 1990, Nature, 344, 43.

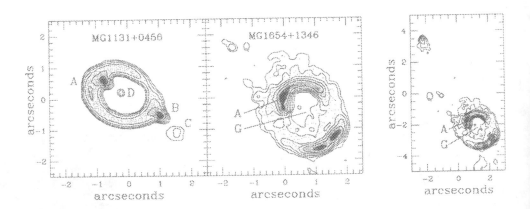

Figure 15. **Left:** 8 GHz VLA map of MG 1131+0456. The contours are $-2, 2, 4, 8, 16, 32, 64, 96\% \times 6.7$ mJy. The grayscale is linear. **Right:** 8 GHz VLA map of MG 1654+1346. The contours are $-2, 2, 4, 8, 16, 32, 64, 96\% \times 4.7$ mJy. The grayscale is linear. **Center:** The lensed radio lobe of MG 1654+1346. The contours and grayscale are the same as in the full image. In all three images, the coordinate origin is arbitrary.

Extended-emission systems

Lens	RA	Dec	Flux
B 1938+666			$F_{5GHz} = 316$ mJy
References: Patnaik et al., 1993, Liège proceedings, p. 208.			
MG 1549+3047 G3	$15^h 47^m 12.56^s$	$+30°56'18.0''$	$m_R = 23.3$
G1	$\Delta\alpha = -6.62''$	$\Delta\delta = 3.39$	$m_R = 16.7$
G2	$-2.48''$	1.78	$m_R = 20.5$
Lensed Radio Lobe			$F_{5GHz} = 185$ mJy
Total			$F_{5GHz} = 257$ mJy

Comments: Positions are B1950. G3 is the quasar core, G1 is the lensing galaxy ($z = 0.111$), and G2 is probably a small galaxy associated with the lensing galaxy. The uncertainty in the absolute position of G3 is 0.3″, and the uncertainty in the relative positions is 0.1″; the magnitude uncertainties are 0.1 mag.

References: Lehár et al., 1993, AJ, 105, 847, also private communication.

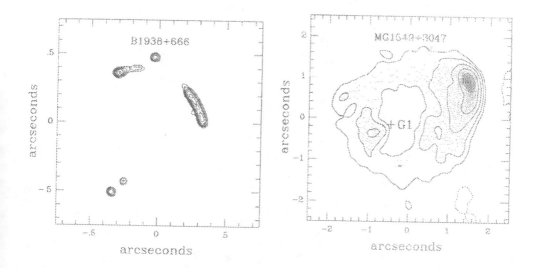

Figure 16. **Left:** 5 GHz MERLIN map of B 1938+666. The contours are −2, 2, 4, 8, 16, 32, 64, 96% × 36 mJy. **Right:** 8.4 GHz VLA map of MG 1549+3047. The contours are −4, 4, 8, 16, 32, 64, 96% × 8.0 mJy. The grayscale is linear. In both images, the coordinate origin is arbitrary.

Extended-emission systems

Lens		RA	Dec	Flux
PKS 1830−211	A	$18^h 33^m 39.94^s$	$-21°03'39.7''$	$F_{2.3\mathrm{GHz}}(\text{total}) \sim 10$ Jy
	B	$\Delta\alpha = -0.7''$	$\Delta\delta = -0.7''$	$F_{8.4\mathrm{GHz}}(\text{total}) \sim 8$ Jy

Comments: Positions are J2000, and the position uncertainty is 0.1″. A and B are compact, bright sources; they sit on top of a low-level ring. A and B are highly variable (see, for example, Lovell et al.). There is evidence for scintillation as the light passes near the galactic center (Jones et al.).

References: Jauncey et al., 1991, Nature, 352, 132; Jones et al., these proceedings; Lovell et al., these proceedings; Subrahmanyan et al., 1990, MNRAS, 246, 263; van Ommen et al., 1995, ApJ, 444, 561.

MG 0751+2716 $F_{5\mathrm{GHz}} = 202$ mJy

Comments: There are four components with mutual separations about 1″. A magnitude $\sim$19 galaxy with $z = 0.351$ is probably the lensing galaxy.

References: Lehár, 1993, Liège proceedings, p. 208, also private communication.

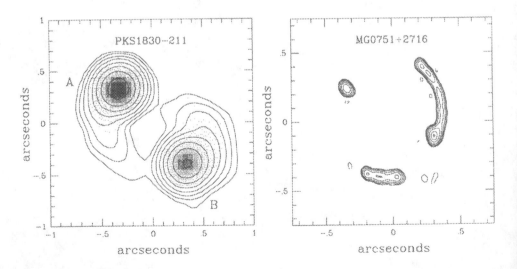

Figure 17. **Left:** 8 GHz VLA map of PKS 1830−211. The contours are −0.5, 0.5, 1, 2, 4, 8, 16, 32, 64, 96% × 3.5 Jy. The grayscale is linear. **Right:** 5 GHz MERLIN map of MG 0751+2716. The contours are −2, 2, 4, 8, 16, 32, 64, 96% × 25 mJy. In both images, the coordinate origin is arbitrary.

Extended-emission systems

Lens		RA	Dec	Flux
B 0218+357	A1	$02^h21^m5.483^s$	$+35°56'13.78''$	$F_{15\text{GHz}}(\text{A}1) = 621 \pm 7$ mJy
	A2−A1	1.07 mas	0.87 mas	$F_{15\text{GHz}}(\text{A}2) = 379 \pm 5$ mJy
	B1−A1	309.2 mas	127.4 mas	$F_{15\text{GHz}}(\text{B}1) = 172 \pm 3$ mJy
	B2−B1	1.47 mas	0.00 mas	$F_{15\text{GHz}}(\text{B}2) = 104 \pm 3$ mJy
				$F_{5\text{GHz}}(\text{total}) = 1209$ mJy

Comments: Positions are J2000. A and B are bright flat-spectrum compact components; the separation of the compact components and the diameter of the ring are 335 mas. The compact components A and B are resolved into two subcomponents each, with separations given above (from Pa95). These observations can be used to derive the magnification matrix (see Pa95, Po95). The bright components are variable, so that the flux ratio changes at the level of ~ 10% (Pa93). The lens redshift of $z_l = 0.6847$ is based on narrow emission and absorption lines (B93) and on HI emission (C93). A source redshift of $z_s = 0.96$ is proposed based on a MgII emission line and associated absorption doublet (L95).

References: B93 (Browne et al., 1993, MNRAS, 263, L32); C93 (Carilli & Rupen, 1993, ApJL, 412, L59); L95 (Lawrence, these proceedings); O92 (O'Dea et al., 1992, AJ, 104, 1320); Pa93 (Patnaik et al., 1993, MNRAS, 261, 435); Pa95 (Patnaik et al., 1995, MNRAS, 274, L5); Po95 (Porcas & Patnaik, these proceedings).

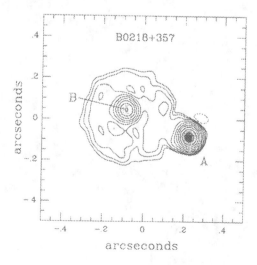

Figure 18. 5 GHz MERLIN map of B 0218+357. The contours are −0.125, 0.125, 0.25, 0.5, 1, 2, 4, 8, 16, 32, 64, 96% × 0.73 Jy. The grayscale is linear. The coordinate origin is arbitrary.

INDEX

442